Intelligent Systems Reference Library

Volume 246

The aim of this series is to publish a Reference Library, including novel advances and developments in all aspects of Intelligent Systems in an easily accessible and well structured form. The series includes reference works, handbooks, compendia, textbooks, well-structured monographs, dictionaries, and encyclopedias. It contains well integrated knowledge and current information in the field of Intelligent Systems. The series covers the theory, applications, and design methods of Intelligent Systems. Virtually all disciplines such as engineering, computer science, avionics, business, e-commerce, environment, healthcare, physics and life science are included. The list of topics spans all the areas of modern intelligent systems such as: Ambient intelligence, Computational intelligence, Social intelligence, Computational neuroscience, Artificial life, Virtual society, Cognitive systems, DNA and immunity-based systems, e-Learning and teaching, Human-centred computing and Machine ethics, Intelligent control, Intelligent data analysis, Knowledge-based paradigms, Knowledge management, Intelligent agents, Intelligent decision making, Intelligent network security, Interactive entertainment, Learning paradigms, Recommender systems, Robotics and Mechatronics including human-machine teaming, Self-organizing and adaptive systems, Soft computing including Neural systems, Fuzzy systems, Evolutionary computing and the Fusion of these paradigms, Perception and Vision, Web intelligence and Multimedia.

Indexed by SCOPUS, DBLP, zbMATH, SCImago.

All books published in the series are submitted for consideration in Web of Science.

Erik Cuevas · Daniel Zaldívar ·
Marco Pérez-Cisneros

New Metaheuristic Schemes: Mechanisms and Applications

Springer

Erik Cuevas
Centro Universitario de Ciencias Exactas e
Ingenierías (CUCEI)
Universidad de Guadalajara
Guadalajara, Jalisco, Mexico

Daniel Zaldívar
Centro Universitario de Ciencias Exactas e
Ingenierías (CUCEI)
University of Guadalajara
Guadalajara, Jalisco, Mexico

Marco Pérez-Cisneros
Centro Universitario de Ciencias Exactas e
Ingenierías (CUCEI)
Universidad de Guadalajara
Guadalajara, Jalisco, Mexico

ISSN 1868-4394 ISSN 1868-4408 (electronic)
Intelligent Systems Reference Library
ISBN 978-3-031-45560-5 ISBN 978-3-031-45561-2 (eBook)
https://doi.org/10.1007/978-3-031-45561-2

This Springer imprint is published by the registered company Springer Nature Switzerland AG
The registered company address is: Gewerbestrasse 11, 6330 Cham, Switzerland

Paper in this product is recyclable.

Preface

Metaheuristic schemes are optimization models of an abstract nature, imitating various biological or social systems. They have demonstrated remarkable success in tackling complex problems where conventional optimization approaches yield unsatisfactory outcomes. These methods begin by generating individuals using random values, with each individual representing a potential solution. An objective function then assesses the quality of each individual's solution. At each iteration, individuals undergo modifications based on rules that govern their behavior within the search space. These rules are abstract models inspired by natural or social mechanisms. The process iterates until a predefined stopping criterion is met. Notably, metaheuristic methods possess two key characteristics that set them apart from other search techniques. Firstly, they differ from traditional optimization methods, as they involve a population of candidate solutions rather than focusing on a single solution. Secondly, metaheuristic methods foster collective cooperation among individuals within the population, encouraging features like adaptation and self-organization.

Over the past decade, an increasing number of metaheuristic techniques have been documented in academic literature, with authors asserting their novelty and impressive optimization capabilities. Despite emulating diverse processes or systems, these schemes share striking similarities in the rules employed to model individual behavior. The underlying concept behind designing many metaheuristic methods is to assemble a collection of successful rules from previous approaches, creating new and effective optimization schemes.

Several key features consistently appear in most metaheuristic approaches. These include the utilization of diversification to explore underexplored regions of the search space, which were seldom visited before, as well as the use of intensification or exploitation to thoroughly investigate promising regions. Another noteworthy aspect is the incorporation of memory to store the best solutions encountered during the optimization process. As a result, metaheuristic methods have swiftly gained popularity among researchers, proving effective in solving a wide range of optimization problems in various domains, from straightforward to highly complex ones.

A vast array of problems across diverse fields, such as science, engineering, economics, and more, can be framed as optimization tasks. While some problems are relatively straightforward and amenable to traditional mathematical analysis-based optimization methods, the majority of practical and significant challenges, like image processing and parameter estimation, involve conflicting conditions that make them exceptionally difficult for classical approaches to handle effectively. In response to these complex formulations, metaheuristic algorithms have emerged as the most viable and promising alternative. Consequently, metaheuristic techniques have gained significant traction as an active and vibrant research area over the past decade. During this period, numerous novel metaheuristic approaches have been introduced and rigorously tested on artificial benchmark problems and real-world applications. Despite being one of the most extensively explored research paradigms in computational intelligence, the field of metaheuristic computation faces various open challenges. These challenges span from addressing issues like premature convergence and the struggle to maintain population diversity to investigating how to effectively combine metaheuristic paradigms with other algorithmic schemes. Furthermore, researchers are continuously seeking to extend existing techniques to tackle increasingly difficult problems as they arise.

Several books have been published, extensively covering well-known metaheuristic methods such as Differential Evolution and Particle Swarm Optimization. However, there is a scarcity of attempts to explore new alternative approaches in these works. Established metaheuristic schemes exhibit certain limitations in their design, including premature convergence and a lack of population diversity maintenance. In recent times, new metaheuristic methods have emerged to address these challenges, generally yielding improved results. These novel approaches introduce fresh models and innovative cooperation strategies that enable effective exploration and exploitation of vast search spaces with a significant number of dimensions. Although most of these new metaheuristic techniques show promising results, they are still in their early stages.

To reach their full potential, these new metaheuristic methods need to be extensively applied to a diverse range of problems and contexts. This will ensure that they not only perform well on the reported optimization problem sets but also excel in tackling new and complex formulations. Achieving this requires disseminating and presenting these methods in various technical fields as valuable optimization tools. In practice, when scientists, engineers, or practitioners encounter a problem as a specific instance of a broader class, they can select from different metaheuristic algorithms that guarantee expected optimization performance. Unfortunately, the current options mainly concentrate on well-established algorithms that have gained popularity and widespread use, leaving behind the potential of newer developments.

The overabundance of publications focused on making simple modifications to popular metaheuristic methods poses a significant drawback: it hinders the exploration of new techniques and procedures that could prove valuable in solving problems posed by academic and industrial communities. Despite the introduction of several promising metaheuristic schemes in recent years, incorporating intriguing concepts and operators, they appear to have been entirely overlooked in the literature. Instead, there seems to be a preference for modifying, hybridizing, or reconfiguring well-known metaheuristic approaches.

The objective of this book is to present advancements that discuss novel alternative metaheuristic developments, which have demonstrated their effectiveness in tackling various complex problems. The book covers different new metaheuristic methods and their practical applications. This approach is of paramount importance to us because we believe it is the most effective way to aid researchers, educators, engineers, and practitioners in addressing their optimization challenges.

To facilitate the readers, each chapter of this book has been structured in a manner that allows it to be read independently from the others. Chapter 1 provides an overview of the main characteristics and properties of metaheuristic and swarm schemes. It delves into the essential concepts of metaheuristics, offering insights into their significance.

Chapter 2 introduces a novel metaheuristic algorithm based on the principles of agent systems. Agent-based modeling represents a relatively recent approach to simulate complex systems comprising agents with behavior defined by simple rules. As agents interact, complex global behavioral patterns emerge without explicit programming. Agent-based systems offer a robust theory and consistent models to characterize global behaviors resulting from collective interactions among individuals governed by basic rules. Within this framework, various agent-based concepts and models that generate highly intricate global search behaviors can be leveraged to develop efficient optimization algorithms or enhance existing ones. The method presented in this chapter is rooted in the "Heroes and Cowards" agent-based model. This model employs a small set of rules to generate two emergent global patterns, which can be interpreted as exploration and exploitation stages in the context of metaheuristic literature. To assess its performance, the algorithm undergoes testing on a selection of representative benchmark functions, encompassing multimodal, unimodal, and hybrid benchmark formulations. The competitive results showcase the promising integration of both paradigms and their potential for producing effective optimization strategies.

Chapter 3 explores a dynamic metaheuristic approach. In recent times, numerous dynamic optimization algorithms based on metaheuristic methods have been proposed. While these schemes efficiently find a single global optimum, they struggle to locate multiple optimal solutions. The primary objective of dynamic multi-modal optimization is to identify multiple optimal solutions for problems where the objective function changes over time. In various applications, it is crucial to find not only the best solution but also other feasible alternatives due to practical limitations. However, the problem of dynamic multi-modal optimization using evolutionary principles has received limited attention in the literature. The algorithm discussed in this chapter utilizes the mean shift scheme to detect global and local optima in dynamic optimization problems. The mean shift is an iterative, nonparametric process for detecting local maxima in a density function represented by a set of samples. It exhibits adaptive characteristics that enable it to find local maxima in dynamic environments. The method presented in this chapter modifies the search strategy of the mean shift to consider not only the density but also the fitness value of candidate solutions. Additionally, a competitive memory and dynamic strategy are integrated to accelerate the convergence process by leveraging crucial information from previous environments.

Consequently, the proposed approach effectively identifies most of the global and local optima in dynamic environments. To showcase the algorithm's performance, a series of comparisons with other well-known dynamic optimization methods have been conducted. The study utilizes the benchmark generator of the CEC competition for dynamic optimization. The experimental results demonstrate the dynamic scheme's competitive accuracy and robustness, validating its effectiveness in solving dynamic multi-modal optimization problems.

Chapter 4 introduces an alternative approach for generating search patterns in metaheuristic schemes. Efficiently exploring a fitness landscape to find the optimal solution can be challenging, especially in the presence of ruggedness and multiple local optima, which can lead to stagnation or premature convergence. Premature convergence occurs when a suboptimal solution is mistakenly identified as the best solution for the optimization problem. The search patterns discussed in this chapter are designed to emulate the response of a second-order system. Depending on the values of their parameters, these patterns exhibit various temporal behaviors, representing search patterns with multiple behaviors and simple configurations. These search patterns are then integrated into a comprehensive search strategy known as the Second-Order Algorithm (SOA), aimed at obtaining global solutions for complex optimization problems. To evaluate the performance of the SOA, it has been compared with a set of representative optimization problems, including multimodal, unimodal, and hybrid benchmark formulations. The competitive results demonstrate the promising potential of the proposed search patterns in effectively solving complex optimization problems and outperforming other approaches.

Chapter 5 introduces a modified version of the Grey Wolf Optimizer (GWO) known as the Group-based Synchronous-Asynchronous Grey Wolf Optimizer (GSAGWO). In this approach, the population is divided into three groups, each containing an equal number of candidate solutions. The optimization process involves individually processing each group in a specific order, employing a control strategy that adjusts the balance between exploration and exploitation. Additionally, the new algorithm incorporates an intriguing mechanism to enhance population diversity by exchanging elements among the groups. To assess the performance of the GSAGWO, it has been subjected to testing on the CEC2017 benchmark, as well as several real-world engineering problems. The results of GSAGWO have been compared against those generated by other state-of-the-art metaheuristic algorithms. The outcomes validate the effectiveness and accuracy of the proposed technique, affirming its capability to tackle optimization problems successfully.

Chapter 6 introduces a novel algorithm for Image Contrast Enhancement (ICE) based on metaheuristic principles. ICE is a critical step in various image processing and computer vision applications, aiming to enhance the visual information quality in processed images. However, the presence of noise and small sets of irrelevant pixels in images can hinder the ICE process, leading to undesirable artifacts, noise amplification, over-saturation, and poor human visual perception in most existing ICE methods. The approach discussed in this chapter addresses these issues by eliminating noisy and irrelevant information to enhance the distribution capacity of significant pixel intensities in the resulting image. It achieves this by removing multiple groups of a small number of pixels that do not represent any essential object or detail in the image. The Mean-shift algorithm is utilized to replace these irrelevant pixel sets in the original histogram with significant pixel densities represented by local maxima. Subsequently, the Moth Swarm Algorithm (MSA) redistributes the pixel intensities in the reduced histogram to maximize the Kullback-Leibler entropy (KL-entropy). To evaluate the proposed approach, various public datasets commonly used in the literature are considered, and the results are compared with those of other well-known ICE techniques. The experimental findings demonstrate that the proposed approach effectively highlights important image details and improves its human visual appearance, overcoming the limitations observed in conventional ICE methods.

Finally, in Chap. 7, a method for anisotropic diffusion filtering is introduced. Anisotropic Diffusion (AD) is a well-established technique used for noise removal in digital images while preserving their edges. However, selecting appropriate parameters for the AD operation significantly impacts the filtering results. Surprisingly, the problem of automatically determining these parameters based on image requirements has received limited attention in the literature. The chapter proposes a novel multi-objective methodology to obtain the AD parameters for achieving effective filtering results. The approach aims to strike a balance between two conflicting scenarios: minimizing the image's noise content while maximizing its contrast. These objectives cannot be improved individually without negatively affecting the other. The proposed scheme employs the Non-dominated Sorting Genetic Algorithm based on Reference Points (NSGA-III), a robust and powerful algorithm for multi-objective optimization, to solve this formulation. The final solution is obtained through an analysis of the Optimal Pareto front, which represents the best possible trade-off between noise reduction and contrast enhancement. Experimental results demonstrate that the proposed method outperforms existing filtering algorithms in terms of both visual quality and standard performance metrics, confirming its effectiveness in obtaining superior filtering results.

As authors, we would like to take this opportunity to extend our heartfelt gratitude to the numerous individuals who contributed to the completion of this book. We express our deepest appreciation to Prof. Lakhmi C. Jain, whose unwavering support and encouragement were instrumental in sustaining and guiding this project. We also want to acknowledge the support of COECYTJAL under the project COECYTJAL 10300_DESIGN OF NEW SUSTAINABLE AND INTELLIGENT VEHICLES IN JALISCO COECYTJAL.

Guadalajara, Mexico

Erik Cuevas
Daniel Zaldívar
Marco Pérez-Cisneros

Contents

Chapter 1
Introduction to Metaheuristic Schemes: Characteristics, Properties, and Importance in Solving Optimization Problems

In this chapter, the primary ideas behind metaheuristic schemes are introduced. The goal of this chapter is to provide an overview of the features and qualities of these approaches. Additionally, an important aim of this chapter is to acknowledge the significance of metaheuristic methods in addressing optimization problems that cannot be effectively solved using conventional techniques.

1.1 Introduction

Most industrial and engineering systems commonly require the implementation of an optimization process to ensure their efficient functioning. These systems necessitate the identification of an optimal solution that minimizes a specified cost function. Broadly speaking, an optimization scheme refers to a search strategy aimed at obtaining the most favorable solution from a pool of potential alternatives. This solution represents the optimal outcome that adequately addresses the optimization problem according to the cost function [1].

To illustrate this concept, let's consider a public transportation system in a specific town. In this scenario, the objective is to determine the "best" route to a designated destination. To evaluate each potential option and subsequently obtain the most optimal solution, a suitable criterion must be taken into consideration. One practical criterion could involve comparing the relative distances among all possible routes. Consequently, a hypothetical optimization scheme would select the option with the shortest distance as the final output. It is important to acknowledge that various evaluation factors are possible, including other important criteria such as the number of transfers, travel time between locations, or ticket price.

The optimization problem can be defined as follows: Given a cost function f, the objective is to find the input value, referred to as the argument, that minimizes the function f.

E. Cuevas et al., *New Metaheuristic Schemes: Mechanisms and Applications*, Intelligent Systems Reference Library 246, https://doi.org/10.1007/978-3-031-45561-2_1

$$x^* = \arg\min_{x \in S} f(\mathbf{x}) \tag{1.1}$$

S denotes the search space, encompassing all conceivable solutions. In broad terms, each potential solution within this space offers a distinct level of effectiveness in addressing the optimization problem. Typically, the unknown components of $\mathbf{x}$ represent the decision variables of the optimization formulation. The cost function f assesses the quality of each potential solution. It quantifies how well a given candidate element $\mathbf{x}$ resolves the optimization problem.

In the case of the public transportation example, S represents the entire set of subway stations, bus lines, and other relevant data available in the transportation system database. The variable $\mathbf{x}$ signifies a potential route connecting the starting location to the final destination. The cost function $f(\mathbf{x})$ evaluates the quality of each feasible path. Furthermore, the problem definition can incorporate additional constraints, such as ticket price or distance to the destination (in some cases, a combination of both criteria is taken into account, depending on individual preferences).

When there are extra constraints to consider, the optimization problem is referred to as constrained optimization, distinguishing it from unconstrained optimization, where such restrictions are not taken into account. In the presence of these conditions, an optimization formulation encompasses the following elements:

- One or multiple decision variables derived from $\mathbf{x}$, incorporating a potential solution
- The cost function $f(\mathbf{x})$, responsible for assessing the merit of each solution $\mathbf{x}$
- The search space S, encompassing the complete range of possible solutions for $f(\mathbf{x})$.
- Constraints that delineate various permissible regions within the search space S.

In practical terms, an optimization strategy aims to find a solution for the cost function $f(\mathbf{x})$. within a given search space S, achieving a balance between computational efficiency and sufficient accuracy. The effectiveness of the optimization method is also contingent upon the nature of the problem formulation. As a result, an optimization problem is considered well-defined when the following conditions are established:

1. There exists at least one solution or a set of solutions that satisfy the optimal values.
2. There is a clear relationship between a solution and its position, such that small changes to the original values result in minor deviations in the objective function $f(\mathbf{x})$.

1.2 Classical Methods for Optimization

Once an engineering problem has been translated into a cost function, the subsequent step is to select an appropriate optimization method. Optimization approaches can be categorized into two groups: classical approaches and metaheuristic methods [2].

In many cases, the cost function $f(\mathbf{x})$ exhibits a nonlinear relationship with its adjustable decision variables x. Classical optimization methods employ iterative algorithms to effectively explore the search space. Among the various approaches described in the literature, methods utilizing derivative-descent principles are the most widely used. These techniques involve determining the new position from the current location by moving in the direction of **d**:

$$x_{k+1} = x_k + \alpha \mathbf{d}, \tag{1.2}$$

where α represents the learning rate, which determines the magnitude of the search step taken in the direction of d. In Eq. 1.2, the direction d is computed assuming the utilization of the gradient (**g**) of the objective function.

Among the classical approaches, one of the most notable methods is the steepest descent scheme. This technique is widely adopted due to its simplicity and efficiency in solving objective functions. Several other derivative-based approaches build upon this scheme as a foundation for the development of more advanced methods. The steepest descent scheme is defined by the following formulation:

$$x_{k+1} = x_k - \alpha \mathbf{g}(f(x)), \tag{1.3}$$

Although classical derivative-based optimization schemes are straightforward, they can still be applied given that the cost function satisfies two crucial constraints:

(I) The cost function is twice differentiable.
(II) The cost function is unimodal, meaning it possesses only one optimal position.

The optimization problem represents a straightforward scenario of a differentiable and unimodal objective function. This function meets the requirements (I) and (II):

$$f(x_1, x_2) = 10 - e^{-(x_1^2 + 3 \cdot x_2^2)} \tag{1.4}$$

Figure 1.1 depicts the function represented by formulation 1.4.

Given the increasing complexity in system design, traditional methods are applicable in only a limited number of cases. The majority of optimization problems involve scenarios that fail to meet the constraints required for the application of gradient-based methods. For instance, combinatorial problems lack a definition of differentiability. Additionally, there are numerous situations where optimization problems cannot be differentiated. A notable example is the "floor" function, which yields the largest integer less than or equal to its argument. When applied in Eq. 1.4, this operation transforms the optimization problem from a differentiable one (Eq. 1.5)

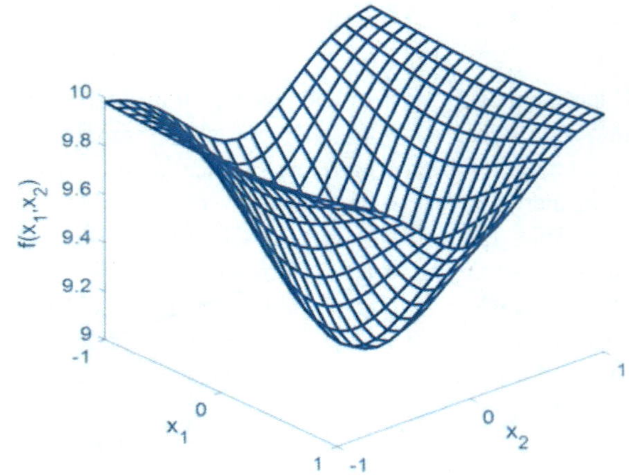

Fig. 1.1 Cost function exhibiting unimodal properties

to a non-differentiable one. This problem can be formulated as follows (Fig. 1.2):

$$f(x_1, x_2) = \text{floor}\left(10 - e^{-(x_1^2 + 3 \cdot x_2^2)}\right) \tag{1.5}$$

While an optimization problem may possess differentiability, there are additional constraints that can restrict the applicability of classical optimization techniques. One such constraint arises when there is a unique optimal solution, meaning the cost function lacks any other significant local optima. To illustrate this point, let's examine the example of minimizing the Griewank function.

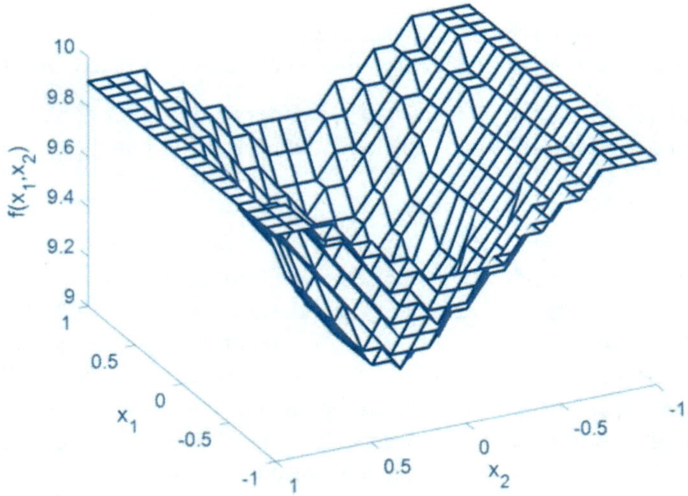

Fig. 1.2 A non-differentiable function is generated by incorporating the floor function.

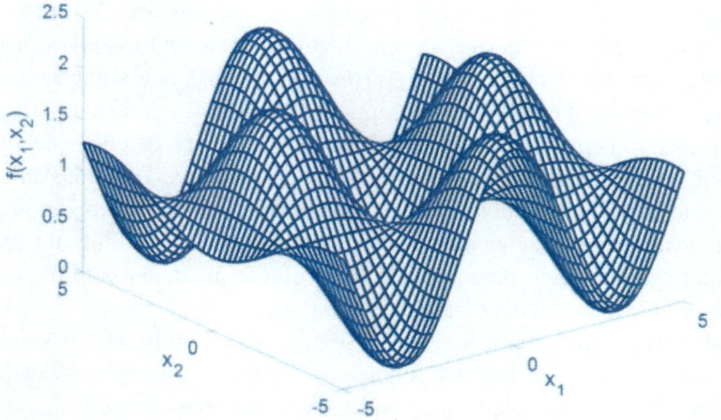

Fig. 1.3 The Griewank function displaying multimodal properties

$$\begin{aligned} \text{minimize} \quad & f(x_1, x_2) = \frac{x_1^2 + x_2^2}{4000} - \cos(x_1)\cos\left(\frac{x_2}{\sqrt{2}}\right) + 1 \\ \text{subject to} \quad & -5 \leq x_1 \leq 5 \\ & -5 \leq x_2 \leq 5 \end{aligned} \tag{1.6}$$

Upon closer examination of the formulation presented in Eq. 1.6, it becomes evident that the optimal global solution is located at $x_1 = x_2 = 0$. Figure 1.3 illustrates the cost function described in Eq. 1.6. As depicted in Fig. 1.3, the cost function exhibits multiple local optimal solutions (multimodal), making it highly probable for gradient-based techniques with randomly generated initial solutions to prematurely converge to one of these local optima.

Considering the constraints associated with gradient-based approaches, their utilization in solving a diverse array of engineering optimization problems becomes challenging. Instead, alternative techniques that do not have such limitations are required. These techniques can be employed effectively across a wide range of problems [3].

1.3 Metaheuristic Algorithms

Metaheuristic schemes [4] are derivative-free methods that do not require the cost function to adhere to the constraints of being twice differentiable or unimodal. These methods, operating as global optimization techniques, are capable of addressing a wide range of optimization problems, including non-convex, nonlinear, and multimodal problems, while accommodating linear or nonlinear constraints and continuous or discrete decision variables.

The field of metaheuristic computation boasts a rich history. As more complex industrial processes emerge, the development of new optimization techniques becomes necessary, ones that do not rely on prior knowledge or assumptions about the optimization problem. This lack of assumptions distinguishes metaheuristic methods from classical gradient-based methods. In fact, a majority of engineering system applications involve highly nonlinear systems or objective functions affected by noise. Additionally, explicit deterministic expressions for the optimization problem are often unavailable. Consequently, evaluating each candidate solution necessitates experimental or simulation processes. In this context, metaheuristic methods have emerged as viable alternatives for optimization.

A metaheuristic approach is a versatile search strategy utilized for solving optimization problems. It treats the cost function in an abstract manner, focusing solely on evaluating the function at specific positions without considering its mathematical properties. Metaheuristic methods do not rely on any assumptions or prior knowledge about the optimization problem or the objective function. They treat the optimization formulation as "black boxes" [5], emphasizing their distinctive and appealing characteristic.

Metaheuristic approaches acquire knowledge about the structure of an optimization problem by leveraging information gathered from all evaluated solutions (i.e., candidate solutions) throughout the optimization process. This knowledge is then utilized to generate new candidate solutions with the expectation that these new solutions exhibit improved quality compared to their predecessors.

Currently, various metaheuristic approaches have been introduced in the literature, yielding favorable results. These methods incorporate models inspired by biological, natural, or social systems, which, from a certain perspective, can be regarded as optimization problems [6].

These approaches involve the cooperative behavior observed in various natural systems, drawing inspiration from references such as the Artificial Bee Colony (ABC) technique [8] which imitates the behavior of bee colonies. The Particle Swarm Optimization (PSO) algorithm [7] is based on the social behavior of bird flocking and fish schooling. The Bat Algorithm (BA) method [10] emulates the behavior of bats, while the Harmony Search (HS) algorithm [9] is inspired by the improvisation process of musicians searching for better states of harmony.

Other metaheuristic methods include the Social Spider Optimization (SSO) algorithm [12] inspired by social-spider behavior, the Firefly (FF) method [11] based on the mating behavior of firefly insects, and the Clonal Selection Algorithm (CSA) [14] which simulates immunological systems. The Collective Animal Behavior algorithm [13] emulates the behavior of animals in groups, while the Differential Evolution (DE) algorithm [16] combines elements of differential and conventional evolution. The Electromagnetism-Like algorithm [15] simulates the phenomenon of electromagnetism, and Genetic Algorithms (GA) [17] draw inspiration from genetic processes in species.

These references serve as inspiration for the development of metaheuristic methods, each offering unique perspectives and strategies for solving optimization problems [18–22].

1.3.1 Typical Scheme of a Metaheuristic Algorithm

Broadly speaking, a metaheuristic scheme can be defined as a search strategy that aims to replicate, from a particular perspective, the behavior of a biological, natural, or social system. A generic metaheuristic method encompasses the following key characteristics:

1. Maintenance of a population of candidate solutions: The method operates with a set of potential solutions, referred to as the population, which undergoes dynamic modifications during the optimization process.
2. Dynamic modification of the population: New solutions are continuously generated and incorporated into the population. This ensures a constant exploration of the solution space and facilitates the discovery of improved solutions.
3. Association of a cost function: A cost function is employed to assess the quality or fitness of each solution in the population. This function determines the solution's ability to survive and potentially reproduce in subsequent iterations.
4. Definition of operations: Various operations or mechanisms are defined to enable effective exploration and exploitation of the solution space. These operations facilitate the production of new promising solutions, allowing the metaheuristic method to navigate the search landscape efficiently.

These characteristics collectively form the foundation of a metaheuristic method, providing a flexible and adaptable framework for solving optimization problems.

Within the metaheuristic methodology, the expectation is that candidate solutions will, on average, improve their quality throughout the evolutionary process. In other words, their effectiveness in solving the optimization problem should increase. As the metaheuristic scheme operates, its defined operators generate new solutions. With each iteration, the quality of these solutions tends to improve. Since the quality of a solution corresponds to its ability to solve the optimization problem, the metaheuristic method guides the population towards the optimal global solution. This powerful mechanism has enabled the application of metaheuristic schemes to address complex engineering problems across various domains.

The majority of metaheuristic schemes [23–26] have been specifically designed to tackle the challenge of finding a global solution for nonlinear optimization problems with box constraints. These problems can be expressed in the following general form:

$$\begin{array}{ll} \text{Maximize/Minimize} & f(\mathbf{x}),\ \mathbf{x} = (x_1, \ldots, x_d) \in \Re^d \\ \text{subject to} & \mathbf{x} \in \mathbf{X} \end{array} \tag{1.7}$$

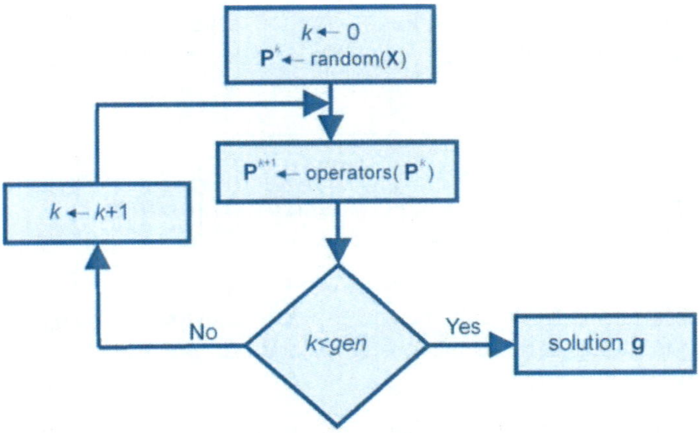

Fig. 1.4 Typical scheme of a metaheuristic algorithm

where $f : \Re^d \rightarrow \Re$ represents a nonlinear function whereas $\mathbf{X} = \left\{\mathbf{x} \in \Re^d | l_i \leq x_i \leq u_i, i = 1, \ldots, d\right\}$ corresponds to the feasible search space, restriced by the lower (l_i) and upper (u_i) bounds.

In order to solve the problem described in Eq. 1.7 using the metaheuristic computation methodology, a population of N possible solutions (individuals) is iteratively modified from an initial point ($k = 0$) to a specified number of generations ($k = gen$). The metaheuristic scheme starts by initializing the population of N candidate solutions with random values that are uniformly distributed within the predefined lower and upper limits.

At each generation, a set of operations is applied to the current population in order to generate a new set of individuals. Each possible solution is represented as a d-dimensional vector, where each element corresponds to a decision variable of the optimization problem being solved. The fitness of each possible solution is evaluated by the cost function which assigns a fitness value. Throughout the evolution process, the best solution encountered so far is preserved as it represents the best solution found up to that point.

Figure 1.4 provides an illustration of the generic procedure employed by a metaheuristic method, depicting the iterative modification of the population and the evaluation of fitness values using the cost function.

References

1. Akay, B., Karaboga, D.: A survey on the applications of artificial bee colony in signal, image, and video processing. SIViP **9**(4), 967–990 (2015)
2. Yang, X.-S.: Engineering Optimization. Wiley (2010)
3. Treiber, M.A.: Optimization for Computer Vision An Introduction to Core Concepts and Methods. Springer (2013)

4. Simon, D.: Evolutionary Optimization Algorithms. Wiley (2013)
5. Blum, C., Roli, A.: Metaheuristics in combinatorial optimization: overview and conceptual comparison. ACM Comput. Surv. (CSUR) **35**(3), 268–308 (2003). https://doi.org/10.1145/937503.937505
6. Nanda, S.J., Panda, G.: A survey on nature inspired metaheuristic algorithms for partitional clustering. Swarm Evol. Comput. **16**, 1–18 (2014)
7. Kennedy, J., Eberhart, R.: Particle swarm optimization. In: Proceedings of the 1995 IEEE International Conference on Neural Networks, vol. 4, pp. 1942–1948 (1995)
8. Karaboga, D.: An idea based on honey bee swarm for numerical optimization. TechnicalReport-TR06. Engineering Faculty, Computer Engineering Department, Erciyes University (2005)
9. Geem, Z.W., Kim, J.H., Loganathan, G.V.: A new heuristic optimization algorithm: harmony search. Simulations **76**, 60–68 (2001)
10. Yang, X.S.: A new metaheuristic bat-inspired algorithm. In: Cruz, C., González, J., Krasnogor, G.T.N., Pelta, D.A. (eds.) Nature Inspired Cooperative Strategies for Optimization (NISCO 2010), Studies in Computational Intelligence, vol. 284, pp. 65–74. Springer, Berlin (2010)
11. Yang, X.S.: Firefly algorithms for multimodal optimization. In: Stochastic Algorithms: Foundations and Applications, SAGA 2009, Lecture Notes in Computer Sciences, vol. 5792, 2009, pp. 169–178
12. Cuevas, E., Cienfuegos, M., Zaldívar, D., Pérez-Cisneros, M.: A swarm optimization algorithm inspired in the behavior of the social-spider. Expert Syst. Appl. **40**(16), 6374–6384 (2013)
13. Cuevas, E., González, M., Zaldivar, D., Pérez-Cisneros, M., García, G.: An algorithm for global optimization inspired by collective animal behaviour. Discrete Dyn. Nat. Soc. art. no. 638275 (2012)
14. de Castro, L.N., von Zuben, F.J.: Learning and optimization using the clon-al selection principle. IEEE Trans. Evol. Comput. **6**(3), 239–251 (2002)
15. Birbil, ŞI., Fang, S.C.: An electromagnetism-like mechanism for global optimization. J. Glob. Optim. **25**(1), 263–282 (2003)
16. Storn, R., Price, K.: Differential Evolution -a simple and efficient adaptive scheme for global optimisation over continuous spaces. Technical Report TR-95–012, ICSI, Berkeley, CA (1995)
17. Goldberg, D.E.: Genetic algorithm in search optimization and machine learning. Addison-Wesley (1989)
18. Cuevas, E., Zaldivar, D., Pérez-Cisneros, M.: Seeking multi-thresholds for image segmentation with learning automata. Mach. Vis. Appl. **22**, 805–818 (2011)
19. Ramírez-Ortegón, M.A., Tapia, E., Ramírez-Ramírez, L.L., Rojas, R., Cuevas, E.: Transition pixel: a concept for binarization based on edge detection and gray-intensity histograms. Pattern Recogn. **43**(4), 1233–1243 (2010)
20. Ibrahim, R.A., Elaziz, M.A., Oliva, D., Cuevas, E., Lu, S.: An opposition-based social spider optimization for feature selection. Soft. Comput. **23**, 13547–13567 (2019)
21. Cuevas, E., González, A., Fausto, F., Zaldívar, D., Pérez-Cisneros, M.: Multithreshold segmentation by using an algorithm based on the behavior of locust swarms. Math. Probl. Eng. (2015)
22. Zaldivar, D., Morales, B., Rodríguez, A., Valdivia-G, A., Cuevas, E., Pérez-Cisneros, M.: A novel bio-inspired optimization model based on Yellow Saddle Goatfish behavior. Biosystems **174**, 1–21 (2018)
23. Hinojosa, S., Dhal, K.G., Abd Elaziz, M., Oliva, D., Cuevas, E.: Entropy-based imagery segmentation for breast histology using the stochastic fractal search. Neurocomputing **321**, 201–215 (2018)
24. Cuevas, E., Wario, F., Osuna-Enciso, V., Zaldivar, D., Pérez-Cisneros, M.: Fast algorithm for multiple-circle detection on images using learning automata. IET Image Proc. **6**(8), 1124–1135 (2012)
25. Cuevas, E., Gálvez, J., Hinojosa, S., Avalos, O., Zaldívar, D., Pérez-Cisneros, M.: A comparison of evolutionary computation techniques for IIR model identification. J. Appl. Math. (2014)
26. Bandyopadhyay, R., Basu, A., Cuevas, E., Sarkar, R.: Harris Hawks optimisation with simulated annealing as a deep feature selection method for screening of COVID-19 CT-scans. Appl. Soft Comput. **111**, 107698 (2021)

Chapter 2
Exploring the Potential of Agent Systems for Metaheuristics

Agent-based modeling is a way of simulating complex systems by creating individual agents that follow simple rules. The interactions between these agents can lead to complex global patterns that were not explicitly programmed. In recent years, there has been a growing interest in using agent-based modeling for optimization problems. This is because agent-based systems can be used to generate very complex search behaviors, which can be used to find optimal solutions to difficult problems. This chapter presents a new metaheuristic algorithm based on agent-based modeling. The presented algorithm is called "Heroes and Cowards," and it involves a small set of rules that can be used to generate two emergent global patterns: exploration and exploitation. This algorithm was evaluated on a set of representative benchmark functions, and it was found to be competitive with other metaheuristic algorithms. These results suggest that agent-based modeling can be a promising approach for optimization problems.

2.1 Introduction

As real-world processes become more interconnected, simple models are no longer sufficient for analysis. The availability of fast computing resources has enabled the development and analysis of more complex models. As a result, a new field of knowledge known as complex systems has emerged. Complex systems research focuses on how systems affect individual behavior, particularly when individuals have the ability to impact those systems. Complex behaviors of higher-level organizations emerge as a result of the collective interaction of individuals who participate in a self-organizing process. Complex systems [1] are systems made up of a large number of interconnected parts that exhibit emergent behavior. Emergent behavior is behavior that arises from the interactions of the parts of a system and cannot be predicted from the behavior of the individual parts. Self-organization [2] is a process by which a

E. Cuevas et al., *New Metaheuristic Schemes: Mechanisms and Applications*, Intelligent Systems Reference Library 246, https://doi.org/10.1007/978-3-031-45561-2_2

system spontaneously organizes itself into a complex structure without any external guidance.

Agent-based modeling (ABM) is a new paradigm in artificial intelligence that uses agents to model complex systems. Agents are autonomous entities that exhibit simple behaviors, and their interactions with each other can give rise to complex global patterns. ABM has been used to model a wide variety of systems, including epidemics, supply chains, the stock market, the immune system, ancient civilizations, consumer purchasing behavior, fire spreading, and segregation phenomena. In ABM, the system is modeled from the individual point of view, but its main properties are visualized from a global perspective. This allows ABM to capture the emergent behavior of complex systems that would be difficult or impossible to model using other methods.

The following are some examples of how ABM has been used:

- To predict the spread of epidemics, such as COVID-19 [4].
- To understand the behavior of supply chains, such as the distribution of food and goods [5].
- To analyze the stock market and predict future trends [6].
- To study the immune system and how it fights disease [7].
- To investigate the collapse of ancient civilizations [8].
- To model consumer purchasing behavior and predict future trends [9].
- To simulate fire spreading and predict how fires will behave [1].
- To study segregation phenomena, such as the separation of different groups of people in a community [10, 11].

ABM is a powerful tool for modeling complex systems, and it has been used to achieve significant results in a wide variety of fields. As ABM continues to develop, it is likely to become an even more valuable tool for understanding and predicting the behavior of complex systems.

In the complex systems community, there is a model known as "Heroes and Cowards" [1, 12–14] that illustrates how simple rules can produce complex collective behaviors. The model produces complex global patterns of concentration and distribution through the interaction of agents that follow simple behavioral rules. In "Heroes and Cowards," each agent selects another agent as its "friend" and another as its "enemy." The model consists of two phases. In the first phase, every agent behaves as "coward." Under this condition, the agent moves so that the friend is always located between it and its enemy (in a cowardly manner, hiding from its enemy behind its friend). During this phase, agents are distributed within the space as a consequence of the escape process from the enemy. In the second phase, each agent behaves as "hero." Therefore, the agent presents a strategy where it is moved in a position between its friend and enemy (protecting the friend from the enemy in a heroic manner). During this stage, agents concentrate around positions marked by the agent distributions.

Metaheuristic schemes are abstract optimization models that emulate biological or social systems. They have been used to solve complex problems where traditional optimization methods are not effective. In metaheuristic methods, a population of candidate solutions is generated randomly. Each solution is evaluated using an objective function. The solutions are then modified according to some rules that define their behavior within the search space. These rules are based on abstract models of natural or social mechanisms, such as [15]. The process is repeated until a stop criterion is reached. Metaheuristic methods have two important characteristics that distinguish them from other search methods. First, they involve a population of candidate solutions, in contrast to other methods that consider a single solution. Second, metaheuristic methods promote adaptation and self-organization through collective cooperation among the individuals in the population. In recent years, several optimization algorithms based on metaheuristic principles have been agent-based. Some of the most popular metaheuristic schemes include:

- Genetic algorithm (GA) [17]
- Artificial bee colony (ABC) [18]
- Differential evolution (DE) [19]
- Differential search (DS) [20]
- Multi-verse optimizer (MVO) [21]
- Sine cosine algorithm (SCA) [22]
- Self-adaptive differential evolution (JADE) [23]
- Covariance matrix adaptation evolution strategies (CMA-ES) [24]
- Moth-flame optimization (MFO) [25].

Agent-based approaches use a set of autonomous agents that interact with an environment to achieve a collective goal. The main characteristic of agent-based algorithms is the continuous interaction among agents and between them and the environment. This interaction process defines the results produced by the agent system. Metaheuristic methods also maintain several properties similar to agent-based systems, such as self-organization, a population of individuals, and adaptation. In metaheuristic approaches, the elements operate collectively to reach a common global objective. The environment with which individuals interact is represented by an objective function and its constraints. According to the literature, there are four ways in which agent-based systems and metaheuristic schemes can be related:

- Assigning to each agent one subtask of the optimization problem to solve.
- Assigning to each agent a candidate solution.
- Assigning to each agent a process of the metaheuristic scheme.
- Assigning a complete metaheuristic scheme to each agent.

One way to associate metaheuristic methods with agent-based systems is to divide the optimization task into different subtasks. Each agent is then assigned to one subtask, and is responsible for solving that subtask under its local perspective. The global solution is then obtained by integrating the results of all the agents.

Some examples of this approach include the following:

- In [29], the decision variables are divided into groups, and each variable group is associated with an agent. Each agent then searches for the optimal value of the variables in its group.

- In [30, 31], the problem is decomposed into a set of subcomponents, and each subcomponent is evolved by an agent. The Differential Evolution (DE) algorithm is then used to obtain the optimal value from the results obtained for all agents.

The most representative association between metaheuristic methods and agent-based systems is when an agent represents a candidate solution. In this view, the population of elements can be seen as a population of agents. Each agent can exchange information or cooperate with other agents through specific operators defined by each metaheuristic method. Each agent aims to increase its chances of finding the optimal solution to an optimization problem by competing or cooperating with other agents in its population.

Several metaheuristic operators can be applied to an agent to achieve cooperation, competition, or self-organization. Cooperation is promoted through information exchange schemes, such as the crossover operator or an attraction mechanism. Competition is generally achieved through a selection process that identifies if an agent or individual has a better quality than others in terms of its ability to solve the optimization problem. Self-organization can be produced through metaheuristic operations that induce movement among agents. Such movements can be explorative, such as long movements (a reset mechanism or strong perturbations), or exploitative, such as a local search.

It is important to note that there is no strict rule to use a specific operation to generate new solutions (or agents) in metaheuristics. In general, any computational strategy can be used to produce new solutions that, considering the information provided by existing solutions, increase the probability of finding the global optimal position. Some illustrative examples of this category are several metaheuristic methods such as:

- Particle Swarm Optimization (PSO) algorithm [36]
- Ant Colony Optimization (ACO) [37]
- Social Spider Optimization (SSO) [38]
- Crow Search Algorithm (CSA) [39, 40]
- Gray Wolf Optimizer (GWO) [41]
- Bat Algorithm (BA) [42]
- Cuckoo Search (CS) [43]

In some cases, metaheuristic methods and agent-based systems can be combined by considering the individual processes within a metaheuristic scheme as agents themselves. These processes may encompass operations or search strategies that form an integral part of the metaheuristic approach [44]. When adopting this association between metaheuristics and agent-based systems, the algorithm's functioning can be interpreted as the cooperative behavior of different agents, whose interactions generate the overall search strategy [45]. Among the notable approaches in this category, the evolutionary multi-agent system (EMAS) stands out [46–48]. In this scheme, evolutionary processes are implemented as agents operating at the population level. The performance evaluation of these agents is carried out through a selection mechanism, aiming to promote the population that best aligns with the global objective of the system. Other noteworthy examples of this combination can be found in the algorithms discussed in [49] and [35].

In another association, each individual metaheuristic method is treated as an autonomous agent. The objective of this approach is to leverage agent-based concepts to enhance the solutions generated by these metaheuristic schemes. In this framework, each agent (or metaheuristic scheme) carries out its specific search process while cooperatively utilizing the solutions produced by other agents to improve its own outcome. As a result, these metaheuristic algorithms (or agents) interact with each other to achieve a global objective, which represents the identification of the optimal value for the given optimization problem. An example of this approach can be found in the work presented in [50], where multiple metaheuristic algorithms such as genetic algorithm (GA), iterated local search (ILS), and variable neighborhood search (VNS) are treated as agents. These algorithms exchange information derived from the search space to refine the overall solution. Other similar approaches, employing different metaheuristic methods, are also documented in [51, 52].

Developing a metaheuristic approach is a complex endeavor, and its performance is primarily influenced by the interactions among its individuals, rather than other factors [53]. Interactions refer to mathematical processes that govern the nature and intensity of communication among individuals. While various metaheuristic schemes simulate interactions based on diverse processes or systems, the employed operators exhibit substantial similarities [15]. Many metaheuristic methods are designed with the idea of assembling a repertoire of proven models from previous approaches to generate new optimization schemes. These common mathematical processes, however, are typically devised without considering the ultimate global outcome resulting from individual interactions. On the other hand, agent-based systems offer a robust theoretical framework and a consistent set of models for characterizing the global behavioral patterns that arise from the collective interaction of individuals governed by simple rules. From this perspective, numerous agent-based concepts and models that produce effective search behaviors can be harnessed to create or enhance efficient optimization algorithms.

A new metaheuristic algorithm based on agent systems principles is presented in this chapter which has two objectives:

- To demonstrate the efficacy of agent-based models as metaheuristic methods.
- To show the promising potential in the combination of both artificial intelligence paradigms.

The agent-based model of "Heroes and Cowards" is implemented as a metaheuristic method to show the capabilities of this association. Under this scheme, candidate solutions from the metaheuristic approach are considered agents while their interactions are characterized following the behaviors of exploration (distribution) and exploitation (concentration) included in the "Heroes and Cowards" model. To evaluate the performance of the algorithm, it has been tested in a set of 23 benchmark functions, including multimodal, unimodal, and hybrid benchmark functions. The competitive results indicate that even though agent-based modeling and metaheuristic schemes refer to distinct scientific communities, metaheuristic methods can increase their capabilities through the incorporation of concepts, formalisms, and models extracted from agent-based techniques.

The structure of this chapter is as follows: Sect. 2.2 provides an introduction to the fundamental principles of agent-based modeling. Section 2.3 examines the Heroes and Cowards model in detail. Section 2.4 presents the agent-based metaheuristic method. Section 2.5 presents the experimental results and a comparative analysis. Lastly, Sect. 2.6 concludes the chapter by summarizing the key findings and drawing overall conclusions.

2.2 Models Through Agent-Based Elements

Agent-based models [3] are computational frameworks utilized to elucidate the behavior of complex systems. These models simulate the actions of individual elements within the system and take into account how these entities influence and are influenced by their environment. Agent-based models are particularly suitable when the interactions among the elements play a significant role in the outcomes.

Agents are artificial entities programmed to perform predetermined actions [54]. While agents carry out their operations based on their individual behaviors, they also engage in cooperation and competition with other elements. The structure of an agent's operations can range from simple decisions (such as yes or no actions) to spatial movements.

Traditional modeling approaches focus solely on aggregate elements, neglecting their interactions. Consequently, they are ill-suited to represent the behavior of complex systems. Even models with elementary elements and basic interactions can produce behaviors that cannot be generated or analyzed from the perspective of classical models that lack interactions [55]. In contrast, agent-based models explicitly incorporate individual interactions, capturing their effects on the system's outcomes.

Many agent-based models do not rely on complex behavioral models or sophisticated architectures. Despite their simplicity, these models can generate a wide range of complex behavioral patterns as a result of interactions among the set of simple agents [56]. Global behavioral patterns manifest as identifiable distributions that represent macroscopic spatial regularities. In agent-based models, virtual elements make decisions based on programmed rules [57]. These rules characterize the behavior of each individual in an abstract manner. Various types of information can be incorporated into the rules, including qualitative information, quantitative data, and expert opinions. When constructing rules, the aim is to strike a balance between accuracy and simplicity. The rules should be simple enough to capture the main theoretical elements of the system [58]. While the focus is on keeping the rules as simple as possible, it is also crucial to ensure they achieve the required level of accuracy. However, excessive detail can be counterproductive, as it hinders the observation of the relationship between the agent and its corresponding behavior.

A generic approach based on agents typically involves the following stages. Initially, a group of A agents, denoted as $\{\mathbf{a}_1, \ldots, \mathbf{a}_A\}$, is initialized. At this stage, each agent is assigned a specific state or location. Subsequently, either randomly or following a specific order, each element $\mathbf{a}_i$ (where i ranges from 1 to A) is selected. This particular agent, $\mathbf{a}_i$, is then subjected to a set of rules that modify its location, state, and relationships with other agents. These rules encompass a collection of conditions and relations determined by local influences (such as neighboring elements) or interactions with other agents. This process continues until a predefined stop criterion is met.

2.3 Example of Agent-Based Model "Heroes and Cowards"

Within the complex system field, there exists an agent-based model referred to as "Heroes and Cowards" [1, 12–14]. This model serves as an illustration of how simple rules can give rise to intricate collective behaviors that are challenging to replicate using traditional modeling approaches. Through the interactions among agents following basic behavioral rules, the model generates complex global patterns of concentration and distribution.

In the Heroes and Cowards model, a set of A agents, denoted as $\{\mathbf{a}_1, \ldots, \mathbf{a}_A\}$, is initialized with random positions in a two-dimensional space. Each agent $\mathbf{a}_i$ selects another agent $\mathbf{a}_p$ as its "friend"$\mathbf{f}_i$, and another agent $\mathbf{a}_q$ as its "enemy"$\mathbf{e}_i$, where i, p, q range from 1 to A and i is not equal to p or q. The association between $\mathbf{a}_i$ and its selected friend $\mathbf{f}_i$, as well as its chosen enemy $\mathbf{e}_i$, remains constant throughout the entire simulation. The model comprises two stages. In the coward stage, each agent behaves as a "coward." In this condition, the agent $\mathbf{a}_i$ moves in a way that keeps its friend $\mathbf{f}_i$ positioned between itself and its enemy $\mathbf{e}_i$, adopting a cowardly strategy of hiding from its enemy behind its friend. The rule governing this behavior is formulated as follows:

$$\mathbf{a}_i(t+1) = (1-\beta)\cdot \mathbf{a}_i(t) + \beta\left(\mathbf{f}_i + \left(\frac{\mathbf{f}_i - \mathbf{e}_i}{2}\right)\right) \tag{2.1}$$

At the current iteration denoted by t, the displacement of agent $\mathbf{a}_i$ is determined by a factor β ($\beta \in [0, 1]$), which controls the velocity of movement. This behavior is visualized in Fig. 2.1a. During the heroical phase, each agent assumes the role of a "hero." Consequently, the agent adopts a strategy where it positions itself between its friend $\mathbf{f}_i$ and enemy $\mathbf{e}_i$, acting heroically to protect the friend from the enemy. The rule that governs this interaction is expressed as follows:

$$\mathbf{a}_i(t+1) = (1-\beta)\cdot \mathbf{a}_i(t) + \beta\left(\frac{\mathbf{f}_i - \mathbf{e}_i}{2}\right) \tag{2.2}$$

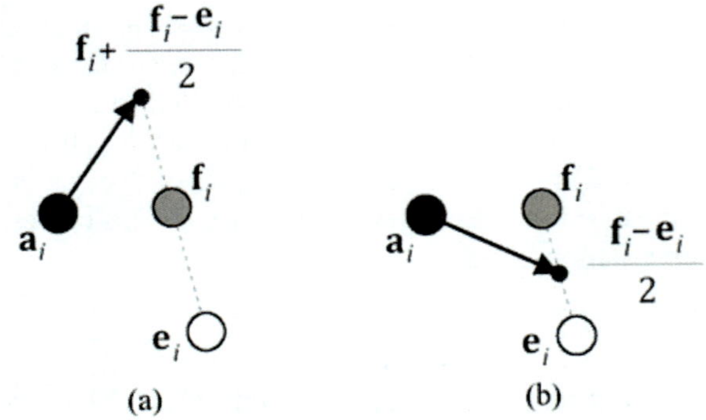

Fig. 2.1 Impact of the rules determining **a** cowardly behavior and **b** heroic behavior

This behavior is depicted in Fig. 2.1b, showcasing the transition between the coward and hero phases. In the Heroes and Cowards model, an artificial moderator determines when the agents should behave as cowards (according to Eq. (2.1)) or heroes (according to Eq. (2.2)) based on a specific number of iterations, denoted as N_t. Consequently, the model alternates between these two phases, with N_t iterations dedicated to the coward phase followed by the subsequent N_t iterations allocated to the heroical phase. This iterative process continues until a predetermined number of phases have been completed.

Figure 2.2 provides illustrations of the resulting global patterns achieved through the operation of the Heroes and Cowards model. The simulation involves a group of 50 agents ($A = 50$). In Fig. 2.2b, the model's coward phase is demonstrated, showcasing the final pattern obtained after 100 iterations using the initial distribution depicted in Fig. 2.2a. Initially concentrated in the center, the agents disperse themselves throughout the space. Figure 2.2d displays the behavioral pattern attained during the heroic phase after 100 iterations, starting with the distribution shown in Fig. 2.2c. Upon examination, it becomes evident that agents initially distributed in the space form semi-concentrations within certain regions of the two-dimensional space.

2.4 Metaheuristic Methods as Agent-Based Models

Although various metaheuristic schemes imitate diverse processes or systems, the operators employed to model individual behavior share similarities. Many metaheuristic methods are designed by reusing a set of rules that have proven successful in previous approaches, without considering the ultimate global pattern resulting from individual interactions.

In contrast, agent-based modeling aims to establish a connection between the collective interaction of individuals and the global behavioral patterns they generate based on the set of behavior-defining rules. This perspective allows for the utilization

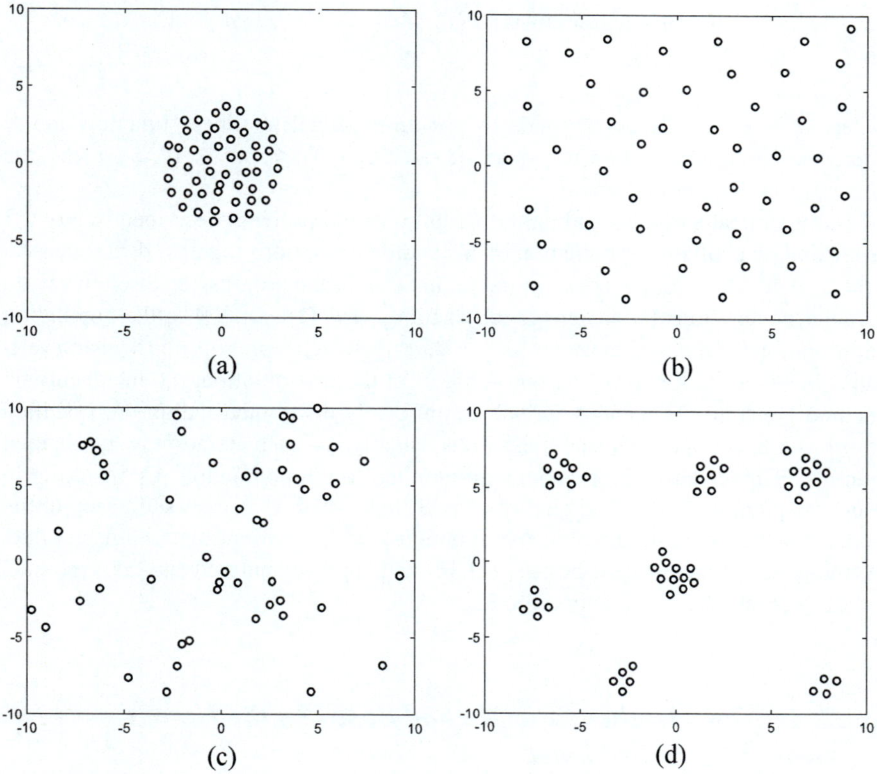

Fig. 2.2 Illustrative instances of the global patterns generated by the Heroes and Cowards model. This includes **a** the initial distribution, **b** the resulting global pattern during the coward phase using the initial configuration depicted in (**a**), **c** the initial distribution, and **d** the final pattern generated by the model during the heroic phase when starting with the agents shown in (**c**)

of various agent-based modeling techniques that produce intricate global search behaviors to enhance and create efficient optimization algorithms.

This chapter explores the relationship between metaheuristic schemes and agent-based modeling, emphasizing the potential of this association. To demonstrate this synergy, the "Heroes and Cowards" agent-based model is implemented as a metaheuristic method. The section is divided into three parts: (4.1) Problem formulation, (4.2) Description of the "Heroes and Cowards" agent-based model as a metaheuristic method, and (4.3) Computational procedure.

2.4.1 Problem Setting

An optimization technique is developed to discover a global solution for a nonlinear problem with box constraints using the formulation provided in [47].

$$\begin{array}{lll} Maximize/\text{minimize} & J(x)x = (x_1, \ldots, x_d) \in \mathbb{R}^d & \\ subject\ to & x \in X & (2.3) \end{array}$$

where $J : \mathbb{R}^d \rightarrow \mathbb{R}$ corresponds to a d-dimensional nonlinear function and $\mathbf{X}$ represents a constrained search space $\left(\mathbf{x} \in \mathbb{R}^d | l_i \leq x_i \leq u_i, i = 1, \ldots, d\right)$ by the lower (l_i) and upper (u_i) bounds.

From a metaheuristic standpoint, the optimization problem described by Eq. 2.3 is tackled by evolving a population of N candidate solutions (agents) denoted as $\mathbf{A}^k$ $\left(\{\mathbf{a}_1^k, \ldots, \mathbf{a}_N^k\}\right)$, starting from an initial point ($k = 1$) and progressing through a fixed number of iterations ($k = gen$). Each agent $\mathbf{a}_i^k$ ($i \in [1, \ldots, N]$) in the population represents a d-dimensional vector $\{a_{i,1}^k, \ldots, a_{i,d}^k\}$, which represents the decision variables involved in the optimization problem. In the first iteration, the metaheuristic method generates N agents with values uniformly distributed within the specified lower (l_i) and upper $((u_i)$ bounds. Subsequently, in each iteration, a predefined number of metaheuristic operations are applied to the population $\mathbf{A}^k$, resulting in a new population $\mathbf{A}^{k+1}$. The quality of each individual $\mathbf{a}_i^k$ is assessed based on its solution with respect to the objective function $J\left(\mathbf{a}_i^k\right)$, representing the fitness value. As the search progresses, the best agent $\mathbf{b}\,\{b_1, \ldots, b_d\}$ encountered thus far is retained as it represents the best solution found.

2.4.2 A New Metaheuristic Method Based on the Heroes and Cowards Model

In this part, we describe the adaptation of the agent-based approach "Heroes and Cowards" to function as a metaheuristic method. The method comprises three key components: Initial phase, Operations, and control stage.

Initial phase

In the first iteration (k=1), the method starts generating a set of A agents $\mathbf{A}^k = \{\mathbf{a}_1^k, \ldots, \mathbf{a}_N^k\}$ with random positions in a d-dimensional space $\left(\mathbf{a}_i^k = \{a_{i,1}^k, \ldots, a_{i,d}^k\}\right)$. In this process, each decision variable $a_{i,j}^k$ ($i \in 1, \ldots, N$; $j = 1, \ldots, d$) that corresponds to the j-th parameter of the i-th agent is set with a numerical value uniformly determined between the defined lower (l_i) and upper (u_i) limits, so that

$$a_{i,j}^k = l_i + \text{rand}(0, 1) \cdot (u_i - l_i) \tag{2.4}$$

Each agent $\mathbf{a}_i^k$ selects another agent $\mathbf{a}_p^k$ as its "friend" $\mathbf{f}_i^k$ and another $\mathbf{a}_q^k$ as its "enemy" $\mathbf{e}_i^k$ where $i, p, q \in (1, \ldots, A)$ and $i \neq p \neq q$. The friend $\mathbf{f}_i^k$ and enemy $\mathbf{e}_i^k$ selected by $\mathbf{a}_i^k$ maintain this association during the complete process.

Operations

The "Heroes and cowards" model consists of two processes: Coward and heroical phases. Under such conditions, the agent-based metaheuristic scheme implements an operator for each phase. Each operator updates the position of an agent $\mathbf{a}_i^k$ in relation to the position of its friend $\mathbf{f}_i^k = \{f_{i,1}^k, \ldots, f_{i,d}^k\}$ and enemy $\mathbf{e}_i^k = \{e_{i,1}^k, \ldots, e_{i,d}^k\}$. Such operations are practically the same as the behavioral rules involved in the original model, with only an adaptation. This modification represents the incorporation of a random number in order to add a stochastic effect in the search strategy. Therefore, the coward operator is defined as follows:

$$a_{i,j}^{k+1} = (1-\beta) \cdot a_{i,j}^k + \beta\left(f_{i,j}^k + \left(\frac{f_{i,j}^k - e_{i,j}^k}{2}\right)\right) + \alpha \cdot \text{rand}(-1, 1) \quad (2.5)$$

where α represents the intensity of the stochastic effect and rand$(-1, 1)$ a function that delivers a random number uniformly distributed between –1 and 1. On the other hand, the heroical operator is modeled as follows:

$$a_{i,j}^{k+1} = (1-\beta) \cdot a_{i,j}^k + \beta \cdot \left(\frac{f_{i,j}^k - e_{i,j}^k}{2}\right) + \alpha \cdot \text{rand}(-1, 1) \quad (2.6)$$

In the adaptation of the heroes and coward model as an optimization scheme, the location of all agents is updated by the coward or heroical operators except for the best element $\mathbf{b}$ of the population $\mathbf{A}^k$. In case of a maximization problem, this agent will be selected in each iteration k so that

$$\mathbf{b} = \underset{\mathbf{a}_i^k \in \mathbf{A}^k}{\arg\max}\, J(\mathbf{a}_i^k) \quad (2.7)$$

Once obtained, this agent $\mathbf{b}$ is not modified by the operators.

Control stage

The position of each agent $\mathbf{a}_i^k$ (except $\mathbf{b}$) is modified iteratively according to one of the operators while the phase of the model has not been changed. Similar to the original model, the phases are performed intercalated. N_t iterations last the coward phase (Eq. (2.5)) while the heroical phase (Eq. (2.6)) considers the next N_t iterations. This process continues until a stop criterion has been reached.

2.4.3 Computational Structure

The adaptation of the model "Heroes and Cowards" has been implemented as an iterative process that involves several steps. Algorithm 2.1 provides a summary of the operations performed throughout the process. The approach requires input parameters such as the number of agents A, displacement velocity β, stochastic effect intensity α, the number of iterations for each phase N_t, , and the maximum number of executions gen (line 1). Similar to other metaheuristic methods, the process begins by generating an initial population $\mathbf{A}^1$ consisting of A agents with positions uniformly distributed within specified limits (line 2). The best agent **b** in terms of fitness value is then selected from $\mathbf{A}^1$ (Line 3). The method starts with N_t iterations of the coward phase (line 4), during which each agent $\mathbf{a}_i^k$, excluding the best agent **b**, is modified using the operator defined in Eq. (2.5) (line 7). Once N_t iterations are completed, a phase change occurs (line 14) to enter the heroical phase. In this phase, each agent $\mathbf{a}_i^k$, excluding the best agent **b**, is modified using the operator defined in Eq. (2.6) (line 10). After each operator application, the best agent **b** in terms of fitness value is selected from $\mathbf{A}^k$ (Line 12). This process continues until the maximum number of generations gen is reached. The final output of the algorithm (line 19) is the last obtained **b**, which represents the final solution.

Algorithm 1. Pseudo-code of "Heroes and cowards"

1 **Input:** A, β, α, N_t, gen
2 $\mathbf{A}^1 \leftarrow$ **Initialize**(A);
3 $\mathbf{b} \leftarrow$ **SelectionBestAgent**($\mathbf{A}^1$);
4 Phase = Coward; m=1;
5 **while** k<=gen **do**
6 **If** (Phase = = Coward)
7 $\mathbf{A}^{k+1} \leftarrow$ **CowardOperation**($\forall \mathbf{a}_i^k \in \mathbf{A}^k$ where $\mathbf{a}_i^k \neq \mathbf{b}$);
8 **end if**
9 **If** (Phase = = Heroical)
10 $\mathbf{A}^{k+1} \leftarrow$ **HeroicalOperation**($\forall \mathbf{a}_i^k \in \mathbf{A}^k$ where $\mathbf{a}_i^k \neq \mathbf{b}$);
11 **end if**
12 $\mathbf{b} \leftarrow$ **SelectionBestAgent**($\mathbf{A}^{k+1}$);
13 m=m+1; $k = k + 1$;
14 **If** (m= = N_t)
15 Phase $\leftarrow$ **ChangeOfPhase**;
16 m=1;
17 **end if**
18 end **while**
19 **Output: b**

Algorithm 2.1 Pseudocode of the "Heroes and cowards" model

To demonstrate the functioning of the "Heroes and Cowards" model, an optimization example is performed. The objective of this example is to identify the maximum value of a two-dimensional objective function, $J(x_1, x_2)$, as defined in Eq. (2.8).

$$J(x_1, x_2) = 3(1 - x_1)^2 e^{-(x_1^2 - x_2^2)} - 10\left(\frac{x_1}{5} - x_1^3 - x_2^5\right)e^{(-x_1^2 - x_2^2)} - 1/3e^{(-(x_1+1)^2 - x_2^2)} \tag{2.8}$$

In the given example, the algorithm is configured with the following parameter values: $A = 20, \beta = 0.7, \alpha = 0.3, N_t = 20, gen = 100$. Using these parameters, the initial agents $\{\mathbf{a}_i^1 = (a_{i,1}^1, a_{i,2}^1)\}$ in $\mathbf{A}^1$ are randomly initialized with values uniformly distributed within the interval of $3 \leq x_1, x_2 \leq 3$. The coward phase of the search strategy encourages the dispersion and exploration of solutions across the search space. Conversely, in the heroical phase, solutions tend to semi-concentrate through attraction movements. As the iterations progress, the exploitation is intensified in the heroical phase to refine the quality of the solutions.

The behavior of the "Heroes and Cowards" model is depicted in Fig. 2.3. Figure 2.3a presents the objective function $J(x_1, x_2)$ to optimize, along with its contour representation. Throughout the algorithm's evolution, five specific points are highlighted to showcase its operation: (b) after 1 iteration (initialization), (c) after 20 iterations (first coward phase), (d) after 40 iterations (first heroical phase), (e) after 60 iterations (second coward phase), and (f) after 80 iterations (second heroical phase). Point (b) represents an early stage where the elements are still randomly distributed. Figures 2.3c-e demonstrate the gradual formation of groups as the iterations progress, leading to the convergence of all elements in Fig. 2.3f towards the global and local maxima.

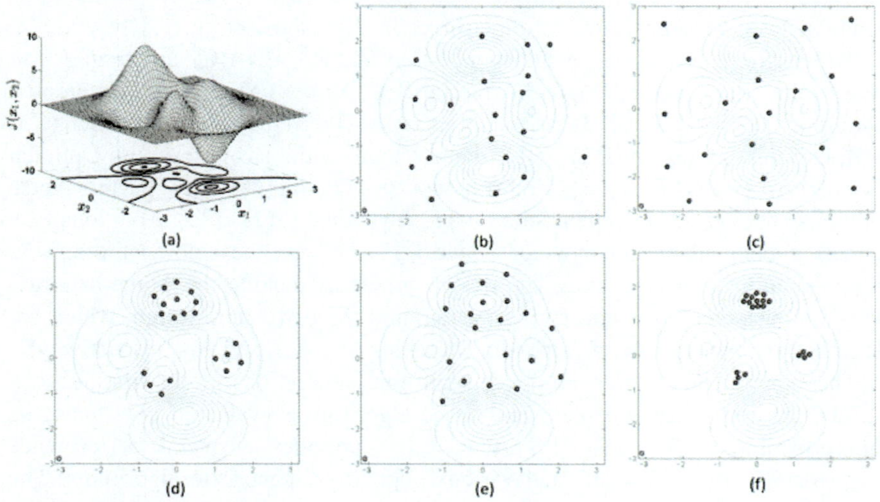

Fig. 2.3 "Heroes and cowards" model behavior considering 100 iterations when it solves the optimization problem formulated by Eq. (2.8). **a** objective function with its respective contour, **b** 1 iteration (initialization), **c** 20 iterations (first coward phase), **d** 40 iterations (First heroical phase), **e** 60 iterations (second coward phase) and **f** 80 iterations (second heroical phase)

2.5 Experiments

Metaheuristic optimization techniques have emerged as stochastic algorithms for solving optimization problems that are challenging for classical methodologies due to the presence of multiple optima. To assess the performance of these metaheuristic schemes, researchers have introduced a standardized set of 23 mathematical benchmark functions [59, 60]. These benchmark functions encompass various levels of complexity, allowing for the evaluation of precision, robustness, and scalability of the metaheuristic methodologies. Furthermore, three engineering design problems commonly encountered in the literature [39, 61] are utilized to evaluate the agent-based method's effectiveness in solving real-world optimization scenarios. Throughout the optimization process, the algorithms are assessed using the maximum number of generations (gen) as a termination criterion, which has been widely employed in the field of metaheuristic optimization.

In this section, we present the numerical results of the agent-based "Heroes and cowards" model, referred to as EA-HC for the sake of identification. As a metaheuristic algorithm, EA-HC is evaluated using a comprehensive set of benchmark functions and engineering design optimization problems. Appendix A, Table A.1 provides a mathematical description of the test functions employed in the performance analysis. The table includes the dimensionality of the vector (n) at which the test functions are evaluated, the optimal value ($f(x^*)$) obtained at the position x^*, and the search space (S) defined by the lower and upper limits. To demonstrate the scalability of the agent-based method, the evaluation for each test function is conducted using search spaces of dimensions 30, 100, and 200.

The performance results of the agent-based method, EA-HC, are compared to nine evolutionary methodologies: Artificial Bee Colony (ABC) [18], Differential Evolution (DE) [19], Particle Swarm Optimization (PSO) [36], Cuckoo Search (CS) [43], Differential Search (DS) [20], Adaptive Differential Evolution With Optional External Archive (JADE) [23], Moth-Flame Optimization (MFO) [25], Multi-Verse Optimizer (MVO) [21], and the Sine Cosine Algorithm (SCA) [22]. Additionally, a numerical comparison is conducted between EA-HC and these competitors using design optimization problems. The design problems include the Three-bar truss design, Tension/compression spring design, and Welded beam design, which are described mathematically in Appendix B (Tables A.2, A.3, and A.4, respectively).

The experimental results are divided into four subsections. In the first Subsect. (2.5.1), the performance of the agent-based algorithm is evaluated in terms of its tuning parameters. The second Subsect. (2.5.2) provides an overall performance comparison between EA-HC and various popular metaheuristic algorithms. The third Subsect. (2.5.3) presents the convergence analysis for each test function using different metaheuristic approaches. Finally, the fourth Subsect. (2.5.4) analyzes the ability of the "heroes and cowards" model to solve engineering problems.

2.5.1 Evaluation of the Performance Considering Its Own Parameters

The influence of the parameters β and α on the performance of the EA-HC scheme is crucial. This subsection analyzes the behavior of the agent-based scheme by examining different parameter configurations. In the test, one parameter is varied while the other parameter is held at a default value. To minimize the stochastic effect, each benchmark function is independently executed 30 times. The termination criterion is set as the maximum number of iterations, which is 1000. Throughout all simulations, the population size is fixed at 50 individuals.

The analysis begins by studying the behavior of the agent-based algorithm with varying values of β while keeping α fixed at 0.3. The values of β range from 0 to 1, and each benchmark function is executed 30 times for each β value. Next, the performance of the agent-based algorithm is evaluated by considering different values of α while keeping β fixed at 0.7. The values of α vary from 0.0 to 0.5, and the experiment is conducted independently for each benchmark function. The results indicate that selecting appropriate parameter values can enhance the performance of the agent-based method and improve solution quality. Based on the experiments, it can be concluded that the optimal parameter set consists of $\beta = 0.7$ and $\alpha = 0.3$. These parameter values are retained for the subsequent experiments.

Please note that the mentioned termination criterion, parameter values, and experimental setup are specific to the context of the study.

2.5.2 Comparison with Other Similar Methods

In this section, the performance of EA-HC is evaluated and numerically compared to nine well-known evolutionary approaches using a set of 23 test functions. The selected test functions encompass uni-modal, multi-modal, and hybrid benchmark functions. To ensure a fair comparison, the evolutionary process of each algorithm is terminated at gen = 1000, which aligns with the termination condition used in related works in the literature. The scalability of the agent-based methods is tested in dimensions $n = 30$, $n = 100$, and $n = 200$, with each experiment conducted for 30 runs. This condition has been chosen to maintain compatibility with the numerical results of most of the related and novel works [56–58] in the literature.

Due to the stochastic nature of metaheuristic methods, statistical validation is employed to eliminate the random effect. The numerical results are validated using the Wilcoxon rank-sum test [62].

The performance of metaheuristic approaches heavily relies on appropriately configuring their parameters to enhance their search capabilities. These configurations are problem-specific, as stated by the No-Free-Lunch (NFL) theorem in optimization theory, which indicates that no single algorithm can solve all optimization problems optimally. Consequently, metaheuristic algorithms often include tuning parameters to increase their efficiency in locating optima.

For the comparison, the parameter configurations for each algorithm are set based on the values reported in Table 2.1, following the guidelines provided by the respective authors. These values represent the configurations where the compared methods achieved their best performance, as stated by the authors themselves. All the algorithms used in the comparisons are obtained from the authors through public repositories.

The experimental study utilized a population size of 50 individuals for each metaheuristic method. The performance results of the numerical comparison are presented in Tables 2.2, 2.3 and 2.4. Table 2.2 displays the results for $n = 30$ dimensions, Table 2.3 for $n = 100$ dimensions, and Table 2.4 for $n = 200$ dimensions. The tables include the best fitness value denoted as f_{Best}, average fitness as $\bar{f}$, standard deviation as σ_f, and worst value as f_{Worst}. The entries with the best performance are highlighted in bold.

According to the results presented in Table 2.2, it is evident that the agent-based EA-HC algorithm surpasses the other metaheuristic methods considered in the comparison study. However, for certain functions such as f_6, f_8, f_9, f_{13}, f_{16}, f_{20}, f_{21}, and f_{22}, EA-HC performs similarly to other evolutionary techniques. In the case of function f_6, ABC, CS, DS, JADE, MFO, and SCA exhibit comparable performance to EA-HC. Similarly, for function f_8, ABC, CS, DS, JADE, MVO, and EA-HC achieve the same best fitness value, although the median fitness value for EA-HC is noticeably higher than the medians of other methods. A similar pattern is observed for function f_9, where ABC, CS, DS, and JADE demonstrate identical fitness values. Furthermore, ABC, DS, JADE, MFO, and SCA achieve the same best fitness value as

Table 2.1 Configuration of parameters for each metaheuristic method employed in the experimental tests

Algorithm	Parameter(s)	References
ABC	The population size has been set as *limit* = 50	[18]
DE	The variant considered is DE/rand/bin, where crossover probability $CR = 0.5$, mutation rate $F = 0.6$ and differential weight $dw = 0.2$	[19]
PSO	The parameters are set to $c_1 = 2$ and $c_2 = 2$ with linearly decreasing weight factor from 0.9 to 0.2	[36]
CS	The discover probability is set to $pa = 0.25$	[43]
DS	The algorithm has been implemented considering the scale parameter $cp = 0.5$ for the Gamma distribution	[20]
JADE	The variant considered is ldquoDE/current-to- p bestrdquo, where crossover probability $cr = 0.5$ and differential weight $dw = 0.6$	[23]
MFO	The source code has been obtained by its reference	[25]
MVO	The source code has been obtained by its reference	[21]
SCA	The implementation considers the guidelines described by the author	[22]
EA-HC	$\beta = 0.7$, $\alpha = 0.3$ and N_t=50	

EA-HC for function f_{13}. One notable characteristic of EA-HC is its ability to dynamically switch between exploration and exploitation stages throughout the optimization process. As a result, the mean fitness value is influenced by this adaptive mechanism. For the remaining functions, ABC and DS directly compete with EA-HC in function f_{16}. Similarly, for functions f_{20} and f_{22}, ABC, DS, SCA, and EA-HC exhibit comparable performance. Finally, in function f_{22}, only the JADE method attains the same results as the agent-based EA-HC algorithm.

Based on the numerical results presented in Table 2.2, it is evident that EA-HC consistently outperforms the compared metaheuristic approaches for the majority of benchmark functions. Although there are a few instances where the agent-based method achieves similar results to some metaheuristic algorithms, the overall performance of EA-HC is remarkable. This superior performance can be attributed to its unique capability of dynamically switching between exploration and exploitation stages throughout the entire optimization process. Typically, metaheuristic optimization algorithms follow a traditional approach where they start with exploration in the search space and then transition to exploitation in the final stages. In contrast, the agent-based EA-HC incorporates a changing frequency that enables both exploration and exploitation throughout the entire process. This dynamic mechanism enhances the search strategy towards the global optimum by flexibly adapting the metaheuristic stages. As a result, the agent-based method consistently achieves the best results for most benchmark functions in the 30-dimensional search space.

To evaluate the scalability of EA-HC in higher-dimensional search spaces, a comparison study was conducted between EA-HC and the other tested metaheuristic methods using a 100-dimensional search space. The performance results of this experiment are presented in Table 2.3. According to the table, it can be concluded that the metaheuristic operators of the agent-based EA-HC method outperform its competitors when evaluating at a higher-dimensional search space. The structure of EA-HC's metaheuristic operators demonstrates a more efficient search strategy compared to the other algorithms, as observed in most benchmark functions listed in Table A.1. However, there are some exceptions, such as functions f_8, f_9 and f_{13}. For functions f_8 and f_9, the JADE algorithm exhibits similar performance to the agent-based EA-HC method. Similarly, only the ABC method achieves a similar objective function value as EA-HC for function f_{13}. When considering the mean and standard deviation metrics, the JADE and ABC methods produce more consistent results compared to EA-HC. This may suggest that EA-HC produces less consistent results. However, this phenomenon is mainly attributed to the changing behavior of EA-HC, which incorporates exploration and exploitation operators throughout the entire optimization process, in contrast to most other metaheuristic methodologies that have fixed exploration and exploitation stages. The key advantage of EA-HC over the tested algorithms lies in its ability to achieve a balanced combination of evolutionary stages throughout the optimization process. EA-HC employs a changing scheme between exploration and exploitation every 50 iterations, which helps maintain population diversity and leads to superior performance.

Table 2.2 Results of minimization for benchmark functions from Table A.1 with $n = 30$

		ABC	DE	PSO	CS	DS	JADE	MFO	MVO	SCA	EA-HC
f_1	f_{Best}	1.05E − 12	7.28E + 03	8.23E + 02	4.78E − 03	1.88E − 06	2.68E − 20	1.94E − 06	1.13E − 01	9.45E − 07	**1.47E − 144**
	f_{Worst}	7.08E − 11	1.74E + 04	3.32E + 03	1.67E − 02	2.40E − 04	5.45E − 06	1.00E + 04	2.92E − 01	9.22E − 02	5.47E − 05
	$\bar{f}$	1.52E − 11	1.23E + 04	1.66E + 03	9.27E − 03	1.99E − 05	4.84E − 07	1.20E + 03	1.84E − 01	6.52E − 03	1.89E − 06
	σ_f	1.72E − 11	2.38E + 03	6.65E + 02	2.85E − 03	4.32E − 05	1.22E − 06	3.28E + 03	4.55E − 02	1.83E − 02	9.98E − 06
f_2	f_{Best}	4.14E − 07	3.99E + 01	2.31E + 01	6.05E − 01	7.66E − 05	2.69E − 05	5.58E − 05	1.98E − 01	2.43E − 09	**3.63E − 77**
	f_{Worst}	2.22E − 06	6.60E + 01	1.03E + 02	5.54E + 00	1.27E − 03	1.66E + 00	8.00E + 01	5.45E − 01	6.03E − 05	4.39E − 02
	$\bar{f}$	1.13E − 06	5.20E + 01	4.16E + 01	1.67E + 00	3.45E − 04	2.37E − 01	2.85E + 01	3.16E − 01	6.33E − 06	1.74E − 03
	σ_f	4.77E − 07	6.63E + 00	1.89E + 01	1.22E + 00	2.78E − 04	5.16E − 01	1.97E + 01	8.69E − 02	1.29E − 05	8.01E − 03
f_3	f_{Best}	6.00E + 03	4.51E + 04	2.13E + 03	2.59E + 02	9.98E + 02	1.21E + 04	2.31E + 02	9.17E + 00	4.36E + 01	**5.88E − 40**
	f_{Worst}	1.56E + 04	7.93E + 04	2.03E + 04	6.77E + 02	4.97E + 03	3.53E + 04	4.00E + 04	3.64E + 01	1.10E + 04	1.44E + 03
	$\bar{f}$	1.14E + 04	5.79E + 04	7.96E + 03	4.37E + 02	2.91E + 03	2.31E + 04	1.56E + 04	1.86E + 01	3.48E + 03	8.31E + 01
	σ_f	2.75E + 03	9.38E + 03	3.64E + 03	9.09E + 01	1.28E + 03	5.87E + 03	1.05E + 04	7.64E + 00	3.42E + 03	3.11E + 02
f_4	f_{Best}	1.83E + 01	5.95E + 01	1.50E + 01	2.21E + 00	2.76E + 00	2.67E − 03	3.25E + 01	3.11E − 01	1.20E + 00	**4.58E − 140**
	f_{Worst}	5.60E + 01	7.65E + 01	3.50E + 01	4.79E + 00	9.96E + 00	5.15E + 01	8.01E + 01	1.14E + 00	2.50E + 01	9.81E − 28
	$\bar{f}$	4.46E + 01	6.86E + 01	2.23E + 01	3.39E + 00	5.67E + 00	6.51E + 00	5.65E + 01	6.28E − 01	1.15E + 01	3.27E − 29
	σ_f	8.00E + 00	3.73E + 00	4.30E + 00	6.72E − 01	1.90E + 00	1.57E + 01	9.87E + 00	2.07E − 01	6.61E + 00	1.79E − 28
f_5	f_{Best}	1.91E + 01	2.75E + 01	1.40E + 02	2.31E + 01	2.24E + 01	1.29E + 01	9.12E + 00	2.34E + 01	2.70E + 01	**5.25E − 03**
	f_{Worst}	2.72E + 01	2.89E + 01	1.05E + 03	2.55E + 01	7.89E + 01	2.63E + 01	9.01E + 04	8.14E + 01	2.87E + 01	2.87E + 01
	$\bar{f}$	2.44E + 01	2.88E + 01	3.68E + 02	2.47E + 01	2.66E + 01	1.97E + 01	1.57E + 04	3.03E + 01	2.79E + 01	2.17E + 01
	σ_f	1.78E + 00	4.66E − 02	1.87E + 02	5.80E − 01	9.92E + 00	5.20E + 00	3.39E + 04	1.34E + 01	4.46E − 01	1.19E + 01

(continued)

Table 2.2 (continued)

		ABC	DE	PSO	CS	DS	JADE	MFO	MVO	SCA	EA-HC
f_6	f_{Best}	**0.00E + 0**	8.45E + 03	7.65E + 02	**0.00E + 0**	**0.00E + 0**	**0.00E + 0**	**0.00E + 0**	1.00E + 00	**0.00E + 0**	**0.00E + 00**
	f_{Worst}	0.00E + 00	1.70E + 04	3.34E + 03	0.00E + 00	0.00E + 00	0.00E + 00	1.00E + 04	1.70E + 01	2.00E + 00	1.00E + 01
	$\bar{f}$	0.00E + 00	1.19E + 04	1.93E + 03	0.00E + 00	0.00E + 00	0.00E + 00	1.60E + 03	6.07E + 00	6.67E − 02	3.33E − 01
	σ_f	0.00E + 00	2.24E + 03	6.78E + 02	0.00E + 00	0.00E + 00	0.00E + 00	3.70E + 03	3.12E + 00	3.65E − 01	1.83E + 00
f_7	f_{Best}	9.31E − 02	3.56E + 00	2.16E + 00	1.59E − 02	1.41E − 02	5.20E − 02	3.01E − 02	6.10E − 03	3.22E − 03	**5.05E − 06**
	f_{Worst}	3.07E − 01	1.57E + 01	1.14E + 01	6.45E − 02	6.47E − 02	2.14E − 01	1.89E + 01	3.71E − 02	1.12E − 01	2.14E − 02
	$\bar{f}$	2.21E − 01	9.27E + 00	5.38E + 00	3.36E − 02	3.77E − 02	1.09E − 01	1.84E + 00	1.38E − 02	2.48E − 02	1.04E − 03
	σ_f	5.28E − 02	2.57E + 00	2.34E + 00	1.10E − 02	1.42E − 02	3.90E − 02	3.91E + 00	6.42E − 03	2.21E − 02	3.89E − 03
f_8	f_{Best}	**2.07E + 0**	5.18E + 00	4.63E + 00	**2.07E + 0**	**2.07E + 0**	**2.07E + 0**	2.08E + 00	**2.07E + 0**	2.23E + 00	**2.07E + 00**
	f_{Worst}	2.07E + 00	9.68E + 00	1.25E + 01	2.07E + 00	2.08E + 00	2.07E + 00	2.13E + 00	2.07E + 00	2.67E + 00	3.25E + 00
	$\bar{f}$	2.07E + 00	7.40E + 00	8.09E + 00	2.07E + 00	2.07E + 00	2.07E + 00	2.10E + 00	2.07E + 00	2.36E + 00	2.12E + 00
	σ_f	2.65E − 08	1.01E + 00	1.93E + 00	4.47E − 06	2.60E − 03	8.50E − 04	1.26E − 02	6.13E − 05	9.78E − 02	2.19E − 01
f_9	f_{Best}	**7.13E − 02**	3.42E + 00	6.05E − 01	**7.13E − 02**	**7.13E − 02**	**7.13E − 02**	8.32E − 02	7.15E − 02	2.16E − 01	**7.13E − 02**
	f_{Worst}	7.13E − 02	6.50E + 00	2.60E + 00	7.13E − 02	7.40E − 02	7.13E − 02	1.33E − 01	7.16E − 02	6.32E − 01	6.51E + 00
	$\bar{f}$	7.13E − 02	4.49E + 00	1.52E + 00	7.13E − 02	7.20E − 02	7.13E − 02	9.99E − 02	7.15E − 02	3.57E − 01	3.83E − 01
	σ_f	1.06E − 10	7.30E − 01	6.29E − 01	6.66E − 07	1.13E − 03	3.86E − 08	1.27E − 02	4.37E − 05	1.12E − 01	1.26E + 00
f_{10}	f_{Best}	8.20E − 09	2.45E + 02	9.56E + 01	6.49E + 01	2.56E + 00	1.27E + 01	6.87E + 01	5.78E + 01	5.95E − 06	**0.00E + 00**
	f_{Worst}	1.02E + 00	3.06E + 02	2.42E + 02	1.08E + 02	1.83E + 01	2.57E + 01	2.42E + 02	2.09E + 02	8.81E + 01	2.53E + 02
	$\bar{f}$	1.67E − 01	2.75E + 02	1.74E + 02	8.52E + 01	8.37E + 00	1.79E + 01	1.43E + 02	1.10E + 02	2.02E + 01	5.41E + 01
	σ_f	3.79E − 01	1.47E + 01	3.12E + 01	1.02E + 01	2.59E + 00	3.16E + 00	4.01E + 01	3.58E + 01	2.75E + 01	9.09E + 01
f_{11}	f_{Best}	2.04E − 06	1.59E + 01	7.60E + 00	2.20E + 00	2.14E − 04	4.28E + 00	5.12E − 04	1.03E − 01	4.16E − 05	**8.88E − 16**
	f_{Worst}	3.54E − 05	1.77E + 01	1.47E + 01	6.92E + 00	4.89E − 03	1.98E + 01	2.00E + 01	2.61E + 00	2.02E + 01	4.67E − 01

(continued)

Table 2.2 (continued)

		ABC	DE	PSO	CS	DS	JADE	MFO	MVO	SCA	EA-HC
	$\bar{f}$	1.55E − 05	1.67E + 01	1.13E + 01	3.91E + 00	1.48E − 03	1.04E + 01	1.39E + 01	9.64E − 01	1.64E + 01	1.56E − 02
	σ_f	8.37E − 06	4.70E − 01	1.66E + 00	1.30E + 00	1.31E − 03	5.79E + 00	8.28E + 00	7.69E − 01	7.50E + 00	8.53E − 02
f_{12}	f_{Best}	5.15E − 11	7.27E + 01	8.75E + 00	4.79E − 02	6.01E − 07	9.61E − 12	5.09E − 06	2.38E − 01	2.26E − 06	**0.00E + 00**
	f_{Worst}	2.80E − 05	1.79E + 02	2.77E + 01	1.81E − 01	3.44E − 02	1.24E − 02	9.09E + 01	6.10E − 01	6.93E − 01	3.15E − 03
	$\bar{f}$	1.20E − 06	1.12E + 02	1.73E + 01	1.02E − 01	7.08E − 03	1.26E − 03	1.45E + 01	4.47E − 01	1.75E − 01	1.05E − 04
	σ_f	5.13E − 06	2.24E + 01	4.68E + 00	2.77E − 02	1.00E − 02	3.14E − 03	3.35E + 01	1.10E − 01	2.00E − 01	5.75E − 04
f_{13}	f_{Best}	**−3.00E + 00**	5.39E − 02	−3.24E − 02	−2.95E + 00	**−3.00E + 00**	**−3.00E + 00**	**−3.00E + 00**	−2.85E + 00	**−3.00E + 00**	**−3.00E + 00**
	f_{Worst}	−3.00E + 00	1.44E + 00	3.92E + 00	−2.42E + 00	−3.00E + 00	−3.00E + 00	−6.00E − 01	−1.08E + 00	−3.00E + 00	−3.00E + 00
	$\bar{f}$	−3.00E + 00	7.96E − 01	1.71E + 00	−2.72E + 00	−3.00E + 00	−3.00E + 00	−2.59E + 00	−2.16E + 00	−3.00E + 00	−3.00E + 00
	σ_f	6.40E − 14	3.31E − 01	1.01E + 00	1.65E − 01	5.82E − 08	1.95E − 05	4.96E − 01	3.98E − 01	2.65E − 05	1.62E−08
f_{14}	f_{Best}	2.75E + 00	3.59E + 01	3.34E + 00	2.75E + 00	2.75E + 00	2.75E + 00	1.97E + 01	2.75E + 00	2.75E + 00	**2.72E + 00**
	f_{Worst}	1.34E − 07	1.27E + 01	8.86E − 01	5.95E − 06	9.84E − 09	2.75E + 00	2.26E + 01	1.05E − 06	2.55E−04	2.75E + 00
	$\bar{f}$	2.75E + 00	1.49E + 01	2.80E + 00	2.75E + 00	2.75E + 00	2.75E + 00	2.75E + 00	2.75E + 00	2.75E + 00	2.72E + 00
	σ_f	2.75E + 00	6.25E + 01	7.68E + 00	2.75E + 00	2.75E + 00	6.79E − 04	9.97E + 01	2.75E + 00	2.75E + 00	5.59E − 03
f_{15}	f_{Best}	9.96E − 16	3.10E + 00	1.13E + 00	1.81E − 06	2.97E − 10	1.44E − 17	7.32E − 10	4.76E − 05	2.93E − 10	**−1.04E + 00**
	f_{Worst}	1.39E − 14	7.40E + 00	3.93E + 00	1.03E − 05	6.53E − 08	6.02E − 09	4.00E + 00	1.69E − 04	3.03E − 06	1.33E − 14
	$\bar{f}$	4.59E − 15	5.05E + 00	2.21E + 00	3.85E − 06	6.86E − 09	3.48E − 10	1.04E + 00	7.72E − 05	3.35E − 07	−3.45E − 02
	σ_f	3.41E − 15	9.97E − 01	7.48E − 01	1.69E − 06	1.22E − 08	1.15E − 09	1.77E + 00	2.54E − 05	7.04E − 07	1.89E − 01

(continued)

Table 2.2 (continued)

		ABC	DE	PSO	CS	DS	JADE	MFO	MVO	SCA	EA-HC
f_{16}	f_{Best}	**−1.25E + 04**	−5.45E + 03	−8.26E + 03	−9.31E + 03	**−1.25E + 04**	−1.06E + 04	−1.05E + 04	−1.00E + 04	−4.83E + 03	**−1.25E + 04**
	f_{Worst}	−1.20E + 04	−3.73E + 03	−4.14E + 03	−8.38E + 03	−1.19E + 04	−8.09E + 03	−7.20E + 03	−5.56E + 03	−3.52E + 03	−3.83E + 03
	$\bar{f}$	−1.22E + 04	−4.40E + 03	−5.81E + 03	−8.65E + 03	−1.23E + 04	−9.44E + 03	−8.68E + 03	−7.92E + 03	−4.07E + 03	−4.95E + 03
	σ_f	1.19E + 02	3.64E + 02	9.44E + 02	2.11E + 02	1.91E + 02	6.93E + 02	7.68E + 02	8.60E + 02	2.85E + 02	2.12E + 03
f_{17}	f_{Best}	8.06E − 14	1.36E + 06	1.31E + 01	5.40E − 01	1.09E − 08	3.53E − 12	4.87E − 07	7.55E − 04	3.03E − 01	**1.57E − 32**
	f_{Worst}	2.12E − 12	4.05E + 07	3.42E + 03	2.18E + 00	1.04E − 01	8.76E − 04	1.98E + 00	4.59E + 00	1.11E + 01	1.12E − 09
	$\bar{f}$	5.04E − 13	1.64E + 07	3.69E + 02	1.15E + 00	3.46E − 03	1.67E − 04	2.93E − 01	9.20E − 01	1.20E + 00	4.17E − 11
	σ_f	3.97E − 13	1.08E + 07	9.05E + 02	4.23E − 01	1.89E − 02	2.26E − 04	5.15E − 01	1.00E + 00	1.94E + 00	2.05E − 10
f_{18}	f_{Best}	5.99E − 13	7.68E + 06	6.98E + 01	5.53E − 02	2.36E − 08	8.89E − 12	2.64E − 05	7.68E − 03	2.10E + 00	**1.35E − 32**
	f_{Worst}	1.33E − 09	1.02E + 08	2.83E + 05	2.95E − 01	9.49E − 05	2.27E − 04	3.60E + 00	1.21E − 01	5.04E + 00	1.56E − 11
	$\bar{f}$	7.02E − 11	4.89E + 07	6.62E + 04	1.59E − 01	6.59E − 06	1.78E − 05	2.03E − 01	3.79E − 02	2.73E + 00	5.20E − 13
	σ_f	2.41E − 10	1.87E + 07	7.98E + 04	6.15E − 02	1.72E − 05	5.30E − 05	6.10E − 01	2.22E − 02	5.80E − 01	2.84E − 12
f_{19}	f_{Best}	2.67E − 07	5.22E + 03	1.21E + 03	6.68E − 01	2.12E − 04	1.63E − 06	4.48E − 04	3.09E − 01	1.09E − 06	**6.49E − 81**
	f_{Worst}	1.70E − 06	1.89E + 04	2.23E + 04	1.51E + 01	2.98E − 03	3.55E + 00	1.00E + 05	7.33E − 01	8.34E − 03	4.34E + 00
	$\bar{f}$	9.72E − 07	1.35E + 04	4.79E + 03	3.64E + 00	9.59E − 04	1.22E − 01	2.53E + 04	5.13E − 01	1.77E − 03	1.45E − 01
	σ_f	4.09E − 07	3.10E + 03	4.07E + 03	3.60E + 00	7.20E − 04	6.48E − 01	2.25E + 04	1.22E − 01	2.11E − 03	7.93E − 01
f_{20}	f_{Best}	**2.90E + 01**	5.15E + 02	2.33E + 02	8.77E + 01	**2.90E + 01**	2.91E + 01	5.30E + 01	9.13E + 01	**2.90E + 01**	**2.90E + 01**
	f_{Worst}	2.90E + 01	7.81E + 02	4.66E + 02	1.23E + 02	4.73E + 01	4.08E + 01	4.01E + 02	1.98E + 02	2.90E + 01	3.04E + 01
	$\bar{f}$	2.90E + 01	6.30E + 02	3.54E + 02	1.09E + 02	3.09E + 01	3.32E + 01	1.29E + 02	1.45E + 02	2.90E + 01	2.90E + 01

(continued)

Table 2.2 (continued)

		ABC	DE	PSO	CS	DS	JADE	MFO	MVO	SCA	EA-HC
	σ_f	1.00E − 05	5.96E + 01	5.38E + 01	8.76E + 00	4.78E + 00	3.04E + 00	8.78E + 01	2.45E + 01	4.52E − 03	2.59E − 01
f_{21}	f_{Best}	2.78E + 02	4.96E + 07	1.12E + 05	6.88E + 01	1.56E + 02	**3.20E + 01**	8.20E + 06	5.13E + 01	3.13E + 02	**3.20E + 01**
	f_{Worst}	3.26E + 01	1.72E + 07	1.64E + 05	8.21E + 00	4.96E + 01	4.78E + 02	5.80E + 07	6.52E + 00	6.28E + 02	3.07E + 02
	$\bar{f}$	1.92E + 02	2.10E + 07	1.84E + 03	5.93E + 01	9.01E + 01	1.81E + 02	4.79E + 01	4.06E + 01	3.20E + 01	4.20E + 01
	σ_f	3.38E + 02	7.77E + 07	6.82E + 05	9.59E + 01	2.63E + 02	1.16E + 02	4.10E + 08	6.80E + 01	3.55E + 03	5.02E + 01
f_{22}	f_{Best}	**2.90E + 01**	5.45E + 02	3.00E + 02	1.27E + 02	**2.90E + 01**	2.92E + 01	4.28E + 01	8.95E + 01	**2.90E + 01**	**2.90E + 01**
	f_{Worst}	2.90E + 01	9.86E + 02	2.97E + 03	1.84E + 02	4.35E + 01	4.22E + 01	2.53E + 03	2.51E + 02	2.90E + 01	2.90E + 01
	$\bar{f}$	2.90E + 01	7.46E + 02	7.10E + 02	1.59E + 02	2.99E + 01	3.38E + 01	9.66E + 02	1.66E + 02	2.90E + 01	2.90E + 01
	σ_f	4.15E − 06	1.07E + 02	5.79E + 02	1.50E + 01	3.56E + 00	2.79E + 00	5.61E + 02	3.47E + 01	3.79E − 03	1.91E − 04
f_{23}	f_{Best}	−8.38E + 01	1.87E + 06	1.40E + 03	−7.29E + 01	−8.31E + 01	−8.34E + 01	−8.27E + 01	−8.17E + 01	−3.23E + 01	**−8.39E + 01**
	f_{Worst}	−8.34E + 01	2.82E + 07	1.56E + 06	−4.76E + 01	−8.19E + 01	−8.18E + 01	1.02E + 09	−1.30E + 01	5.25E + 02	−8.10E + 01
	$\bar{f}$	−8.36E + 01	9.86E + 06	5.81E + 04	−6.34E + 01	−8.26E + 01	−8.25E + 01	1.84E + 08	−7.68E + 01	6.67E + 00	−8.38E + 01
	σ_f	1.17E − 01	6.42E + 06	2.83E + 05	7.02E + 00	3.69E − 01	3.55E − 01	2.74E + 08	1.65E + 01	9.99E + 01	5.39E − 01

Table 2.3 Results of minimization for benchmark functions from Table A.1 with $n = 100$

		ABC	DE	PSO	CS	DS	JADE	MFO	MVO	SCA	EA-HC
f_1	f_{Best}	3.95E − 03	8.61E + 04	9.25E + 03	1.34E + 02	8.15E + 01	1.78E − 06	2.27E + 03	1.69E + 01	2.23E + 02	**1.05E − 105**
	f_{Worst}	1.11E − 01	1.18E + 05	2.19E + 04	3.10E + 02	5.41E + 02	2.80E − 03	5.01E + 04	3.45E + 01	1.25E + 04	1.45E + 01
	$\bar{f}$	2.25E − 02	1.03E + 05	1.55E + 04	2.15E + 02	1.95E + 02	1.58E − 04	2.37E + 04	2.32E + 01	3.39E + 03	4.83E − 01
	σ_f	2.20E − 02	6.98E + 03	3.05E + 03	4.36E + 01	1.16E + 02	5.36E − 04	1.28E + 04	4.28E + 00	3.17E + 03	2.65E + 00
f_2	f_{Best}	6.04E − 02	1.35E + 07	9.20E + 01	1.00E + 10	3.26E + 00	1.01E − 03	8.46E + 01	3.19E + 02	9.23E − 03	**2.15E − 91**
	f_{Worst}	1.61E − 01	2.82E + 19	1.52E + 02	1.00E + 10	1.18E + 01	4.55E + 01	3.33E + 02	2.36E + 16	6.47E + 00	3.06E + 00
	$\bar{f}$	9.62E − 02	1.08E + 18	1.17E + 02	1.00E + 10	6.60E + 00	3.43E + 00	1.59E + 02	1.43E + 15	8.98E − 01	1.74E − 01
	σ_f	2.53E − 02	5.15E + 18	1.58E + 01	0.00E + 00	2.04E + 00	8.40E + 00	4.98E + 01	5.47E + 15	1.36E + 00	6.68E − 01
f_3	f_{Best}	1.49E + 05	3.82E + 05	4.30E + 04	3.32E + 04	4.66E + 04	1.41E + 04	1.02E + 05	2.30E + 04	9.75E + 04	**6.89E − 149**
	f_{Worst}	2.31E + 05	7.80E + 05	1.45E + 05	6.17E + 04	1.13E + 05	4.09E + 05	3.09E + 05	4.55E + 04	2.49E + 05	5.45E + 04
	$\bar{f}$	1.85E + 05	5.52E + 05	7.65E + 04	4.81E + 04	7.19E + 04	2.72E + 05	1.79E + 05	3.15E + 04	1.69E + 05	7.16E + 03
	σ_f	1.93E + 04	9.79E + 04	2.72E + 04	7.64E + 03	1.86E + 04	9.31E + 04	5.21E + 04	5.18E + 03	4.46E + 04	1.40E + 04
f_4	f_{Best}	8.68E + 01	8.16E + 01	2.81E + 01	1.38E + 01	2.85E + 01	1.30E + 01	8.80E + 01	3.11E + 01	7.66E + 01	**301E − 133**
	f_{Worst}	9.26E + 01	8.90E + 01	4.38E + 01	2.16E + 01	5.53E + 01	9.32E + 01	9.57E + 01	5.64E + 01	9.11E + 01	**0.2**.33E − 51
	$\bar{f}$	9.01E + 01	8.53E + 01	3.61E + 01	1.72E + 01	4.34E + 01	8.58E + 01	9.22E + 01	4.23E + 01	8.50E + 01	8.77E − 53
	σ_f	1.72E + 00	1.52E + 00	3.95E + 00	2.01E + 00	6.01E + 00	1.40E + 01	1.96E + 00	6.57E + 00	3.29E + 00	4.27E − 52
f_5	f_{Best}	9.66E + 01	1.26E + 03	3.29E + 03	9.96E + 01	9.73E + 01	8.93E + 01	2.86E + 02	9.16E + 01	1.21E + 02	**1.75E − 01**
	f_{Worst}	1.18E + 02	1.70E + 03	6.04E + 03	1.03E + 02	1.64E + 02	9.87E + 01	6.98E + 02	1.86E + 02	4.16E + 02	9.80E + 01
	$\bar{f}$	1.04E + 02	1.44E + 03	4.62E + 03	1.01E + 02	1.16E + 02	9.24E + 01	4.86E + 02	1.05E + 02	2.34E + 02	7.43E + 01
	σ_f	6.05E + 00	1.11E + 02	6.65E + 02	1.06E + 00	1.97E + 01	2.94E + 00	1.13E + 02	2.24E + 01	8.39E + 01	4.02E + 01
f_6	f_{Best}	4.00E + 00	8.90E + 04	1.15E + 04	3.37E + 02	1.24E + 02	1.00E + 00	3.96E + 03	8.50E + 01	4.81E + 02	**0.00E + 00**

(continued)

Table 2.3 (continued)

		ABC	DE	PSO	CS	DS	JADE	MFO	MVO	SCA	EA-HC
	f_{Worst}	2.80E + 01	1.19E + 05	2.14E + 04	5.87E + 02	9.14E + 02	1.10E + 02	4.51E + 04	2.34E + 02	1.18E + 04	0.00E + 00
	$\bar{f}$	1.28E + 01	1.03E + 05	1.64E + 04	4.48E + 02	3.20E + 02	1.97E + 01	2.77E + 04	1.53E + 02	4.20E + 03	0.00E + 00
	σ_f	6.09E + 00	7.81E + 03	2.41E + 03	7.26E + 01	1.87E + 02	2.21E + 01	1.06E + 04	4.12E + 01	3.19E + 03	0.00E + 00
f_7	f_{Best}	1.38E + 00	2.32E + 02	3.19E + 02	3.25E − 01	8.16E − 01	5.06E − 01	4.70E + 00	1.36E − 01	4.70E + 00	**6.98E − 06**
	f_{Worst}	2.79E + 00	4.55E + 02	7.56E + 02	8.35E − 01	2.35E + 00	1.39E + 00	4.16E + 02	3.88E − 01	1.13E + 02	2.52E − 03
	$\bar{f}$	2.26E + 00	3.58E + 02	5.01E + 02	5.14E − 01	1.30E + 00	8.80E − 01	1.41E + 02	2.43E − 01	4.86E + 01	3.19E − 04
	σ_f	3.25E − 01	5.24E + 01	1.01E + 02	1.22E − 01	4.04E − 01	2.06E − 01	1.04E + 02	6.36E − 02	3.03E + 01	5.18E − 04
f_8	f_{Best}	2.03E + 00	5.72E + 01	3.12E + 01	5.19E + 00	2.03E + 00	**2.02E + 00**	2.04E + 00	2.03E + 00	2.11E + 00	**2.02E + 00**
	f_{Worst}	2.10E + 00	6.69E + 01	4.47E + 01	8.87E + 00	2.04E + 00	2.03E + 00	6.38E + 00	2.88E + 00	2.29E + 00	3.69E + 01
	$\bar{f}$	2.05E + 00	6.22E + 01	3.77E + 01	7.03E + 00	2.03E + 00	2.02E + 00	3.18E + 00	2.23E + 00	2.18E + 00	3.53E + 00
	σ_f	1.47E − 02	2.40E + 00	3.37E + 00	8.48E − 01	3.31E − 03	2.78E − 03	1.35E + 00	2.10E − 01	3.57E − 02	6.42E + 00
f_9	f_{Best}	2.34E − 02	4.85E + 01	4.05E + 01	1.59E − 01	3.67E − 02	**2.04E − 02**	4.54E − 02	2.40E − 02	9.74E − 02	**2.04E − 02**
	f_{Worst}	3.65E − 02	6.56E + 01	5.90E + 01	4.77E − 01	1.15E − 01	2.05E − 02	8.11E + 00	2.64E − 02	4.80E − 01	7.35E + 00
	$\bar{f}$	2.68E − 02	5.66E + 01	5.00E + 01	3.20E − 01	7.09E − 02	2.04E − 02	1.70E + 00	2.53E − 02	1.87E − 01	4.99E − 01
	σ_f	3.05E − 03	3.96E + 00	4.03E + 00	9.18E − 02	2.33E − 02	1.00E − 05	2.75E + 00	4.96E − 04	7.26E − 02	1.82E + 00
f_{10}	f_{Best}	4.38E + 01	1.14E + 03	6.49E + 02	4.08E + 02	1.62E + 02	2.40E + 02	5.76E + 02	4.43E + 02	1.05E + 01	**0.00E + 00**
	f_{Worst}	8.54E + 01	1.27E + 03	9.56E + 02	5.41E + 02	3.54E + 02	3.23E + 02	8.57E + 02	7.29E + 02	5.13E + 02	9.43E + 02
	$\bar{f}$	6.60E + 01	1.21E + 03	8.19E + 02	4.73E + 02	2.31E + 02	2.77E + 02	7.12E + 02	5.91E + 02	2.08E + 02	4.64E + 01
	σ_f	1.06E + 01	2.88E + 01	7.23E + 01	3.38E + 01	4.46E + 01	1.90E + 01	6.04E + 01	7.69E + 01	1.05E + 02	1.77E + 02
f_{11}	f_{Best}	2.25E + 00	1.88E + 01	1.21E + 01	6.63E + 00	3.28E + 00	1.69E + 01	1.90E + 01	2.91E + 00	6.52E + 00	**8.88E − 16**
	f_{Worst}	3.73E + 00	1.96E + 01	1.46E + 01	1.70E + 01	1.99E + 01	1.99E + 01	2.00E + 01	1.99E + 01	2.06E + 01	4.85E − 07
	$\bar{f}$	3.03E + 00	1.93E + 01	1.34E + 01	1.09E + 01	5.22E + 00	1.92E + 01	1.98E + 01	5.23E + 00	1.85E + 01	1.99E − 08

(continued)

Table 2.3 (continued)

		ABC	DE	PSO	CS	DS	JADE	MFO	MVO	SCA	EA-HC
	σ_f	4.06E − 01	1.81E − 01	6.22E − 01	2.67E + 00	3.13E + 00	9.37E − 01	2.71E − 01	4.98E + 00	4.63E + 00	8.93E − 08
f_{12}	f_{Best}	1.74E − 02	7.90E + 02	8.74E + 01	2.26E + 00	1.64E + 00	8.48E − 07	4.26E + 01	1.14E + 00	6.96E + 00	**0.00E + 00**
	f_{Worst}	5.92E − 01	1.03E + 03	1.89E + 02	3.95E + 00	9.16E + 00	1.53E − 01	5.65E + 02	1.24E + 00	1.31E + 02	1.02E − 09
	$\bar{f}$	1.96E − 01	9.16E + 02	1.41E + 02	2.94E + 00	3.05E + 00	1.01E − 02	2.32E + 02	1.20E + 00	4.14E + 01	3.39E − 11
	σ_f	1.55E − 01	6.06E + 01	2.70E + 01	4.24E − 01	1.77E + 00	2.95E − 02	1.27E + 02	2.17E − 02	3.45E + 01	1.85E − 10
f_{13}	f_{Best}	**−1.00E + 01**	6.69E + 00	1.64E + 01	−8.65E + 00	−9.54E + 00	−8.96E + 00	−6.52E + 00	−6.72E + 0	−9.97E + 00	**−1.00E + 01**
	f_{Worst}	−9.70E + 00	1.13E + 01	2.66E + 01	−6.25E + 00	−7.32E + 00	−7.04E + 00	−6.82E − 01	−3.32E + 00	−7.39E + 00	−1.00E + 01
	$\bar{f}$	−9.90E + 00	9.86E + 00	2.09E + 01	−7.23E + 00	−8.63E + 00	−8.24E + 00	−3.64E + 00	−5.23E + 00	−9.09E + 00	−1.00E + 01
	σ_f	9.79E − 02	9.98E − 01	2.64E + 00	5.83E − 01	5.02E − 01	4.93E − 01	1.54E + 00	8.49E − 01	8.30E − 01	8.66E − 10
f_{14}	f_{Best}	2.73E + 00	6.02E + 02	1.50E + 01	3.22E + 00	2.76E + 00	2.73E + 00	6.05E + 01	2.73E + 00	5.19E + 00	**2.72E + 00**
	f_{Worst}	5.48E + 00	1.23E + 03	5.94E + 01	4.13E + 00	5.49E + 00	6.00E + 00	6.84E + 02	2.73E + 00	2.42E + 02	2.73E + 00
	$\bar{f}$	3.40E + 00	8.62E + 02	2.50E + 01	3.69E + 00	3.18E + 00	2.84E + 00	2.33E + 02	2.73E + 00	5.77E + 01	2.72E + 00
	σ_f	6.22E − 01	1.42E + 02	9.54E + 00	2.70E − 01	5.71E − 01	5.97E − 01	1.46E + 02	1.41E − 05	6.60E + 01	3.34E − 03
f_{15}	f_{Best}	1.67E − 04	3.23E + 01	1.76E + 01	5.88E − 02	3.03E − 02	3.97E − 10	8.47E − 01	5.81E − 03	1.11E − 01	**1.06E − 165**
	f_{Worst}	4.82E − 03	4.67E + 01	3.00E + 01	1.34E − 01	2.44E − 01	1.95E − 04	2.06E + 01	1.13E − 02	5.85E + 00	1.13E − 01
	$\bar{f}$	9.76E − 04	3.94E + 01	2.37E + 01	8.51E − 02	8.08E − 02	1.38E − 05	1.08E + 01	8.66E − 03	1.47E + 00	3.76E − 03
	σ_f	9.26E − 04	3.14E + 00	3.72E + 00	1.96E − 02	4.47E − 02	4.45E − 05	5.57E + 00	1.29E − 03	1.42E + 00	2.06E − 02
f_{16}	f_{Best}	−3.64E + 04	−8.99E + 03	−9.27E + 03	−2.33E + 04	−3.14E + 04	−2.57E + 04	−2.90E + 04	−2.74E + 04	−8.19E + 03	**−4.19E + 04**

(continued)

Table 2.3 (continued)

		ABC	DE	PSO	CS	DS	JADE	MFO	MVO	SCA	EA-HC
	f_{Worst}	−3.43E + 04	−6.61E + 03	−5.72E + 03	−2.11E + 04	−2.74E + 04	−1.93E + 04	−2.03E + 04	−2.20E + 04	−6.58E + 03	−7.35E + 03
	$\bar{f}$	−3.52E + 04	−7.86E + 03	−7.61E + 03	−2.19E + 04	−2.91E + 04	−2.19E + 04	−2.38E + 04	−2.44E + 04	−7.34E + 03	−1.81E + 04
	σ_f	6.17E + 02	6.87E + 02	1.05E + 03	4.63E + 02	1.06E + 03	2.19E + 03	2.62E + 03	1.56E + 03	3.99E + 02	1.47E + 04
f_{17}	f_{Best}	3.28E − 05	2.47E + 08	2.58E + 04	4.40E + 00	2.74E + 00	2.22E + 00	3.93E + 04	2.58E + 04	2.51E + 07	**4.71E − 33**
	f_{Worst}	3.11E − 02	6.70E + 08	9.29E + 05	8.46E + 00	1.42E + 01	7.81E + 00	5.22E + 08	9.29E + 05	3.11E + 08	4.20E − 09
	$\bar{f}$	1.26E − 03	4.48E + 08	2.84E + 05	5.67E + 00	6.58E + 00	4.91E + 00	1.08E + 08	2.84E + 05	1.16E + 08	1.40E − 10
	σ_f	5.65E − 03	9.17E + 07	2.72E + 05	7.80E − 01	2.43E + 00	1.38E + 00	1.61E + 08	2.72E + 05	6.55E + 07	7.66E − 10
f_{18}	f_{Best}	5.75E − 04	6.37E + 08	1.89E + 06	6.13E + 01	4.02E + 01	9.12E + 00	9.25E + 06	1.89E + 06	4.98E + 07	**1.35E − 32**
	f_{Worst}	5.15E − 02	1.21E + 09	2.22E + 07	5.88E + 02	3.17E + 02	6.09E + 01	4.57E + 08	2.22E + 07	5.70E + 08	3.71E − 12
	$\bar{f}$	4.43E − 03	9.42E + 08	7.18E + 06	1.27E + 02	8.65E + 01	2.39E + 01	1.89E + 08	7.18E + 06	2.48E + 08	1.49E − 13
	σ_f	9.47E − 03	1.64E + 08	4.60E + 06	9.95E + 01	4.93E + 01	9.50E + 00	1.80E + 08	4.60E + 06	1.23E + 08	6.80E − 13
f_{19}	f_{Best}	3.82E − 02	3.53E + 12	1.44E + 04	1.00E + 10	7.59E + 01	1.33E − 03	7.40E + 04	1.44E + 04	2.14E + 02	**2.00E − 104**
	f_{Worst}	1.83E − 01	1.12E + 20	6.63E + 04	1.00E + 10	4.75E + 02	4.37E + 03	2.32E + 05	6.63E + 04	1.23E + 04	2.88E + 00
	$\bar{f}$	7.61E − 02	9.05E + 18	2.20E + 04	1.00E + 10	2.14E + 02	2.01E + 02	1.37E + 05	2.20E + 04	4.57E + 03	9.60E − 02
	σ_f	2.84E − 02	2.48E + 19	9.22E + 03	0.00E + 00	9.09E + 01	8.16E + 02	4.78E + 04	9.22E + 03	3.31E + 03	5.25E − 01
f_{20}	f_{Best}	1.47E + 02	3.06E + 03	1.27E + 03	5.34E + 02	3.25E + 02	2.86E + 02	9.81E + 02	1.27E + 03	1.41E + 02	**9.90E + 01**
	f_{Worst}	2.12E + 02	4.18E + 03	1.99E + 03	6.61E + 02	5.18E + 02	4.96E + 02	3.04E + 03	1.99E + 03	9.59E + 02	9.90E + 01
	$\bar{f}$	1.81E + 02	3.66E + 03	1.44E + 03	6.01E + 02	4.03E + 02	3.81E + 02	1.86E + 03	1.44E + 03	5.03E + 02	9.90E + 01
	σ_f	1.70E + 01	2.56E + 02	1.39E + 02	2.73E + 01	4.79E + 01	5.37E + 01	5.34E + 02	1.39E + 02	2.06E + 02	4.40E − 04
f_{21}	f_{Best}	2.81E + 03	7.38E + 08	2.03E + 06	2.41E + 03	6.78E + 03	1.49E + 03	6.63E + 06	2.03E + 06	3.61E + 07	**1.09E + 02**

(continued)

Table 2.3 (continued)

		ABC	DE	PSO	CS	DS	JADE	MFO	MVO	SCA	EA-HC
	f_{Worst}	5.74E + 03	1.33E + 09	1.83E + 07	7.10E + 03	1.16E + 05	4.40E + 03	1.27E + 09	1.83E + 07	7.69E + 08	3.10E + 03
	$\bar{f}$	3.86E + 03	1.00E + 09	6.53E + 06	3.78E + 03	1.72E + 04	2.14E + 03	2.25E + 08	6.53E + 06	2.56E + 08	3.92E + 02
	σ_f	6.81E + 02	1.48E + 08	4.25E + 06	8.91E + 02	1.97E + 04	6.70E + 02	3.10E + 08	4.25E + 06	1.47E + 08	7.86E + 02
f_{22}	f_{Best}	9.95E + 01	5.62E + 09	1.43E + 03	1.00E + 10	2.34E + 02	2.99E + 02	1.88E + 03	1.43E + 03	1.26E + 02	**9.90E + 01**
	f_{Worst}	1.54E + 02	8.00E + 18	2.92E + 03	1.00E + 10	4.27E + 02	5.73E + 02	7.43E + 03	2.92E + 03	1.02E + 03	9.90E + 01
	$\bar{f}$	1.13E + 02	3.12E + 17	1.70E + 03	1.00E + 10	3.15E + 02	4.38E + 02	4.48E + 03	1.70E + 03	3.66E + 02	9.90E + 01
	σ_f	1.40E + 01	1.47E + 18	3.07E + 02	0.00E + 00	4.54E + 01	6.27E + 01	1.23E + 03	3.07E + 02	2.31E + 02	4.02E − 03
f_{23}	f_{Best}	−2.95E + 02	1.82E + 10	1.62E + 04	1.00E + 10	−1.24E + 02	−2.87E + 02	2.66E + 08	1.62E + 04	4.90E + 05	**−2.98E + 02**
	f_{Worst}	−2.91E + 02	7.53E + 20	2.23E + 06	1.00E + 10	1.37E + 03	3.07E + 03	6.40E + 09	2.23E + 06	1.43E + 08	−2.87E + 02
	$\bar{f}$	−2.94E + 02	3.22E + 19	4.53E + 05	1.00E + 10	5.15E + 02	−4.16E + 00	2.81E + 09	4.53E + 05	2.67E + 07	−2.97E + 02
	σ_f	7.66E − 01	1.38E + 20	5.90E + 05	0.00E + 00	4.13E + 02	6.15E + 02	1.52E + 09	5.90E + 05	3.13E + 07	1.88E + 00

Table 2.4 Results of minimization for benchmark functions from Table A.1 with $n = 200$

		ABC	DE	PSO	CS	DS	JADE	MFO	MVO	SCA	EA-HC
f_1	f_{Best}	3.95E − 03	8.61E + 04	9.25E + 03	1.34E + 02	8.15E + 01	1.78E − 06	2.27E + 03	1.69E + 01	2.23E + 02	**1.05E − 105**
	f_{Worst}	1.11E − 01	1.18E + 05	2.19E + 04	3.10E + 02	5.41E + 02	2.80E − 03	5.01E + 04	3.45E + 01	1.25E + 04	1.45E + 01
	$\bar{f}$	2.25E − 02	1.03E + 05	1.55E + 04	2.15E + 02	1.95E + 02	1.58E − 04	2.37E + 04	2.32E + 01	3.39E + 03	4.83E − 01
	σ_f	2.20E − 02	6.98E + 03	3.05E + 03	4.36E + 01	1.16E + 02	5.36E − 04	1.28E + 04	4.28E + 00	3.17E + 03	2.65E + 00
f_2	f_{Best}	6.04E − 02	1.35E + 07	9.20E + 01	1.00E + 10	3.26E + 00	1.01E − 03	8.46E + 01	3.19E + 02	9.23E − 03	**2.15E − 91**
	f_{Worst}	1.61E − 01	2.82E + 19	1.52E + 02	1.00E + 10	1.18E + 01	4.55E + 01	3.33E + 02	2.36E + 16	6.47E + 00	3.06E + 00
	$\bar{f}$	9.62E − 02	1.08E + 18	1.17E + 02	1.00E + 10	6.60E + 00	3.43E + 00	1.59E + 02	1.43E + 15	8.98E − 01	1.74E − 01
	σ_f	2.53E − 02	5.15E + 18	1.58E + 01	0.00E + 00	2.04E + 00	8.40E + 00	4.98E + 01	5.47E + 15	1.36E + 00	6.68E − 01
f_3	f_{Best}	1.49E + 05	3.82E + 05	4.30E + 04	3.32E + 04	4.66E + 04	1.41E + 04	1.02E + 05	2.30E + 04	9.75E + 04	**6.89E − 149**
	f_{Worst}	2.31E + 05	7.80E + 05	1.45E + 05	6.17E + 04	1.13E + 05	4.09E + 05	3.09E + 05	4.55E + 04	2.49E + 05	5.45E + 04
	$\bar{f}$	1.85E + 05	5.52E + 05	7.65E + 04	4.81E + 04	7.19E + 04	2.72E + 05	1.79E + 05	3.15E + 04	1.69E + 05	7.16E + 03
	σ_f	1.93E + 04	9.79E + 04	2.72E + 04	7.64E + 03	1.86E + 04	9.31E + 04	5.21E + 04	5.18E + 03	4.46E + 04	1.40E + 04
f_4	f_{Best}	8.68E + 01	8.16E + 01	2.81E + 01	1.38E + 01	2.85E + 01	1.30E + 01	8.80E + 01	3.11E + 01	7.66E + 01	**301E − 133**
	f_{Worst}	9.26E + 01	8.90E + 01	4.38E + 01	2.16E + 01	5.53E + 01	9.32E + 01	9.57E + 01	5.64E + 01	9.11E + 01	**0.2**.33E − 51
	$\bar{f}$	9.01E + 01	8.53E + 01	3.61E + 01	1.72E + 01	4.34E + 01	8.58E + 01	9.22E + 01	4.23E + 01	8.50E + 01	8.77E − 53
	σ_f	1.72E + 00	1.52E + 00	3.95E + 00	2.01E + 00	6.01E + 00	1.40E + 01	1.96E + 00	6.57E + 00	3.29E + 00	4.27E − 52
f_5	f_{Best}	9.66E + 01	1.26E + 03	3.29E + 03	9.96E + 01	9.73E + 01	8.93E + 01	2.86E + 02	9.16E + 01	1.21E + 02	**1.75E − 01**
	f_{Worst}	1.18E + 02	1.70E + 03	6.04E + 03	1.03E + 02	1.64E + 02	9.87E + 01	6.98E + 02	1.86E + 02	4.16E + 02	9.80E + 01
	$\bar{f}$	1.04E + 02	1.44E + 03	4.62E + 03	1.01E + 02	1.16E + 02	9.24E + 01	4.86E + 02	1.05E + 02	2.34E + 02	7.43E + 01
	σ_f	6.05E + 00	1.11E + 02	6.65E + 02	1.06E + 00	1.97E + 01	2.94E + 00	1.13E + 02	2.24E + 01	8.39E + 01	4.02E + 01
f_6	f_{Best}	4.00E + 00	8.90E + 04	1.15E + 04	3.37E + 02	1.24E + 02	1.00E + 00	3.96E + 03	8.50E + 01	4.81E + 02	**0.00E + 00**

(continued)

Table 2.4 (continued)

		ABC	DE	PSO	CS	DS	JADE	MFO	MVO	SCA	EA-HC
	f_{Worst}	2.80E + 01	1.19E + 05	2.14E + 04	5.87E + 02	9.14E + 02	1.10E + 02	4.51E + 04	2.34E + 02	1.18E + 04	0.00E + 00
	$\bar{f}$	1.28E + 01	1.03E + 05	1.64E + 04	4.48E + 02	3.20E + 02	1.97E + 01	2.77E + 04	1.53E + 02	4.20E + 03	0.00E + 00
	σ_f	6.09E + 00	7.81E + 03	2.41E + 03	7.26E + 01	1.87E + 02	2.21E + 01	1.06E + 04	4.12E + 01	3.19E + 03	0.00E + 00
f_7	f_{Best}	1.38E + 00	2.32E + 02	3.19E + 02	3.25E − 01	8.16E − 01	5.06E − 01	4.70E + 00	1.36E − 01	4.70E + 00	**6.98E − 06**
	f_{Worst}	2.79E + 00	4.55E + 02	7.56E + 02	8.35E − 01	2.35E + 00	1.39E + 00	4.16E + 02	3.88E − 01	1.13E + 02	2.52E − 03
	$\bar{f}$	2.26E + 00	3.58E + 02	5.01E + 02	5.14E − 01	1.30E + 00	8.80E − 01	1.41E + 02	2.43E − 01	4.86E + 01	3.19E − 04
	σ_f	3.25E − 01	5.24E + 01	1.01E + 02	1.22E − 01	4.04E − 01	2.06E − 01	1.04E + 02	6.36E − 02	3.03E + 01	5.18E − 04
f_8	f_{Best}	2.03E + 00	5.72E + 01	3.12E + 01	5.19E + 00	2.03E + 00	**2.02E + 00**	2.04E + 00	2.03E + 00	2.11E + 00	**2.02E + 00**
	f_{Worst}	2.10E + 00	6.69E + 01	4.47E + 01	8.87E + 00	2.04E + 00	2.03E + 00	6.38E + 00	2.88E + 00	2.29E + 00	3.69E + 01
	$\bar{f}$	2.05E + 00	6.22E + 01	3.77E + 01	7.03E + 00	2.03E + 00	2.02E + 00	3.18E + 00	2.23E + 00	2.18E + 00	3.53E + 00
	σ_f	1.47E − 02	2.40E + 00	3.37E + 00	8.48E − 01	3.31E − 03	2.78E − 03	1.35E + 00	2.10E − 01	3.57E − 02	6.42E + 00
f_9	f_{Best}	2.34E − 02	4.85E + 01	4.05E + 01	1.59E − 01	3.67E − 02	**2.04E − 02**	4.54E − 02	2.40E − 02	9.74E − 02	**2.04E − 02**
	f_{Worst}	3.65E − 02	6.56E + 01	5.90E + 01	4.77E − 01	1.15E − 01	2.05E − 02	8.11E + 00	2.64E − 02	4.80E − 01	7.35E + 00
	$\bar{f}$	2.68E − 02	5.66E + 01	5.00E + 01	3.20E − 01	7.09E − 02	2.04E − 02	1.70E + 00	2.53E − 02	1.87E − 01	4.99E − 01
	σ_f	3.05E − 03	3.96E + 00	4.03E + 00	9.18E − 02	2.33E − 02	1.00E − 05	2.75E + 00	4.96E − 04	7.26E − 02	1.82E + 00
f_{10}	f_{Best}	4.38E + 01	1.14E + 03	6.49E + 02	4.08E + 02	1.62E + 02	2.40E + 02	5.76E + 02	4.43E + 02	1.05E + 01	**0.00E + 00**
	f_{Worst}	8.54E + 01	1.27E + 03	9.56E + 02	5.41E + 02	3.54E + 02	3.23E + 02	8.57E + 02	7.29E + 02	5.13E + 02	9.43E + 02
	$\bar{f}$	6.60E + 01	1.21E + 03	8.19E + 02	4.73E + 02	2.31E + 02	2.77E + 02	7.12E + 02	5.91E + 02	2.08E + 02	4.64E + 01
	σ_f	1.06E + 01	2.88E + 01	7.23E + 01	3.38E + 01	4.46E + 01	1.90E + 01	6.04E + 01	7.69E + 01	1.05E + 02	1.77E + 02
f_{11}	f_{Best}	2.25E + 00	1.88E + 01	1.21E + 01	6.63E + 00	3.28E + 00	1.69E + 01	1.90E + 01	2.91E + 00	6.52E + 00	**8.88E − 16**
	f_{Worst}	3.73E + 00	1.96E + 01	1.46E + 01	1.70E + 01	1.99E + 01	1.99E + 01	2.00E + 01	1.99E + 01	2.06E + 01	4.85E − 07
	$\bar{f}$	3.03E + 00	1.93E + 01	1.34E + 01	1.09E + 01	5.22E + 00	1.92E + 01	1.98E + 01	5.23E + 00	1.85E + 01	1.99E − 08

(continued)

Table 2.4 (continued)

		ABC	DE	PSO	CS	DS	JADE	MFO	MVO	SCA	EA-HC
	σ_f	4.06E − 01	1.81E − 01	6.22E − 01	2.67E + 00	3.13E + 00	9.37E − 01	2.71E − 01	4.98E + 00	4.63E + 00	8.93E − 08
f_{12}	f_{Best}	1.74E − 02	7.90E + 02	8.74E + 01	2.26E + 00	1.64E + 00	8.48E − 07	4.26E + 01	1.14E + 00	6.96E + 00	**0.00E + 00**
	f_{Worst}	5.92E − 01	1.03E + 03	1.89E + 02	3.95E + 00	9.16E + 00	1.53E − 01	5.65E + 02	1.24E + 00	1.31E + 02	1.02E − 09
	$\bar{f}$	1.96E − 01	9.16E + 02	1.41E + 02	2.94E + 00	3.05E + 00	1.01E − 02	2.32E + 02	1.20E + 00	4.14E + 01	3.39E − 11
	σ_f	1.55E − 01	6.06E + 01	2.70E + 01	4.24E − 01	1.77E + 00	2.95E − 02	1.27E + 02	2.17E − 02	3.45E + 01	1.85E − 10
f_{13}	f_{Best}	**−1.00E + 01**	6.69E + 00	1.64E + 01	−8.65E + 00	−9.54E + 00	−8.96E + 00	−6.52E + 00	−6.72E + 0	−9.97E + 00	**−1.00E + 01**
	f_{Worst}	−9.70E + 00	1.13E + 01	2.66E + 01	−6.25E + 00	−7.32E + 00	−7.04E + 00	−6.82E − 01	−3.32E + 00	−7.39E + 00	−1.00E + 01
	$\bar{f}$	−9.90E + 00	9.86E + 00	2.09E + 01	−7.23E + 00	−8.63E + 00	−8.24E + 00	−3.64E + 00	−5.23E + 00	−9.09E + 00	−1.00E + 01
	σ_f	9.79E − 02	9.98E − 01	2.64E + 00	5.83E − 01	5.02E − 01	4.93E − 01	1.54E + 00	8.49E − 01	8.30E − 01	8.66E − 10
f_{14}	f_{Best}	2.73E + 00	6.02E + 02	1.50E + 01	3.22E + 00	2.76E + 00	2.73E + 00	6.05E + 01	2.73E + 00	5.19E + 00	**2.72E + 00**
	f_{Worst}	5.48E + 00	1.23E + 03	5.94E + 01	4.13E + 00	5.49E + 00	6.00E + 00	6.84E + 02	2.73E + 00	2.42E + 02	2.73E + 00
	$\bar{f}$	3.40E + 00	8.62E + 02	2.50E + 01	3.69E + 00	3.18E + 00	2.84E + 00	2.33E + 02	2.73E + 00	5.77E + 01	2.72E + 00
	σ_f	6.22E − 01	1.42E + 02	9.54E + 00	2.70E − 01	5.71E − 01	5.97E − 01	1.46E + 02	1.41E − 05	6.60E + 01	3.34E − 03
f_{15}	f_{Best}	1.67E − 04	3.23E + 01	1.76E + 01	5.88E − 02	3.03E − 02	3.97E − 10	8.47E − 01	5.81E − 03	1.11E − 01	**1.06E − 165**
	f_{Worst}	4.82E − 03	4.67E + 01	3.00E + 01	1.34E − 01	2.44E − 01	1.95E − 04	2.06E + 01	1.13E − 02	5.85E + 00	1.13E − 01
	$\bar{f}$	9.76E − 04	3.94E + 01	2.37E + 01	8.51E − 02	8.08E − 02	1.38E − 05	1.08E + 01	8.66E − 03	1.47E + 00	3.76E − 03
	σ_f	9.26E − 04	3.14E + 00	3.72E + 00	1.96E − 02	4.47E − 02	4.45E − 05	5.57E + 00	1.29E − 03	1.42E + 00	2.06E − 02
f_{16}	f_{Best}	−3.64E + 04	−8.99E + 03	−9.27E + 03	−2.33E + 04	−3.14E + 04	−2.57E + 04	−2.90E + 04	−2.74E + 04	−8.19E + 03	**−4.19E + 04**

(continued)

Table 2.4 (continued)

		ABC	DE	PSO	CS	DS	JADE	MFO	MVO	SCA	EA-HC
	f_{Worst}	−3.43E + 04	−6.61E + 03	−5.72E + 03	−2.11E + 04	−2.74E + 04	−1.93E + 04	−2.03E + 04	−2.20E + 04	−6.58E + 03	−7.35E + 03
	$\bar{f}$	−3.52E + 04	−7.86E + 03	−7.61E + 03	−2.19E + 04	−2.91E + 04	−2.19E + 04	−2.38E + 04	−2.44E + 04	−7.34E + 03	−1.81E + 04
	σ_f	6.17E + 02	6.87E + 02	1.05E + 03	4.63E + 02	1.06E + 03	2.19E + 03	2.62E + 03	1.56E + 03	3.99E + 02	1.47E + 04
f_{17}	f_{Best}	3.28E − 05	2.47E + 08	2.58E + 04	4.40E + 00	2.74E + 00	2.22E + 00	3.93E + 04	2.58E + 04	2.51E + 07	**4.71E − 33**
	f_{Worst}	3.11E − 02	6.70E + 08	9.29E + 05	8.46E + 00	1.42E + 01	7.81E + 00	5.22E + 08	9.29E + 05	3.11E + 08	4.20E − 09
	$\bar{f}$	1.26E − 03	4.48E + 08	2.84E + 05	5.67E + 00	6.58E + 00	4.91E + 00	1.08E + 08	2.84E + 05	1.16E + 08	1.40E − 10
	σ_f	5.65E − 03	9.17E + 07	2.72E + 05	7.80E − 01	2.43E + 00	1.38E + 00	1.61E + 08	2.72E + 05	6.55E + 07	7.66E − 10
f_{18}	f_{Best}	5.75E − 04	6.37E + 08	1.89E + 06	6.13E + 01	4.02E + 01	9.12E + 00	9.25E + 06	1.89E + 06	4.98E + 07	**1.35E − 32**
	f_{Worst}	5.15E − 02	1.21E + 09	2.22E + 07	5.88E + 02	3.17E + 02	6.09E + 01	4.57E + 08	2.22E + 07	5.70E + 08	3.71E − 12
	$\bar{f}$	4.43E − 03	9.42E + 08	7.18E + 06	1.27E + 02	8.65E + 01	2.39E + 01	1.89E + 08	7.18E + 06	2.48E + 08	1.49E − 13
	σ_f	9.47E − 03	1.64E + 08	4.60E + 06	9.95E + 01	4.93E + 01	9.50E + 00	1.80E + 08	4.60E + 06	1.23E + 08	6.80E − 13
f_{19}	f_{Best}	3.82E − 02	3.53E + 12	1.44E + 04	1.00E + 10	7.59E + 01	1.33E − 03	7.40E + 04	1.44E + 04	2.14E + 02	**2.00E − 104**
	f_{Worst}	1.83E − 01	1.12E + 20	6.63E + 04	1.00E + 10	4.75E + 02	4.37E + 03	2.32E + 05	6.63E + 04	1.23E + 04	2.88E + 00
	$\bar{f}$	7.61E − 02	9.05E + 18	2.20E + 04	1.00E + 10	2.14E + 02	2.01E + 02	1.37E + 05	2.20E + 04	4.57E + 03	9.60E − 02
	σ_f	2.84E − 02	2.48E + 19	9.22E + 03	0.00E + 00	9.09E + 01	8.16E + 02	4.78E + 04	9.22E + 03	3.31E + 03	5.25E − 01
f_{20}	f_{Best}	1.47E + 02	3.06E + 03	1.27E + 03	5.34E + 02	3.25E + 02	2.86E + 02	9.81E + 02	1.27E + 03	1.41E + 02	**9.90E + 01**
	f_{Worst}	2.12E + 02	4.18E + 03	1.99E + 03	6.61E + 02	5.18E + 02	4.96E + 02	3.04E + 03	1.99E + 03	9.59E + 02	9.90E + 01
	$\bar{f}$	1.81E + 02	3.66E + 03	1.44E + 03	6.01E + 02	4.03E + 02	3.81E + 02	1.86E + 03	1.44E + 03	5.03E + 02	9.90E + 01
	σ_f	1.70E + 01	2.56E + 02	1.39E + 02	2.73E + 01	4.79E + 01	5.37E + 01	5.34E + 02	1.39E + 02	2.06E + 02	4.40E − 04
f_{21}	f_{Best}	2.81E + 03	7.38E + 08	2.03E + 06	2.41E + 03	6.78E + 03	1.49E + 03	6.63E + 06	2.03E + 06	3.61E + 07	**1.09E + 02**

(continued)

Table 2.4 (continued)

		ABC	DE	PSO	CS	DS	JADE	MFO	MVO	SCA	EA-HC
	f_{Worst}	5.74E + 03	1.33E + 09	1.83E + 07	7.10E + 03	1.16E + 05	4.40E + 03	1.27E + 09	1.83E + 07	7.69E + 08	3.10E + 03
	$\bar{f}$	3.86E + 03	1.00E + 09	6.53E + 06	3.78E + 03	1.72E + 04	2.14E + 03	2.25E + 08	6.53E + 06	2.56E + 08	3.92E + 02
	σ_f	6.81E + 02	1.48E + 08	4.25E + 06	8.91E + 02	1.97E + 04	6.70E + 02	3.10E + 08	4.25E + 06	1.47E + 08	7.86E + 02
f_{22}	f_{Best}	9.95E + 01	5.62E + 09	1.43E + 03	1.00E + 10	2.34E + 02	2.99E + 02	1.88E + 03	1.43E + 03	1.26E + 02	**9.90E + 01**
	f_{Worst}	1.54E + 02	8.00E + 18	2.92E + 03	1.00E + 10	4.27E + 02	5.73E + 02	7.43E + 03	2.92E + 03	1.02E + 03	9.90E + 01
	$\bar{f}$	1.13E + 02	3.12E + 17	1.70E + 03	1.00E + 10	3.15E + 02	4.38E + 02	4.48E + 03	1.70E + 03	3.66E + 02	9.90E + 01
	σ_f	1.40E + 01	1.47E + 18	3.07E + 02	0.00E + 00	4.54E + 01	6.27E + 01	1.23E + 03	3.07E + 02	2.31E + 02	4.02E − 03
f_{23}	f_{Best}	−2.95E + 02	1.82E + 10	1.62E + 04	1.00E + 10	−1.24E + 02	−2.87E + 02	2.66E + 08	1.62E + 04	4.90E + 05	**−2.98E + 02**
	f_{Worst}	−2.91E + 02	7.53E + 20	2.23E + 06	1.00E + 10	1.37E + 03	3.07E + 03	6.40E + 09	2.23E + 06	1.43E + 08	−2.87E + 02
	$\bar{f}$	−2.94E + 02	3.22E + 19	4.53E + 05	1.00E + 10	5.15E + 02	−4.16E + 00	2.81E + 09	4.53E + 05	2.67E + 07	−2.97E + 02
	σ_f	7.66E − 01	1.38E + 20	5.90E + 05	0.00E + 00	4.13E + 02	6.15E + 02	1.52E + 09	5.90E + 05	3.13E + 07	1.88E + 00

Furthermore, to further explore higher dimensionality in the optimization process, Table 2.4 presents the performance results considering a 200-dimensional search space ($n = 200$).

The objective of this experiment is to assess the capabilities of the agent-based metaheuristic operators when applied to optimization processes in higher-dimensional search spaces. As evidenced by Table 2.4, the findings indicate that the agent-based EA-HC method achieves superior performance compared to the other evolutionary methodologies tested for most benchmark functions. The only exception to this is the case of the JADE method in function f_8, where EA-HC and JADE yield the same fitness value.

Based on the numerical results obtained from the comparison of evolutionary methods, it is clear that the agent-based EA-HC method demonstrates outstanding performance in higher-dimensional search spaces. The structure of EA-HC's evolutionary mechanism consistently delivers better outcomes than its competitors. The numerical results obtained in higher-dimensional search spaces strongly suggest that the changing nature of EA-HC exhibits a high level of scalability, as it surpasses the performance of the other tested algorithms across the majority of benchmark functions.

To validate the performance results obtained from Tables 2.2, 2.3 and 2.4, a non-parametric statistical test is conducted. The Wilcoxon rank-sum test [62] is employed in this study to assess the significance of the results. A significance level of 0.05 is considered, and the test is applied to the 30 independent runs for each benchmark function. Table 2.5 presents the p-values resulting from the Wilcoxon test applied to the performance results of Table 2.2 ($n = 30$). Similarly, Table 2.6 displays the p-values obtained from the rank-sum test for the performance results of Table 2.3 ($n = 100$). Furthermore, Table 2.7 provides the rank-sum test values for the performance results of Table 2.4 ($n = 200$). Tables 2.5, 2.6 and 2.7 present pair-wise comparisons between EA-HC and the other tested algorithms. The study encompasses nine groups: EA-HC versus ABC, EA-HC versus DE, EA-HC versus PSO, EA-HC versus CS, EA-HC versus DS, EA-HC versus JADE, EA-HC versus MFO, EA-HC versus MVO, and EA-HC versus SCA. These tables provide insights into the statistical significance of the performance differences between EA-HC and the competing algorithms.

In the statistical analysis, the null hypothesis (H_0) assumes no significant difference within a group, while the alternative hypothesis (H_1) suggests a significant difference. Tables 2.5, 2.6 and 2.7 utilize symbols ▲, ▼, and ▶ for better visualization of the results. The symbol ▲ indicates that EA-HC achieves significantly better results than a specific competitor, ▼ represents that EA-HC produces worse results, and ▶ signifies situations where the Wilcoxon test cannot distinguish between the numerical results. Based on the p-values in Table 2.5 ($n = 30$), it is evident that for function f_6, the rank-sum test cannot differentiate among the results for the groups EA-HC versus ABC, EA-HC versus CS, EA-HC versus DS, EA-HC versus JADE, EA-HC versus MFO, and EA-HC versus SCA. Similarly, for function f_8, the groups EA-HC versus ABC, EA-HC versus CS, EA-HC versus DS, EA-HC versus JADE, and EA-HC versus MVO show similar performance to the agent-based approach.

Table 2.5 The p-values resulting from the Wilcoxon rank sum test, which compares EA-HC against ABC, DE, PSO, CS, DS, JADE, MFO, MVO, and SCA in terms of the average fitness value $\bar{f}$ for each function in Table 2.2

EA-HC versus	ABC	DE	PSO	CS	DS	JADE	MFO	MVO	SCA
f_1	1.066E − 07▲	1.558E − 08▲	2.249E − 07▲	6.765E − 05▲	1.491E − 06▲	2.028E − 07▲	8.476E − 05▲	6.765E − 05▲	1.635E − 05▲
f_2	9.958E − 05▲	3.019E − 11▲	3.019E − 11▲	3.010E − 07▲	3.037E − 06▲	4.713E − 04▲	3.061E − 11▲	1.193E − 06▲	6.463E − 10▲
f_3	6.184E − 10▲	3.019E − 11▲	5.462E − 06▲	3.019E − 11▲	3.019E − 11▲	3.196E − 09▲	5.183E − 10▲	3.019E − 11▲	7.380E − 10▲
f_4	6.045E − 07▲	3.019E − 11▲	1.173E − 09▲	3.019E − 11▲	3.019E − 11▲	8.841E − 07▲	2.534E − 12▲	3.019E − 11▲	3.338E − 11▲
f_5	3.019E − 11▲	3.019E − 11▲	3.689E − 11▲	3.019E − 11▲	6.695E − 11▲	3.019E − 11▲	1.357E − 07▲	1.613E − 10▲	3.019E − 11▲
f_6	0.0000E00►	4.645E − 08▲	2.701E − 03▲	0.0000E00►	0.0000E00►	0.0000E00►	0.0000E00►	1.825E − 04▲	0.0000E00►
f_7	3.019E − 11▲	6.065E − 11▲	7.043E − 07▲	3.019E − 11▲	3.019E − 11▲	3.019E − 11▲	1.374E − 05▲	3.019E − 11▲	3.019E − 11▲
f_8	0.0000E00►	2.388E − 04▲	5.561E − 04▲	0.0000E00►	0.0000E00►	0.0000E00►	9.424E − 14▲	0.0000E00►	3.338E − 11▲
f_9	0.0000E00►	3.019E − 11▲	3.019E − 11▲	0.0000E00►	0.0000E00►	0.0000E00►	9.434E − 14▲	3.019E − 11▲	3.019E − 11▲
f_{10}	4.975E − 11▲	6.518E − 09▲	5.434E − 08▲	3.279E − 02▲	3.497E − 09▲	1.698E − 08▲	4.342E − 05▲	7.341E − 07▲	7.695E − 08▲

(continued)

Table 2.5 (continued)

EA-HC versus	ABC	DE	PSO	CS	DS	JADE	MFO	MVO	SCA
f_{11}	6.749E − 05▲	1.422E − 08▲	1.335E − 02▲	3.980E − 04▲	6.749E − 05▲	5.394E − 03▲	4.262E − 04▲	3.980E − 06▲	3.559E − 06▲
f_{12}	4.998E − 09▲	1.473E − 07▲	1.127E − 05▲	1.107E − 06▲	1.156E − 07▲	2.601E − 08▲	1.199E − 05▲	1.107E − 06▲	4.112E − 07▲
f_{13}	0.0000E00▶	2.262E − 07▲	2.504E − 11▲	3.019E − 11▲	0.0000E00▶	0.0000E00▶	0.0000E00▶	1.042E − 07▲	0.0000E00▶
f_{14}	3.019E − 11▲	1.776E − 10▲	4.077E − 11▲	3.019E − 11▲	3.019E − 11▲	3.016E − 11▲	1.964E − 11▲	3.019E − 11▲	3.019E − 11▲
f_{15}	5.572E − 10▲	1.010E − 08▲	6.876E − 09▲	1.066E − 07▲	8.484E − 09▲	8.891E − 10▲	5.139E − 06▲	1.156E − 07▲	1.429E − 08▲
f_{16}	0.0000E00▶	3.231E − 05▲	6.736E − 06▲	3.019E − 11▲	0.0000E00▶	3.019E − 11▲	9.434E − 14▲	4.077E − 11▲	1.359E − 07▲
f_{17}	3.019E − 11▲	3.019E − 11▲	6.121E − 10▲	3.158E − 10▲	3.019E − 11▲	3.019E − 11▲	1.594E − 13▲	1.205E − 10▲	1.205E − 10▲
f_{18}	3.019E − 11▲	8.152E − 11▲	5.572E − 10▲	3.019E − 11▲	3.019E − 11▲	3.019E − 11▲	1.017E − 13▲	3.019E − 11▲	1.205E − 11▲
f_{19}	1.107E − 06▲	2.227E − 09▲	1.621E − 05▲	6.765E − 05▲	1.107E − 06▲	2.491E − 06▲	4.913E − 05▲	6.765E − 05▲	1.491E − 06▲
f_{20}	0.0000E00▶	2.154E − 10▲	5.497E − 06▲	8.484E − 09▲	0.0000E00▶	3.019E − 11▲	4.670E − 09▲	7.690E − 08▲	0.0000E00▶
f_{21}	3.019E − 11▲	3.019E − 11▲	3.689E − 11▲	3.019E − 11▲	3.019E − 11▲	0.0000E00▶	8.649E − 13▲	3.019E − 11▲	3.019E − 11▲

(continued)

Table 2.5 (continued)

EA-HC versus	ABC	DE	PSO	CS	DS	JADE	MFO	MVO	SCA
f_{22}	0.0000E00▶	5.490E − 11▲	3.874E − 03▲	2.596E − 05▲	0.0000E00▶	1.107E − 06▲	2.564E − 07▲	3.156E − 05▲	0.0000E00▶
f_{23}	3.019E − 11▲	3.019E − 11▲	5.607E − 05▲	3.019E − 11▲	3.019E − 11▲	3.019E − 11▲	9.986E − 09▲	3.019E − 11▲	6.722E − 10▲
▲	16	23	23	20	16	18	21	22	19
▼	0	0	0	0	0	0	0	0	0
▶	7	0	0	3	7	5	2	1	4

Table 2.6 The p-values resulting from the Wilcoxon rank sum test, which compares EA-HC against ABC, DE, PSO, CS, DS, JADE, MFO, MVO, and SCA in terms of the average fitness value $\bar{f}$ for each function in Table 2.3

EA-HC versus	ABC	DE	PSO	CS	DS	JADE	MFO	MVO	SCA
f_1	3.019E − 11▲	3.019E − 11▲	3.019E − 11▲	3.019E − 11▲	3.019E − 11▲	3.019E − 11▲	2.194E − 08▲	3.019E − 11▲	3.019E − 11▲
f_2	1.066E − 07▲	3.019E − 11▲	2.572E − 07▲	1.211E − 12▲	2.530E − 04▲	4.744E − 06▲	5.461E − 09▲	3.019E − 11▲	6.526E − 07▲
f_3	3.081E − 08▲	3.019E − 11▲	1.193E − 06▲	3.019E − 11▲	3.645E − 08▲	7.695E − 08▲	1.681E − 04▲	3.019E − 11▲	8.120E − 04▲
f_4	3.019E − 11▲	3.019E − 11▲	6.282E − 06▲	3.019E − 11▲	3.019E − 11▲	5.572E − 10▲	3.019E − 11▲	6.736E − 06▲	3.019E − 11▲
f_5	3.019E − 11▲	3.019E − 11▲	3.019E − 11▲	3.019E − 11▲	3.019E − 11▲	3.019E − 11▲	7.233E − 09▲	3.019E − 11▲	3.019E − 11▲
f_6	2.982E − 11▲	3.019E − 11▲	3.019E − 11▲	3.014E − 11▲	3.019E − 11▲	2.993E − 11▲	1.354E − 07▲	3.010E − 11▲	3.019E − 11▲
f_7	3.019E − 11▲	3.019E − 11▲	3.019E − 11▲	3.019E − 11▲	3.019E − 11▲	3.019E − 11▲	6.207E − 04▲	3.019E − 11▲	4.329E − 05▲
f_8	3.019E − 11▲	2.138E − 06▲	8.352E − 08▲	3.019E − 11▲	3.019E − 11▲	0.0000E00▶	3.099E − 11▲	3.019E − 11▲	3.019E − 11▲
f_9	3.019E − 11▲	2.028E − 07▲	4.185E − 09▲	3.019E − 11▲	3.019E − 11▲	0.0000E00▶	3.019E − 11▲	3.019E − 11▲	3.019E − 11▲
f_{10}	3.019E − 11▲	3.019E − 11▲	7.064E − 09▲	3.624E − 11▲	4.440E − 07▲	8.165E − 05▲	9.524E − 08▲	4.186E − 05▲	6.526E − 07▲

(continued)

Table 2.6 (continued)

EA-HC versus	ABC	DE	PSO	CS	DS	JADE	MFO	MVO	SCA
f_{11}	5.572E − 10▲	3.019E − 11▲	6.120E − 10▲	8.484E − 09▲	7.772E − 09▲	3.820E − 10▲	3.019E − 11▲	8.841E − 07▲	6.736E − 06▲
f_{12}	3.019E − 11▲	3.019E − 11▲	2.371E − 10▲	3.019E − 11▲	3.019E − 11▲	3.019E − 11▲	2.159E − 07▲	3.019E − 11▲	3.338E − 11▲
f_{13}	0.0000E00►	3.019E − 11▲	3.019E − 11▲	5.572E − 10▲	3.822E − 10▲	5.572E − 10▲	3.099E − 06▲	2.227E − 09▲	9.918E − 11▲
f_{14}	3.019E − 11▲	3.019E − 11▲	5.494E − 11▲	3.019E − 11▲	3.019E − 11▲	3.019E − 11▲	1.243E − 05▲	3.099E − 11▲	1.529E − 05▲
f_{15}	5.572E − 10▲	3.019E − 11▲	3.830E − 05▲	5.572E − 10▲	5.572E − 10▲	3.3384–11▲	1.018E − 05▲	5.572E − 10▲	8.810E − 10▲
f_{16}	3.099E − 11▲	1.105E − 04▲	1.047E − 04▲	3.099E − 11▲	3.099E − 11▲	3.019E − 11▲	3.099E − 11▲	3.019E − 11▲	4.182E − 09▲
f_{17}	3.099E − 11▲	3.019E − 11▲	2.015E − 08▲	3.019E − 11▲	3.019E − 11▲	3.019E − 11▲	4.981E − 04▲	3.019E − 11▲	3.019E − 11▲
f_{18}	3.019E − 11▲	3.019E − 11▲	3.689E − 11▲	3.019E − 11▲	3.019E − 11▲	3.019E − 11▲	2.342E − 10▲	3.019E − 11▲	2.609E − 10▲
f_{19}	3.019E − 11▲	3.019E − 11▲	2.194E − 08▲	1.211E − 12▲	3.019E − 11▲	3.019E − 11▲	3.019E − 11▲	3.019E − 11▲	3.019E − 11▲

(continued)

Table 2.6 (continued)

EA-HC versus	ABC	DE	PSO	CS	DS	JADE	MFO	MVO	SCA
f_{20}	3.019E – 11▲	3.019E – 11▲	2.665E – 09▲	3.099E – 11▲	3.019E – 11▲	3.019E – 11▲	1.356E – 06▲	3.384E – 11▲	3.099E – 11▲
f_{21}	3.019E – 11▲	3.019E – 11▲	1.270E – 09▲	3.019E – 11▲	3.199E – 11▲	3.019E – 11▲	9.753E – 05▲	3.019E – 11▲	8.114E – 10▲
f_{22}	3.019E – 11▲	3.019E – 11▲	5.835 E – 11▲	1.218E – 12▲	3.820E – 10▲	5.572E – 10▲	7.391E – 11▲	3.019E – 11▲	1.632E – 10▲
f_{23}	3.019E – 11▲	3.019E – 11▲	1.328E – 10▲	1.218E – 12▲	3.019E – 11▲	3.099E – 11▲	3.019E – 11▲	5.527E – 10▲	1.846E – 06▲
▲	22	23	23	23	23	21	23	23	23
▼	0	0	0	0	0	0	0	0	0
►	1	0	0	0	0	2	0	0	0

Table 2.7 The p – values obtained from the Wilcoxon rank – sum test, which compares EA – HC with ABC, DE, PSO, CS, DS, JADE, MFO, MVO, and SCA based on the average fitness value $\bar{f}$ for each function in Table 2.4

EA – HC versus	ABC	DE	PSO	CS	DS	JADE	MFO	MVO	SCA
f_1	2.371E – 10▲	3.099E – 11▲	3.099E – 11▲	4.503E – 11▲	3.019E – 11▲	2.605E – 08▲	3.019E – 11▲	5.577E – 10▲	3.019E – 11▲
f_2	2.499E – 03▲	3.019E – 11▲	3.472E – 10▲	1.218E – 12▲	2.678E – 06▲	1.056E – 03▲	3.019E – 11▲	3.019E – 11▲	2.380E – 03▲
f_3	1.442E – 02▲	6.720E – 10▲	3.157E – 05▲	4.118E – 06▲	1.167E – 05▲	1.335E – 03▲	6.372E – 03▲	1.498E – 06▲	3.031E – 02▲
f_4	3.019E – 11▲	3.099E – 11▲	3.019E – 11▲	3.019E – 11▲	3.019E – 11▲	3.019E – 11▲	3.019E – 11▲	3.019E – 11▲	3.019E – 11▲
f_5	9.916E – 11▲	3.019E – 11▲	3.019E – 11▲	9.519E – 06▲	1.3289E – 10▲	5.9969E – 02▲	3.0199E – 11▲	3.988E – 04▲	3.334E – 11▲
f_6	1.871E – 10▲	2.365E – 12▲	2.365E – 12▲	3.554E – 11▲	2.367E – 12▲	1.238E – 09▲	2.365E – 12▲	1.234E – 09▲	2.367E – 12▲
f_7	5.577E – 10▲	3.019E – 11▲	3.019E – 11▲	5.577E – 10▲	3.158E – 10▲	5.572E – 10▲	3.019E – 11▲	5.527E – 10▲	3.019E – 11▲
f_8	6.627E – 02▲	5.572E – 10▲	1.232E – 02▲	3.957E – 02▲	6.843E – 02▲	0.0000E00►	6.952E – 02▲	6.673E – 02▲	1.877E – 01▲
f_9	6.623E – 02▲	3.099E – 11▲	2.609E – 10▲	6.200E – 02▲	6.627E – 02▲	5.593E – 02▲	1.452E – 02▲	6.623E – 02▲	6.623E – 01▲
f_{10}	7.694E – 02▲	2.721E – 11▲	4.494E – 11▲	6.415E – 06▲	8.789E – 04▲	7.376E – 05▲	4.061E – 11▲	1.554E – 09▲	4.625E – 02▲

(continued)

Table 2.7 (continued)

EA − HC versus	ABC	DE	PSO	CS	DS	JADE	MFO	MVO	SCA
f_{11}	5.468E − 10▲	2.954E − 11▲	1.301E − 10▲	3.744E − 10▲	1.433E − 10▲	2.924E − 11▲	2.924E − 11▲	6.518E − 10▲	4.864E − 11▲
f_{12}	1.796E − 08▲	1.795E − 11▲	3.528E − 10▲	3.521E − 10▲	3.581E − 10▲	8.046E − 06▲	3.581E − 10▲	8.324E − 07▲	3.521E − 10▲
f_{13}	4.414E − 09▲	1.271E − 11▲	1.271E − 11▲	4.44E − 09▲	4.414E − 09▲	4.411E − 09▲	1.410E − 11▲	4.414E − 09▲	4.414E − 09▲
f_{14}	3.023E − 11▲	3.123E − 11▲	3.013E − 11▲	3.013E − 11▲	3.012E − 11▲	3.012E − 11▲	3.012E − 11▲	3.012E − 11▲	3.013E − 11▲
f_{15}	4.351E − 05▲	3.099E − 11▲	3.019E − 11▲	6.750E − 05▲	4.117E − 06▲	1.875E − 03▲	3.019E − 11▲	6.650E − 05▲	1.5581E − 0▲
f_{16}	1.337E − 05▲	2.385E − 04▲	1.249E − 05▲	3.981E − 04▲	3.981E − 04▲	3.981E − 04▲	3.988E − 04▲	3.988E − 04▲	7.040E − 07▲
f_{17}	4.686E − 08▲	3.019E − 11▲	4.197E − 10▲	8.488E − 09▲	5.527E − 10▲	8.848E − 09▲	3.099E − 11▲	8.448E − 09▲	3.099E − 11▲
f_{18}	2.1959E − 07▲	3.0199E − 11▲	4.975E − 11▲	8.4848E − 09▲	2.9215E − 09▲	8.4848E − 09▲	3.019E − 11▲	8.484E − 09▲	3.019E − 11▲
f_{19}	7.959E − 03▲	3.0199E − 11	5.577E − 10▲	1.211E − 12▲	7.186E − 09▲	7.284E − 03▲	3.019E − 11▲	3.019E − 11▲	3.849E − 09▲

(continued)

Table 2.7 (continued)

EA − HC versus	ABC	DE	PSO	CS	DS	JADE	MFO	MVO	SCA
f_{20}	9.539E − 06▲	3.019E − 11▲	1.680E − 08▲	9.519E − 06▲	9.519E − 06▲	9.539E − 06▲	3.019E − 11▲	7.220E − 06▲	2.493E − 06▲
f_{21}	8.649E − 03▲	3.019E − 11▲	1.286E − 06▲	6.200E − 02▲	9.797E − 05▲	3.253E − 02▲	3.019E − 11▲	3.794E − 02▲	3.019E − 11▲
f_{22}	3.961E − 04▲	2.953E − 11▲	3.406E − 10▲	1.177E − 12▲	1.808E − 05▲	7.594E − 05▲	2.953E − 11▲	2.953E − 11▲	7.147E − 05▲
f_{23}	7.487E − 03▲	3.019E − 11▲	1.205E − 10▲	1.218E − 12▲	5.072E − 10▲	2.002E − 06▲	3.019E − 11▲	3.019E − 11▲	3.099E − 11▲
▲	23	23	23	23	23	22	23	23	23
▼	0	0	0	0	0	0	0	0	0
▶	0	0	0	0	0	1	0	0	0

The p-values for functions f_9 and f_{13} suggest that EA-HC produces similar results to ABC, CS, DS, JADE, MFO, and SCA. In function f_{16}, the rank-sum test cannot distinguish between the numerical results for the groups EA-HC versus ABC and EA-HC versus DS. Additionally, the groups EA-HC versus ABC, EA-HC versus DS, and EA-HC versus SCA exhibit similar performance in functions f_{20} and f_{22}. The group EA-HC versus JADE statistically confirms the results from Table 2.2 for function f_{21}. Referring to the p-values in Table 2.6 ($n = 100$), it is evident that the agent-based approach outperforms each metaheuristic algorithm considered in the experimental study. Notably, in functions f_8 and f_9, the Wilcoxon test indicates that EA-HC and JADE perform similarly. Additionally, for function f_{13}, the ABC and EA-HC methods achieve the best fitness value compared to other evolutionary methodologies. In the remaining entries, the EA-HC method outperforms the competitors in the 100-dimensional search space for each benchmark function.

Based on the p-value in Table 2.7 ($n = 200$), it is evident that the EA-HC method directly competes with the JADE algorithm as both algorithms yield similar results. However, when considering the overall experiment, the scalability of the evolutionary operators in EA-HC enables a balance between exploration and exploitation stages due to its adaptive switching mechanism between these stages.

The statistical findings presented in Tables 2.5, 2.6 and 2.7 indicate that the dynamic nature of the agent-based approach's evolutionary structure yields more resilient and adaptable results compared to other metaheuristic algorithms. This is achieved by preserving population diversity and maintaining a balance between exploration and exploitation throughout the optimization process, leading to enhanced robustness and scalability.

2.5.3 *Study of Convergence*

This section focuses on the convergence analysis of both the agent-based approach and the tested metaheuristic algorithms. While Tables 2.2, 2.3 and 2.4 present the performance comparison in terms of fitness values, it is essential to include a convergence study in line with existing literature. This study evaluates the rate at which metaheuristic approaches converge during the optimization process for each benchmark function. For the convergence experiment, data was selected based on average fitness values, specifically considering the evaluation conducted in 200-dimensional search spaces as shown in Table 2.4.

Figure 2.4 demonstrates that the agent-based method exhibits the highest convergence rate compared to the tested algorithms. This indicates that EA-HC achieves faster convergence during the optimization process. The exceptional performance of EA-HC can be attributed to its unique changing characteristic, which effectively preserves population diversity throughout the entire process, leading to more reliable results.

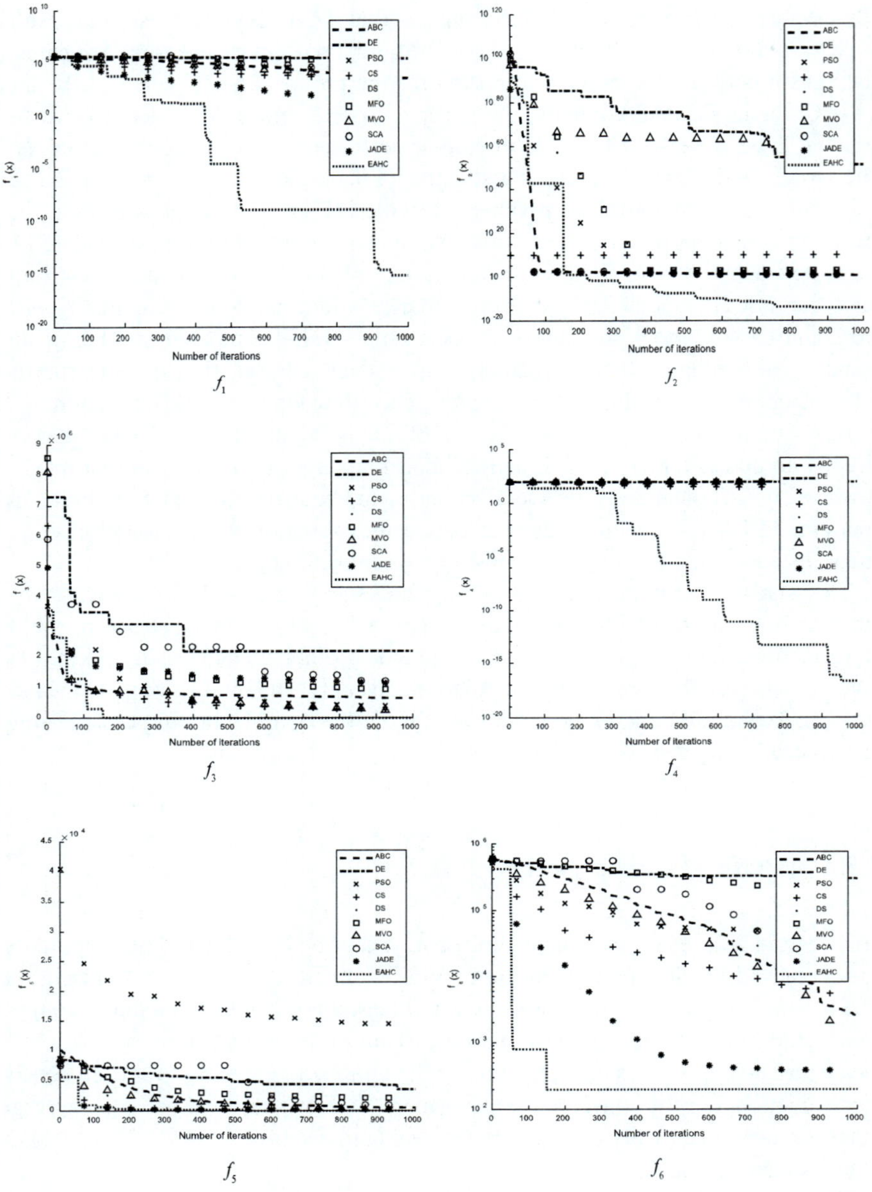

Fig. 2.4 Graphs that show the convergence results from Table 2.4

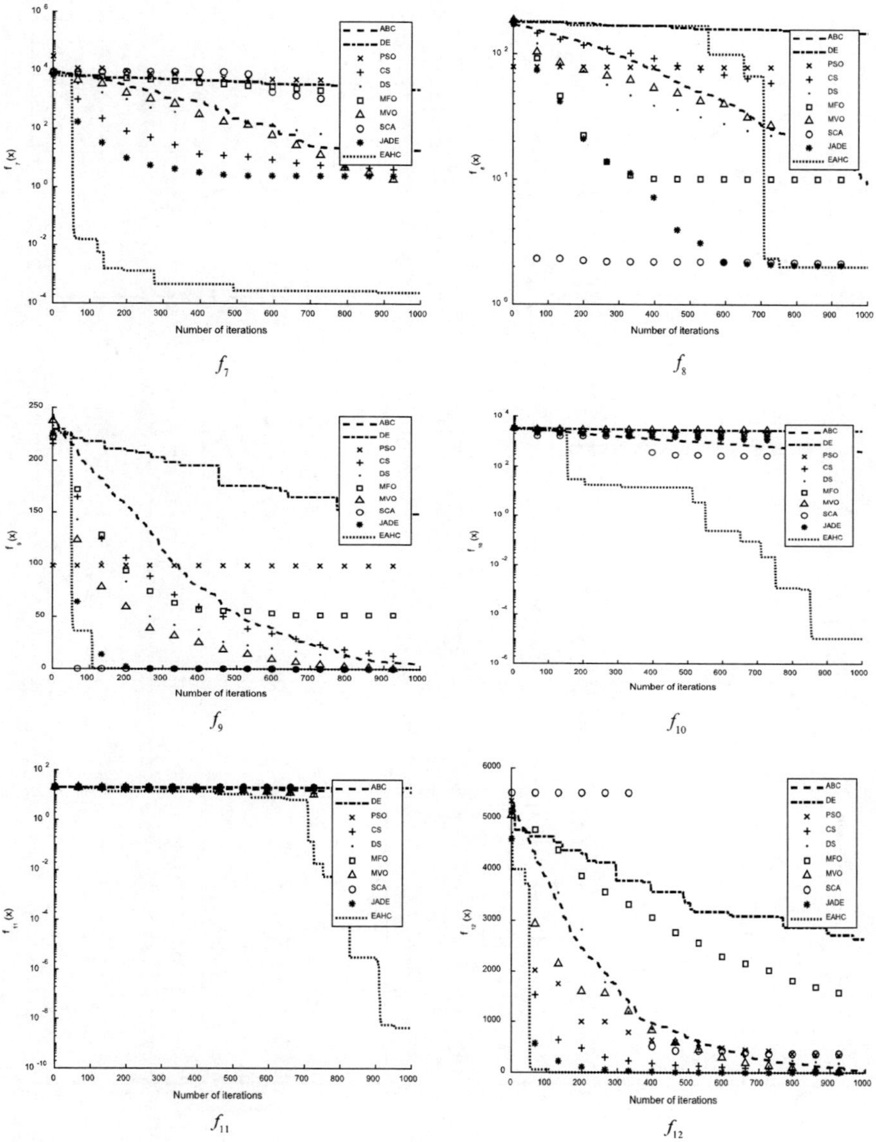

Fig. 2.4 (continued)

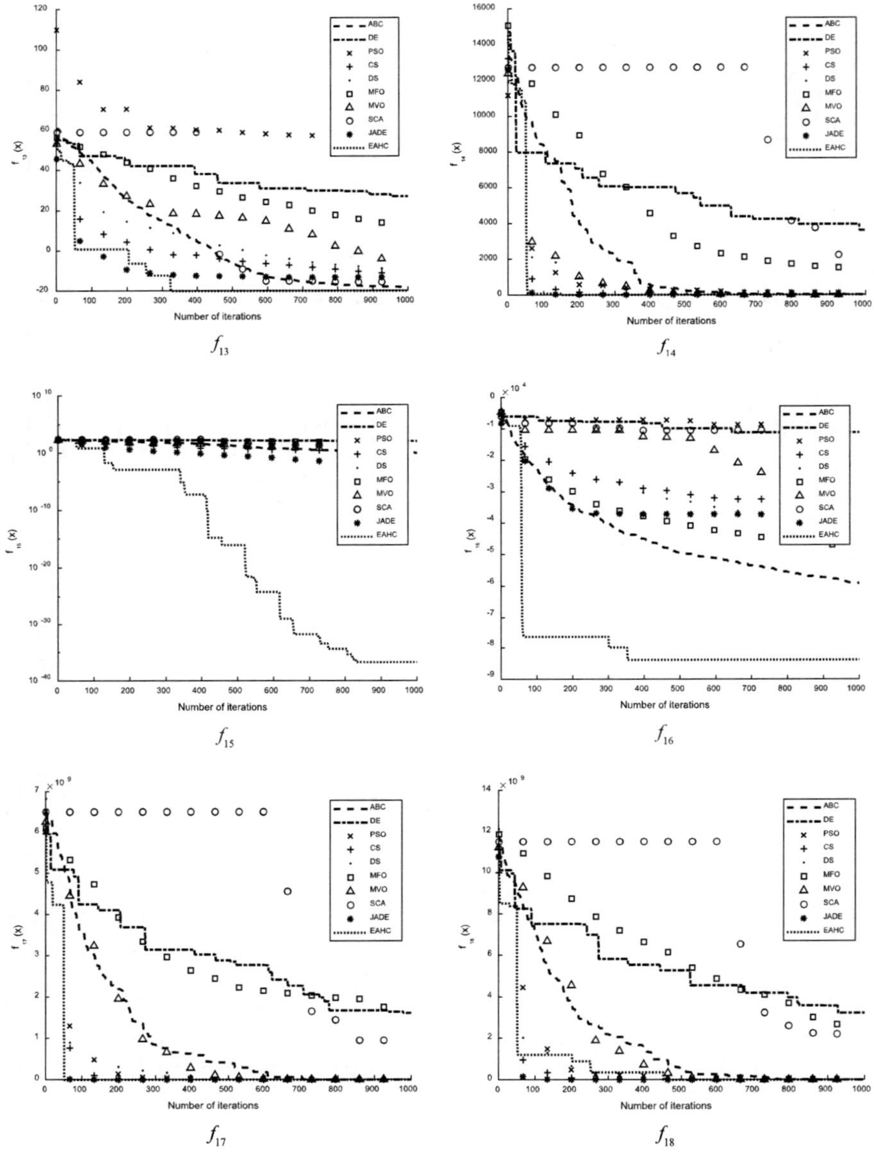

Fig. 2.4 (continued)

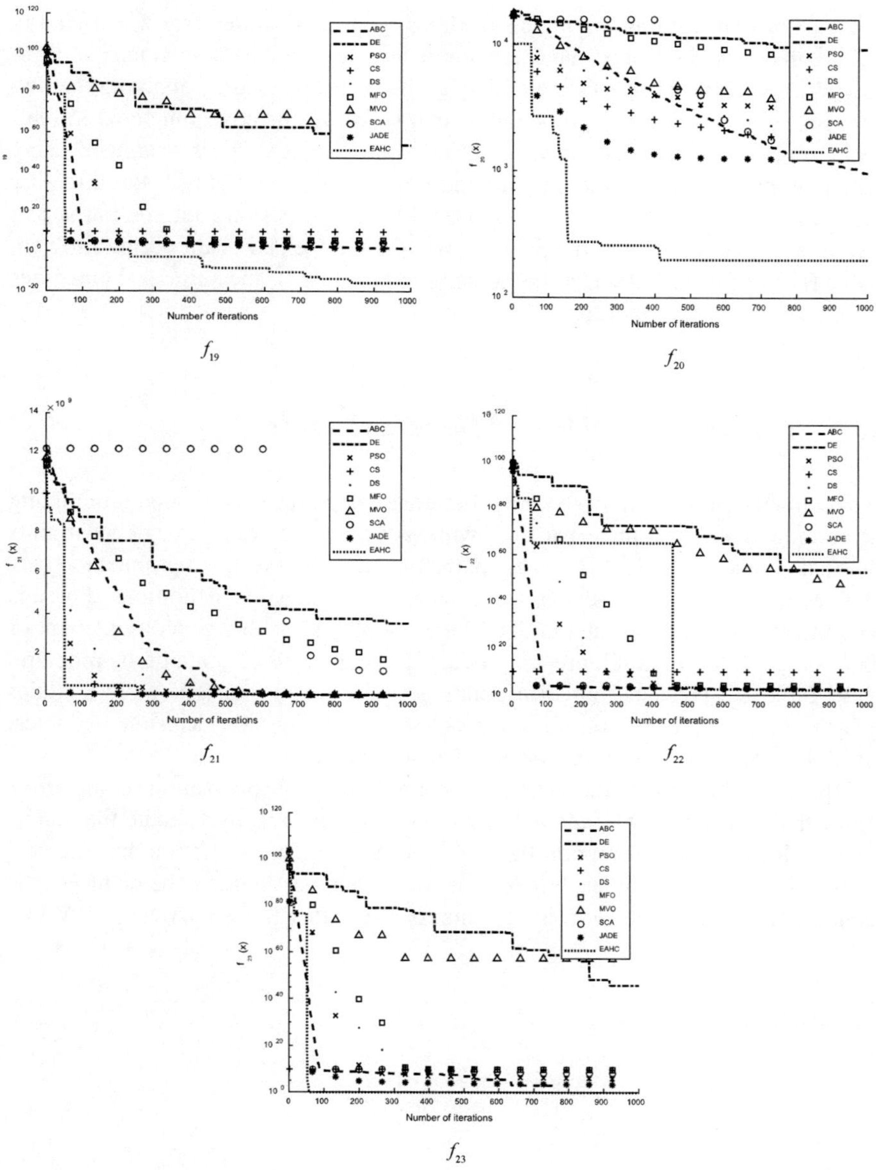

Fig. 2.4 (continued)

2.5.4 Applications

Optimization problems serve as mathematical representations of various real-world issues and aim to find the optimal solution for a given objective function. As numerous fields, including engineering, medicine, and economics, formulate their problems in terms of minimizing or maximizing certain parameters, it is crucial to evaluate recently developed metaheuristic techniques within these contexts. Typically, the performance of metaheuristic algorithms is assessed by applying them to various engineering design problems, which serve as objective functions in real-world applications. In this study, the effectiveness of EA-HC is examined using three commonly encountered engineering design problems [64.65]: the three-bar truss design problem (Table A.2 in Appendix B), the tension/compression spring design problem (Table A.3 in Appendix B), and the welded beam design problem (Table B3 in Appendix B). The subsequent section presents the numerical outcomes obtained from this evaluation.

2.5.5 *The Design Problem of Three-Bar Truss*

The optimization problem of the three-bar truss design revolves around minimizing the volume of a loaded three-bar truss while adhering to various design constraints imposed on each truss [63–67]. The mathematical model of this optimization task involves a 2-dimensional search space with three inequality conditions. For this experiment, the setup outlined in Sect. 5.1 is employed, with a generation count of 1000 for each algorithm. Figure 2.5 provides a visual representation of the problem, while Table 2.8 presents the corresponding numerical results. The table includes the parameters of the optimization problem, specifying the decision variables, constraints, and fitness value associated with each entry.

Table 2.8 indicates that the agent-based method is capable of obtaining competitive results than CS, DS, JADE and MVO methods. To statistically validate the results from Tables 2.8 and 2.9 reports the worst, mean, standard deviation and the best fitness values achieved by each metaheuristic technique. As it can be deduced, the agent-based EA-HC achieves similar fitness value to CS, DS, JADE and MVO.

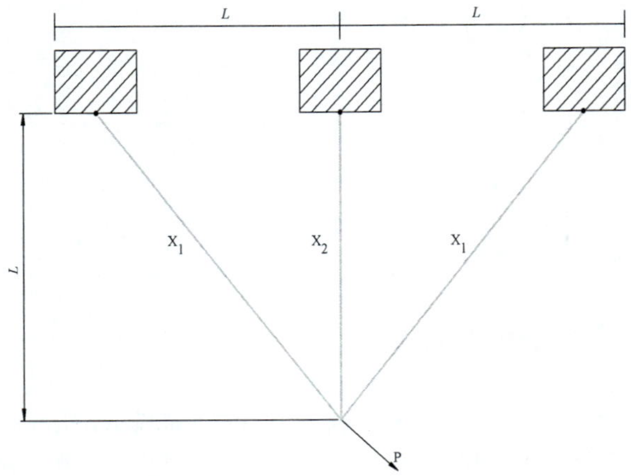

Fig. 2.5 Graphical Description of the three-bar truss problem

Table 2.8 Obtained results of the three-bar truss problem

	ABC	DE	PSO	CS	DS	JADE	MFO	MVO	SCA	EA-HC
x_1	0.8643	0.8708	0.8573	0.8698	0.8700	0.8837	0.8702	0.8697	0.8730	0.8522
x_2	0.2278	0.2149	0.2710	0.2166	0.2162	0.1945	0.2157	0.2167	0.2101	0.2528
$g_1(x)$	−1.49E−04	−3.60E−04	−0.0275	−2.2204e−16	−6.2774e−10	−4.7746e−05	0	−2.5611e−07	−5.7400e−05	0.0001
$g_2(x)$	−1.6858	−1.7029	−1.6396	−1.7005	−1.7010	−1.7295	−1.7018	−1.7005	−1.7091	−1.6532
$g_3(x)$	−0.3143	−0.2975	−0.3878	−0.2995	−0.2990	−0.2704	−0.2982	−0.2995	−0.2909	0.3467
$f(x)$	279.7452	279.7472	280.5426	**279.7245**	**279.7245**	**279.7245**	279.7246	**279.7245**	279.7312	**279.7245**

Table 2.9 Results of the statistical analysis

Algorithm	Worst	Mean	Std	Best
ABC	279.7849	279.7372	0.0125	279.7452
DE	280.1867	279.8096	0.0902	279.7472
PSO	282.8226	93.2079	29.9639	280.5426
CS	279.7245	279.7245	1.3843e−13	**279.7245**
DS	280.0971	279.7484	0.0677	**279.7245**
JADE	282.8427	279.8500	0.5675	**279.7245**
MFO	280.1945	279.7763	0.1028	279.7246
MVO	279.7270	279.7251	5.7728e−04	**279.7245**
SCA	282.8427	281.4059	1.5624	279.7312
EA–HC	279.9114	279.7491	0.0411	**279.7245**

However, it presents higher mean and standard deviation values due to its changing characteristics in the evolutionary structure in CS, DS and MVO.

2.5.6 *Tension/Compression of a Spring as Design Problem*

The spring design optimization task serves as a benchmark problem to evaluate the effectiveness of metaheuristic methods in minimizing the weight of a tension/compression spring. This problem involves a 3-dimensional search space [67–71], with variables representing the wire diameter $W(x_1)$, mean coil diameter $d(x_2)$, and the number of active coils $L(x_3)$. Additionally, there are three non-linear inequality constraints to be considered. The experimental results for this task can be found in Table 2.10. A schematic of the tension/compression spring problem is provided in Fig. 2.6.

Based on the results presented in Table 2.10, it is evident that EA-HC achieves a lower fitness value compared to CS, DS, JADE, MFO, MVO, and SCA. While these algorithms outperform EA-HC, it should be noted that EA-HC still produces competitive results when compared to state-of-the-art algorithms. This can be attributed to the efficient performance achieved by the dynamic balance between the exploration and exploitation operators of EA-HC. The statistical results for this experiment are reported in Table 2.11.

Problem of the welded beam design.

The optimization problem of the welded beam design is a challenging engineering task aimed at minimizing the cost of a welded beam. The design problem is visually depicted in Fig. 2.7. It involves a 4-dimensional search space, including the width $h(x_1)$, length $l(x_2)$, depth $t(x_3)$, and thickness $b(x_4)$ of the beam. Additionally, there are 7 constraints associated with this optimization problem, which are

Table 2.10 Obtained results for the problem

	ABC	DE	PSO	CS	DS	JADE	MFO	MVO	SCA	EA-HC
w	0.0500	0.0500	0.0516	0.0500	0.0500	0.0500	0.0500	0.0500	0.0500	0.0518
d	0.4532	0.4278	0.3560	0.4800	0.4800	0.3712	0.4800	0.4800	0.4796	0.4156
L	4.8539	5.8778	11.3295	4.0563	4.0577	14.352	4.0563	4.0580	4.0806	9.4088
$g_1(x)$	−0.0070	−0.0257	−0.000006	−1.577e−13	−4.834e−06	−0.6374	−0.0025	−6.6620e−05	−0.0035	−0.3079
$g_2(x)$	−0.1001	−0.1896	−0.000013	−2.061e−14	−2.155e−04	−0.3702	−1.6653e−15	−2.1897e−04	−0.0015	−0.2361
$g_3(x)$	−6.0440	−5.5282	−4.0523	−6.5135	−6.5126	−2.5491	−6.4950	−6.5121	−6.4810	−3.4755
$g_4(x)$	−0.6645	−0.6815	−0.7282	−0.6467	−0.6467	−0.7191	−0.6467	−0.6467	−0.6469	−0.6884
$f(x)$	0.0078	0.0084	0.0127	**0.0073**	**0.0073**	**0.0073**	**0.0073**	**0.0073**	**0.0073**	0.0076

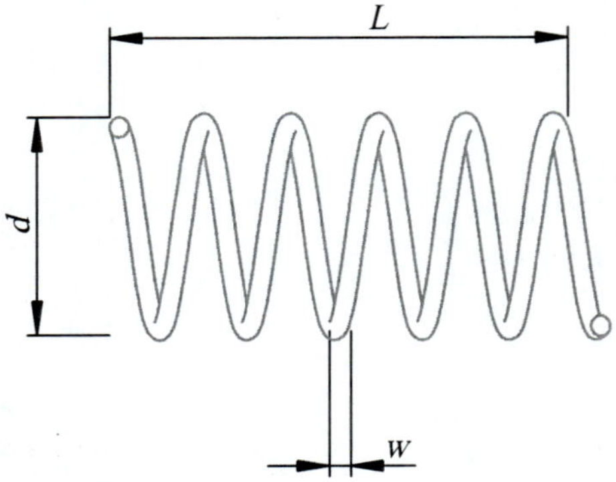

Fig. 2.6 Illustration of the spring design problem

Table 2.11 Results of the statistical analysis

Algorithm	Worst	Mean	Std	Best
ABC	0.0084	0.0076	0.0002	0.0078
DE	1.00E + 06	2.00E + 05	4.07E + 05	0.0084
PSO	1.00E + 06	1.67E + 05	3.79E + 05	0.0127
CS	0.0073	0.0073	2.3681e − 12	**0.0073**
DS	0.0081	0.0074	2.2511e − 04	**0.0073**
JADE	1.00E + 06	3.333E + 04	1.8257E + 05	**0.0073**
MFO	0.0073	0.0073	1.5149e − 07	**0.0073**
MVO	1.00E + 06	4.0000e + 05	4.9827e + 05	**0.0073**
SCA	0.0077	0.0074	9.4466e − 05	**0.0073**
EA − HC	0.0237	0.0129	0.0031	0.0076

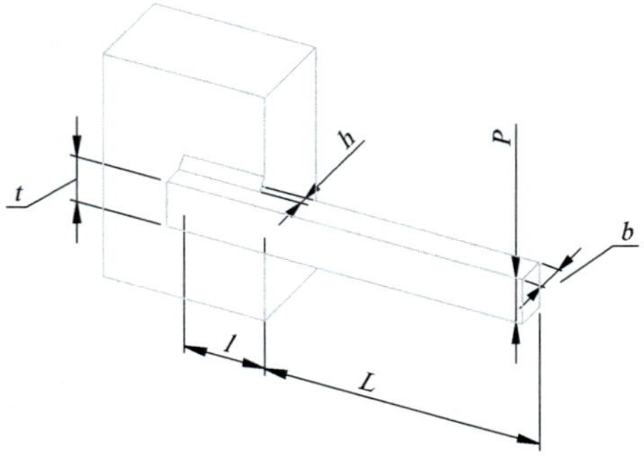

Fig. 2.7 Illustration of the Welded beam problem

detailed in Table A.4 in Appendix 2. The numerical results for this optimization task are presented in Table 2.12.

Table 2.12 Obtained results for the welded beam problem

	ABC	DE	PSO	CS	DS	JADE	MFO	MVO	SCA	EA-HC
h	0.1993	0.1626	0.8120	0.2057	0.2057	0.2739	0.2057	0.2045	0.2004	0.2057
l	3.8115	6.8239	6.4319	3.4705	3.4706	3.4241	3.4705	3.5040	3.7893	3.4705
t	8.6442	8.7650	9.2837	9.0366	9.0385	9.0331	9.0366	9.0375	9.0748	9.0366
b	0.2259	0.2346	0.7008	0.2057	0.2057	0.2869	0.2057	0.2058	0.2062	0.2057
$g_1(x)$	−81.6604	−2977.1639	−1.16E + 04	−0.0062	−0.0322	−3.330e + 03	−0.0062	−23.2984	−616.4397	−0.0062
$g_2(x)$	−135.7300	−2035.9101	−2.17E + 04	−0.1334	−14.1393	−8.778e + 03	−0.1334	−9.186	−315.5598	−0.1334
$g_3(x)$	−0.0265	−0.0720	0.1112	−1.286e − 06	−5.025e − 05	−0.0130	−1.286e − 06	−0.0012	−0.0058	−1.286e - 06
$g_4(x)$	−3.3229	−2.9372	1.4640	−3.4330	−3.4326	−2.8189	−3.4406	−3.4297	−3.3945	−3.4306
$g_5(x)$	−0.0743	−0.0376	−0.6870	−0.0807	−0.0807	−0.1489	−0.0807	−0.0795	−0.0754	−0.0807
$g_6(x)$	−0.2350	−0.2361	−0.2461	−0.2355	−0.2356	−0.2396	−0.2355	−0.2355	−0.2358	−0.2355
$g_7(x)$	−1706.7356	−2718.3738	−2.35E + 05	−0.0157	−1.9242	−1.023e + 04	−0.0157	−2.5351	−55.4759	−0.0157
$f(x)$	1.8403	2.2593	2.0118	**1.7249**	1.7252	1.7577	**1.7249**	1.7278	1.7694	**1.7249**

Table 2.13 Results of the statistical analysis

Algorithm	Worst	Mean	Std	Best
ABC	2.0848	1.9269	0.0849	1.8403
DE	3.6161	2.6481	0.3445	2.2593
PSO	3.3931	2.1544	0.3594	2.0118
CS	1.7250	1.7249	2.2364e − 05	**1.7249**
DS	2.2552	1.8165	0.1205	1.7252
JADE	2.9267	2.2761	0.3097	1.7578
MFO	2.3171	1.7854	0.1282	**1.7249**
MVO	1.8169	1.7412	0.0182	1.7278
SCA	1.9711	1.8519	0.0493	1.7694
EA-HC	2.3171	1.9850	0.2046	**1.7249**

From Table 2.12, it can be demonstrated the EA-HC is capable of achieving quite similar results than CS and MFO. To corroborate the experimental results from Tables 2.12 and 2.13 reports that EA-HC produces results with greater mean and standard deviation. However, this is not a limitation of the agent-based approach, since it uses changing framework overexploitation and exploration stages during the entire evolutionary process.

2.6 Conclusions

The performance of metaheuristic approaches is primarily influenced by the interactions among individuals rather than other factors. Although different metaheuristic schemes model interactions based on various processes or systems, their operators share common mathematical processes that overlook the final global result of individual interactions. In contrast, agent-based systems offer a solid theoretical foundation and consistent models to characterize collective behavioral patterns resulting from the interactions of individuals governed by simple rules. This perspective enables the utilization of agent-based concepts and models to develop efficient optimization algorithms or enhance existing ones.

In this chapter, a novel metaheuristic technique is presented, incorporating agent-based principles. The agent-based approach adopts the "Heroes and Cowards" agent-based model, where candidate solutions are treated as agents and their interactions follow a small set of rules, giving rise to two emergent search processes: Coward and Heroic. These processes correspond to the exploration and exploitation stages commonly found in metaheuristics. During the Coward process, agents disperse across the search space as a result of evading enemies. Conversely, during the Heroic process, agents semi-converge around positions influenced by agent distributions. The model also employs a moderator to dynamically transition between these two phases.

The performance of the agent-based method is evaluated by comparing it numerically against several state-of-the-art metaheuristic schemes on 23 benchmark functions with varying complexities. The experimental results demonstrate that the agent-based metaheuristic approach outperforms its competitors in terms of accuracy and scalability. This remarkable performance stems from the dynamic switching between exploration and exploitation throughout the entire optimization process.

The chapter aims to achieve two main objectives: (I) to demonstrate the effectiveness of agent-based models as metaheuristic methods, and (II) to highlight the promising potential of combining both artificial intelligence paradigms. The results clearly indicate that the use of agent-based concepts and models enables the development of effective metaheuristic schemes that maintain superior innovative search patterns by incorporating a well-established methodology to implement complex interactions among agents.

Appendix 1 Functions Used in the Experiments

See Table A.1.

Appendix 2 Engineering Design Problems

See Tables A.2, A.3 and A.4.

Table A.1 Functions used in the experimental analysis

Function	S	Dimension	Minimum
$f_1(\mathrm{x}) = \sum_{i=1}^{n} x_i^2$	$[-100, 100]^n$	$n = 30$ $n = 100$ $n = 200$	$x^* = (0, \ldots, 0)$; $f(x^*) = 0$
$f_2(\mathrm{x}) = \sum_{i=1}^{n} \lvert x_i \rvert + \prod_{i=1}^{n} \lvert x_i \rvert$	$[-100, 100]^n$	$n = 30$ $n = 100$ $n = 200$	$x^* = (0, \ldots, 0)$; $f(x^*) = 0$
$f_3(\mathrm{x}) = \sum_{i=1}^{n} \left(\sum_{j=1}^{i} x_i^2 \right)^2$	$[-100, 100]^n$	$n = 30$ $n = 100$ $n = 200$	$x^* = (0, \ldots, 0)$; $f(x^*) = 0$
$f_4(\mathrm{x}) = \max\{\lvert x_i \rvert, 1 \le i \le n\}$	$[-100, 100]^n$	$n = 30$ $n = 100$ $n = 200$	$x^* = (0, \ldots, 0)$; $f(x^*) = 0$
$f_5(\mathrm{x}) = \sum_{i=1}^{n} \left[100\left(x_{i+1} - x_i^2\right)^2 + (x_i - 1)^2 \right]$	$[-30, 30]^n$	$n = 30$ $n = 100$ $n = 200$	$x^* = (1, \ldots, 1)$; $f(x^*) = n$
$f_6(\mathrm{x}) = \sum_{i=1}^{n} ([x_i + 0.5])^2$	$[-100, 100]^n$	$n = 30$ $n = 100$ $n = 200$	$x^* = (0, \ldots, 0)$; $f(x^*) = 0$
$f_7(\mathrm{x}) = \sum_{i=1}^{n} i x_i^4 + random(0, 1)$	$[-1.28, 1.28]^n$	$n = 30$ $n = 100$ $n = 200$	$x^* = (0, \ldots, 0)$; $f(x^*) = 0$
$f_8(\mathrm{x}) = \sum_{i=1}^{n} x_i + \sum_{i=1}^{n} \frac{x_i}{\left(-x_i + \sum_{i1=1}^{n} x_{i1}\right)^2}$	$\left[10e^{-6}, 2\right]^n$	$n = 30$ $n = 100$ $n = 200$	$\mathrm{x}^* =$ $\left(\frac{1}{n-1}, \ldots, \frac{1}{n-1}\right)$; $f(\mathbf{x}^*) = \frac{2n}{n-1}$

(continued)

Table A.1 (continued)

Function	S	Dimension	Minimum
$f_9(\mathrm{x}) = \sum_{i=1}^{n} x_i^2 + \sum_{i=1}^{n} \frac{x_i{}^2}{\left(-x_i + \sum_{i1=1}^{n} x_{i1}\right)^4}$	$\left[10e^{-6}, 2\right]^n$	$n = 30$ $n = 100$ $n = 200$	$\mathrm{x}^* = \left(\frac{1}{n-1}, \ldots, \frac{1}{n-1}\right)$; $f(\mathbf{x}^*) = \frac{2n}{(n-1)^2}$
$f_{10}(\mathrm{x}) = \sum_{i=1}^{n}\left[x_i^2 - 10\cos(2\pi x_i) + 10\right]$	$[-5.12, 5.12]^n$	$n = 30$ $n = 100$ $n = 200$	$\mathrm{x}^* = (0, \ldots, 0)$; $f(\mathrm{x}^*) = 0$
$f_{11}(\mathrm{x}) = -20\exp\left(-0.2\sqrt{\frac{1}{n}\sum_{i=1}^{n} x_i^2}\right) - \exp\left(\frac{1}{n}\sum_{i=1}^{n}\cos(2\pi x_i)\right) + 20 + e$	$[-32, 32]^n$	$n = 30$ $n = 100$ $n = 200$	$\mathrm{x}^* = (0, \ldots, 0)$; $f(\mathrm{x}^*) = 0$
$f_{12}(\mathrm{x}) = -0.1\sum_{i=1}^{n}\cos(5\pi x_i) - \sum_{i=1}^{n} x_i{}^2$	$[-1, 1]^n$	$n = 30$ $n = 100$ $n = 200$	$\mathrm{x}^* = (0, \ldots, 0)$; $f(\mathrm{x}^*) = -0.1 \times n$
$f_{13}(\mathrm{x}) = -0.1\sum_{i=1}^{n}\cos(5\pi x_i) - \sum_{i=1}^{n} x_i{}^2$	$[-1, 1]^n$	$n = 30$ $n = 100$ $n = 200$	$\mathrm{x}^* = (0, \ldots, 0)$; $f(\mathrm{x}^*) = -0.1 \times n$
$f_{14}(\mathrm{x}) = \sum_{i=1}^{n}\left[x_i^2\left(2x_i^2 + x_{i+1} + 2\right) - x_i x_{i-1}(3x_i + 3x_{i-1} - x_{i+1})\right]^2$ $+\frac{1}{n^2}\sum_{i=1}^{n}(x_i - 1)^2 + e;\ x_{n+1} = x_1,\ x_0 = x_n$	$[-1, 2]^n$	$n = 30$ $n = 100$ $n = 200$	$\mathrm{x}^* = (1, \ldots, 1)$; $f(\mathrm{x}^*) = 1$
$f_{15}(\mathrm{x}) = -(n+1)e^{-10\sqrt{n}\left[\sum_{i=1}^{n}(x_i-1)^2\right]^{1/2}} + \sum_{i=1}^{n} x_i{}^2$	$[-2, 2]^n$	$n = 30$ $n = 100$ $n = 200$	$\mathrm{x}^* = (1, \ldots, 1)$; $f(\mathrm{x}^*) = -1$

(continued)

Table A.1 (continued)

Function	S	Dimension	Minimum
$f_{16}(\mathrm{x}) = \sum_{i=1}^{n} -x_i \sin\left(\sqrt{\lvert x_i \rvert}\right)$	$[-500, 500]^n$	$n = 30$ $n = 100$ $n = 200$	$\mathrm{x}^* = (420, ..., 420)$; $f(\mathrm{x}^*) =$ $-418.9829 \times n$
$f_{17}(\mathrm{x}) = \frac{\pi}{n}\left\{10sin(\pi y_i - 1) + \sum_{i=1}^{n-1}(y_i - 1)^2\left[1 + 10sin^2(\pi y_{i+1}) + (y_n - 1)^2\right]\right\}$ $+ \sum_{i=1}^{n} u(x_i, 10, 100, 4)$; $y_i = 1 + \frac{x_i+1}{4}$; $u(x_i, a, k, m) = \begin{cases} k(x_i - a)^m, x_i > a \\ 0, -a < x_i < a \\ k(-x_i - a)^m, x_i < -a \end{cases}$	$[-50, 50]^n$	$n = 30$ $n = 100$ $n = 200$	$\mathrm{x}^* = (0, ..., 0)$; $f(\mathrm{x}^*) = 0$
$f_{18}(\mathrm{x}) = 0.1\left\{sin^2(3\pi x_i) + \sum_{i=1}^{n}(x_i - 1)^2\left[1 + sin^2(3\pi x_1 + 1)\right] + \left((x_n - 1)^2\left[1 + sin^2(2\pi x_n)\right]\right)\right\} +$ $u(x_i, 5, 100, 4)$	$[-50, 50]^n$	$n = 30$ $n = 100$ $n = 200$	$\mathrm{x}^* = (1, ..., 1)$; $f(\mathrm{x}^*) = 0$
$f_{19}(\mathrm{x}) = f_1(\mathrm{x}) + f_2(\mathrm{x}) + f_{10}(\mathrm{x})$	$[-100, 100]^n$	$n = 30$ $n = 100$ $n = 200$	$\mathrm{x}^* = (0, ..., 0)$; $f(\mathrm{x}^*) = 0$
$f_{20}(\mathrm{x}) = f_5(\mathrm{x}) + f_{10}(\mathrm{x}) + f_{12}(\mathrm{x})$	$[-100, 100]^n$	$n = 30$ $n = 100$ $n = 200$	$\mathrm{x}^* = (0, ..., 0)$; $f(\mathrm{x}^*) = n - 1$

(continued)

Table A.1 (continued)

Function	S	Dimension	Minimum
$f_{21}(\mathrm{x}) = f_3(\mathrm{x}) + f_5(\mathrm{x}) + f_{11}(\mathrm{x}) + f_{18}(\mathrm{x})$	$[-100, 100]^n$	$n = 30$ $n = 100$ $n = 200$	$\mathrm{x}^* = (0, ..., 0)$; $f(\mathrm{x}^*) = (1.1 \times n) - 1$
$f_{22}(\mathrm{x}) = f_2(\mathrm{x}) + f_5(\mathrm{x}) + f_{10}(\mathrm{x}) + f_{11}(\mathrm{x}) + f_{12}(\mathrm{x})$	$[-100, 100]^n$	$n = 30$ $n = 100$ $n = 200$	$\mathrm{x}^* = (0, ..., 0)$; $f(\mathrm{x}^*) = n - 1$
$f_{23}(\mathrm{x}) = f_1(\mathrm{x}) + f_2(\mathrm{x}) + f_{11}(\mathrm{x}) + f_{16}(\mathrm{x}) + f_{17}(\mathrm{x})$	$[-100, 100]^n$	$n = 30$ $n = 100$ $n = 200$	$\mathrm{x}^* = (1, ..., 1)$; $f(\mathrm{x}^*) = -3 \times n$

Table A.2 Three-bar truss design problem

Function	S	Dimension	Constraints
$f(\mathrm{x}) = \left(2\sqrt{2}x_1 + x_2\right) \times L$ $L = 100cm,\ P = 2kN/cm^2, \sigma = 2kN/cm^2$	$0 \le x_i \le 1, i = 1, 2$	$n = 2$	$g_1(\mathrm{x}) = \frac{\sqrt{2}x_1 + x_2}{\sqrt{2}x_1{}^2 + 2x_1x_2} P - \sigma \le 0$ $g_2(\mathrm{x}) = \frac{x_2}{\sqrt{2}x_1{}^2 + 2x_1x_2} P - \sigma \le 0$ $g_3(\mathrm{x}) = \frac{1}{\sqrt{2}x_2 + x_1} P - \sigma \le 0$

Table A.3 Tension/compression spring design

Function	S	Dimension	Constraints
$f(\mathrm{x}) = (x_3 + 2)x_2x_1{}^2$	$0.05 \le x_1 \le 2$ $0.25 \le x_2 \le 1.3$ $2 \le x_3 \le 15$	$n = 3$	$g_1(\mathrm{x}) = 1 - \frac{x_2{}^3x_3}{71{,}785x_1{}^4} \le 0$ $g_2(\mathrm{x}) = \frac{4x_2{}^2 - x_1x_2}{12{,}566\left(x_2x_1{}^3 - x_1{}^4\right)} + \frac{1}{5108x_1{}^2} - 1 \le 0$ $g_3(\mathrm{x}) = 1 - \frac{140.45x_1}{x_2{}^2x_3} \le 0$ $g_4(\mathrm{x}) = \frac{x_1 + x_2}{1.5}$ -1 ≤ 0

Table A.4 Welded beam design problem

Function	S	Dimension	Constraints
$f(\mathrm{x}) = 1.10471x_1{}^2x_2 + 0.04811x_3x_4(14 + x_2)$	$0.1 \le x_i \le 2$, $i = 1, 4$ $0.1 \le x_i \le 10$, $i = 2, 3$	$n = 4$	$g_1(\mathrm{x}) = \tau(x) - \tau_{max}$ $g_2(\mathrm{x}) = \sigma(x) - \sigma_{max}$ $g_3(\mathrm{x}) = x_1 - x_4 \le 0$ $g_4(\mathrm{x}) =$ $0.10471x_1{}^2 + 0.04811x_3x_4(14 + x_2) - 5 \le 0$ $g_5(\mathrm{x}) = 0.125 - x_1 \le 0$ $g_6(\mathrm{x}) = \delta(x) - \delta_{max}$ $g_7(\mathrm{x}) = P - P_c(x) \le 0;$ $\tau(x) = \sqrt{(\tau')^2 + 2\tau'\tau''\frac{x_2}{2R} + (\tau'')^2},$ $\tau' = \frac{P}{\sqrt{2}x_1x_2}, \tau'' = \frac{MR}{J}, M = P\left(L + \frac{x_2}{2}\right),$ $R = \sqrt{\frac{x_2{}^2}{2} + \left(\frac{x_1 + x_3}{2}\right)^2}, \delta(x) = \frac{4PL^3}{Ex_3{}^3x_4},$ $J = 2\left[\sqrt{2}x_1x_2\left\{\frac{x_2{}^2}{12} + \left(\frac{x_1 + x_3}{2}\right)^2\right\}\right], \sigma(x) = \frac{6PL}{x_4x_3{}^2},$ $P_c(x) = \frac{4.013E\sqrt{\frac{x_3{}^2x_4{}^6}{36}}}{L^2}\left(1 - \frac{x_3}{2L}\sqrt{\frac{E}{4G}}\right),$

References

1. Wilensky, Uri, Rand, William.: An introduction to agent-based modeling modeling natural, Social, and Engineered Complex Systems with NetLogo, MIT Press (2010)
2. Banisch, Sven: Markov Chain Aggregation for Agent-Based Models, Springer Verlag (2016)
3. Macall, C., North, M.: Tutorial on agent-based modelling and simulation. J. Simul. **4**, 151–162 (2010)
4. Cuevas, E.: An agent-based model to evaluate the COVID-19 transmission risks in facilities. Comput. Biol. Med. **121**, 103827 (2020)
5. Macal, C. M.: Emergent structures from trust relationships in supply chains. In: Macal, C., Sallach, D., North, M. (eds.). Proceedings of Agent 2004: Conference on Social Dynamics: Interaction, Reflexivity and Emergence. Argonne National Laboratory: Chicago, IL, 7–9 October, pp 743–760
6. Arthur, W. B., Durlauf, S. N., Lane, D. A. (eds.).: The economy as an evolving complex system II, SFI Studies in the Sciences of Complexity. Addison-Wesley: Reading, MA (1997)
7. Folcik, V. A. An, G. C., Orosz, C. G.: The basic immune simulator: an agent-based model to study the interactions between innate and adaptive immunity. Theoret Biol. Med. Model. **4**(39) (2007). http://www.tbiomed.com/content/4/1/39
8. Kohler, T.A., Gumerman, G.J., Reynolds, R.G.: Simulating ancient societies. Sci. Amer **293**(1), 77–84 (2005)
9. North, M., et al.: Multi-scale agent-based consumer market modeling. Complexity **15**, 37–47 (2009)
10. Schelling, T.C.: Dynamic models of segregation. J. Math. Sociol. **1**(2), 143–186 (1971)
11. Sayama, H.: Introduction to the modeling and analysis of complex systems. Open SUNY Textbooks, Milne Library (2015)
12. Bonabeau, E.: (2012). http://www.icosystem.com/labsdemos/the-game/
13. Sweeney, L.B., Meadows, D.: The systems thinking playbook: exercises to stretch and build learning and systems thinking capabilities. Chelsea Green Publishing, White River Junction, VT (2010)
14. Bonabeau, E., Meyer, C.: Swarm intelligence: a whole new way to think about business. Harv. Bus. Rev. **5**, 107–114 (2001)
15. Sörensen, K.: Metaheuristics—the metaphor exposed. Int. Trans. Oper. Res. **22**(1), 3–18 (2015)
16. Sarker, Ruhul A., Ray, Tapabrata (Eds.): Agent-based evolutionary search. Adaptation, Learning, and Optimization, Springer Verlag (2010)
17. Tang, K.S., Man, K.F., Kwong, S., He, Q.: Genetic algorithms and their applications. IEEE Signal Process. Mag. **13**, 22–37 (1996). https://doi.org/10.1109/79.543973
18. Karaboga, D.: An idea based on honey bee swarm for numerical optimization. TechnicalReport-TR06. Engineering Faculty, Computer Engineering Department, Erciyes University (2005)
19. Storn, R., Price, K.: Differential evolution—a simple and efficient heuristic for global optimization over continuous spaces. J. Glob. Optim. **11**, 341–359 (1997)
20. Civicioglu, P.: Transforming geocentric cartesian coordinates to geodetic coordinates by using differential search algorithm. Comput. Geosci. **46**, 229–247 (2012)
21. Mirjalili, S., Mirjalili, S.M., Hatamlou, A.: Multi-verse optimizer: a nature-inspired algorithm for global optimization. Neural Comput. Appl. **27**(2), 495–513 (2016)
22. Mirjalili, S.: SCA: a sine cosine algorithm for solving optimization problems. Knowledge-Based Syst. **96**, 120–133 (2016)
23. Zhang, J., Sanderson, A. C.: JADE: self-adaptive differential evolution with fast and reliable convergence performance, 2007 IEEE Congr. Evol. Comput. CEC (2007)
24. Hansen, N.: The CMA evolution strategy: a comparing review, Towards a new evolutionary computation. Advances on estimation of distribution algorithms, Springer, pp. 1769–1776 (2006)
25. Mirjalili, S.: Moth-flame optimization algorithm: a novel nature-inspired heuristic paradigm. Knowledge-Based Syst. **89**, 228–249 (2015)

26. Barbati, M., Bruno, G., Genovese, A.: Applications of agent-based models for optimization problems: a literature review. Expert Syst. Appl. **39**, 6020–6028 (2012)
27. Byrski, Aleksander, Kisiel-Dorohinicki, Marek: Evolutionary multi-agent systems, from inspirations to applications. Studies in Computational Intelligence, Springer Verlag (2010)
28. Silva, M.A.L.: Sérgio Ricardo de Souza, Marcone Jamilson Freitas Souza, Moacir Felizardo de Franca Filho, Hybrid metaheuristics and multi-agent systems for solving, optimization problems: a review of frameworks and a comparative análisis. Appl. Soft Comput. **71**, 433–459 (2018)
29. Jin, X., Liu, J.: Multiagent SAT (MASSAT): autonomous pattern search inconstrained domains, In: H. Yin, N. Allinson, R. Freeman, J. Keane, S. Hubbard (Eds.), Proceedings of the Third International Conference on Intelligent DataEngineering and Automated Learning (IDEAL'02), Springer, Berlin, London,UK, 2002, pp. 318–328
30. Liu, J., Sycara, K.: Distributed problem solving through coordination in asociety of agents, Proceedings of the 13th International Workshop onDistributed Artificial Intelligence, pp. 169–185 (1994)
31. Lotfi, N., Acan, A.: Learning-based multi-agent system for solvingcombinatorial optimization problems: a new architecture. In: Onieva E., Santos, I., Osaba, E., Quintián, H., Corchado, E. (Eds.), –24, 2015, Proceedings, S, pp. 319–332 (2015)
32. Liu, J., Zhong, W., Jiao, L.: A multiagent evolutionary algorithm for combinatorial optimization, problems. IEEE Trans. Syst. Man Cybern. B Cybern. **40**(1), 229–240 (2010)
33. Zhang, J., Liang, C., Huang, Y., Wu, J., Yang, S.: An effective multiagent evolutionary algorithm integrating a novel roulette inversion operator for engineering optimization. Appl. Math. Comput. **211**, 392–416 (2009)
34. Drezewski, R., Siwik, L.: Agent-based co-operative co-evolutionary algorithm for multi-objective optimization. In: Rutkowski, L., Tadeusiewicz, R., Zadeh, L.A., Zurada, J.M. (eds.) ICAISC 2008. LNCS (LNAI), vol. 5097, pp. 388–397. Springer, Heidelberg (2008)
35. Milano, M., Roli, A.: MAGMA: a multiagent architecture for metaheuristics. IEEE Trans. Syst. Man, Cybern. Part B **34**(2), 925–941 (2004)
36. Kennedy, J., Eberhart, R.: Particle swarm optimization, Neural Networks, 1995. Proceedings., IEEE Int. Conf. **4**, 1942–1948 (1995)
37. Dorigo, M., Maniezzo, V., Colorni, A.: Ant system: optimization by a colony of cooperating agents. IEEE Trans. Syst. Man Cybern. Part B **26**(1), 29–41 (1996)
38. Cuevas, E., Cienfuegos, M., Zaldívar, D., Pérez-cisneros, M.: A swarm optimization algorithm inspired in the behavior of the social-spider. Expert Syst. Appl. **40**, 6374–6384 (2013)
39. Askarzadeh, A.: A novel metaheuristic method for solving constrained engineering optimization problems: crow search algorithm. Comput. Struct. **169**, 1–12 (2016). https://doi.org/10.1016/j.compstruc.2016.03.001
40. Díaz, P., Pérez, M., Cuevas, E., Avalos, O., Gálvez, J., Hinojosa, S., Zaldivar, D.: An Improved Crow Search Algorithm Applied to Energy Problems, pp. 1–23 (2018). https://doi.org/10.3390/en11030571
41. Mirjalili, S., Mirjalili, S.M., Lewis, A.: Grey Wolf optimizer. Adv. Eng. Softw. **69**, 46–61 (2014)
42. Yang, X.-S.: A new metaheuristic Bat-inspired algorithm. Stud. Comput. Intell. **284**, 65–74 (2010)
43. Yang, X. S., Deb, S.: Cuckoo search via Lévy flights, in: 2009 World Congr. Nat. Biol. Inspired Comput. NABIC 2009 - Proc., pp. 210–214 (2009)
44. Hippolyte, J. -L., Bloch, C., Chatonnay, P., Espanet, C., Chamagne, D.: A self-adaptive multi-agent evolutionary algorithm for electrical machine design. In: Proceedings of the 9th annual conference on Genetic and evolutionary computation, pp. 1250–1255 (2007)
45. Giardini, G., Kalmar-Nagy, T.: Genetic algorithm for multi-agent space exploration. In: 2007 AIAA InfoTech at Aerospace Conference, vol. 2, pp. 1146–1160 (2007)
46. Kisiel-Dorohinicki, M., Dobrowolski, G., Nawarecki, E.: Agent populations as computational intelligence. In L. Rutkowski, J. Kacprzyk (eds), Neural Networks and Soft Computing, pp. 608–614. Physica Verlag (2002)

47. Cetnarowicz, K.: Agent oriented technology based on the m-agent architecture. In Proceedings of international conference on Intelligent techniques in robotics, control and decision making, pp. 16–31. Polish-Japanese Institute of Information Technology, Warsaw (1999)
48. Cetnarowicz, K., Kisiel-Dorohinicki, M., Nawarecki, E.: The application of evolution process in multi-agent world (MAW) to the prediction system. In M. Tokoro (ed.), Proc. Of the 2nd Int. Conf. on Multi-Agent Systems (ICMAS'96). AAAI Press (1996)
49. Aydin, M. E.: Agentification of individuals: a multi-agent approach to metaheuristics. J. Comp. Sci. Syst. Biol. **6**(5) (2013)
50. Fernandes, F. C., de Souza, S. R., Silva, M. A. L., Borges, H. E., Ribeiro, F. F.: A multi-agent architecture for solving combinatorial optimization problems through metaheuristics, Proceedings of the 2009 IEEE International Conference on Systems, Man and Cybernetics (SMC 2009), pp. 3071–3076 (2009)
51. Silva, M. A. L., de Souza, S. R., de Oliveira, S. M., Souza, M. J. F.: An agent-based metaheuristic approach applied to the vehicle routing problem with time-windows, in: Proceedings of the 2014 Brazilian Conference on Intelligent Systems - Enc. Nac. de Inteligência Artificial e Computacional (BRACIS-ENIAC 2014), São Carlos, SP, Brazil (2014)
52. Silva, M. A. L., de Souza, S. R., Souza, M. J. F., de Oliveira, S. M.: A multi-agent metaheuristic optimization framework with cooperation, in: Proceedings of the 2015 Brazilian Conference on Intelligent Systems (BRACIS), Natal, Brazil, pp. 104–109 (2015)
53. Morales-Castañeda, B., Zaldívar, D., Cuevas, E., Fausto, F., Rodríguez, A.: A better balance in metaheuristic algorithms: does it exist? Swarm Evol. Comput. **54**, 100671 (2020)
54. Chan, V., Son, Y., Macal, C., Agent-based simulation tutorial - simulation of emergent behavior and differences between agent-based simulation and discrete-event simulation, Proceedings of the 2010 Winter Simulation Conference B. Johansson, S. Jain, J. Montoya-Torres, J. Hugan, and E. Yücesan, eds.
55. Railsback, S. F., Grimm, V.: Agent-based and individual-based modeling: a practical introduction. Princeton University Press, New Jersey (2011)
56. Bruch, E., Atwell, J.: Agent-based models in empirical social research. Soc. Methods Res. **44**(2), 186–221 (2015)
57. Dorin, A., Geard, N.: The practice of agent-based model visualization. Artif. Life **20**(2), 271–289 (2014)
58. Badhama, J., Chattoe-Brown, E., Gilbert, N., Chalabi, Z., Keea, F., Hunter, R.F.: Developing agent-based models of complex health behaviour. Health Place **54**, 170–177 (2018)
59. Anescu, Further scalable test functions for multidimensional continuous optimization (2017)
60. Li, M.D., Zhao, H., Weng, X.W., Han, T.: A novel nature-inspired algorithm for optimization: Virus colony search. Adv. Eng. Softw. **92**, 65–88 (2016)
61. Yang, X. -S., Wiley InterScience (Online service): Engineering optimization : an introduction with metaheuristic applications. John Wiley (2010)
62. Wilcoxon, F.: Individual comparisons by ranking methods. Biometrics, pp. 80–83 (1945)
63. Cuevas, E., González, A., Fausto, F., Zaldívar, D., Pérez-Cisneros, M.: Multithreshold segmentation by using an algorithm based on the behavior of locust swarms. Mathem. Probl. Eng. (2015)
64. Hinojosa, S., Dhal, K.G., Abd Elaziz, M., Oliva, D., Cuevas, E.: Entropy-based imagery segmentation for breast histology using the stochastic fractal search. Neurocomputing **321**, 201–215 (2018)
65. Cuevas, E., Gálvez, J., Hinojosa, S., Avalos, O., Zaldívar, D., Pérez-Cisneros, M.: A comparison of evolutionary computation techniques for IIR model identification. J. Appl. Mathem. (2014)
66. Bandyopadhyay, R., Basu, A., Cuevas, E., Sarkar, R.: Harris Hawks optimisation with Simulated Annealing as a deep feature selection method for screening of COVID-19 CT-scans. Appl. Soft Comput. **111**, 107698 (2021)
67. Cuevas, E., Zaldivar, D., Pérez-Cisneros, M.: Seeking multi-thresholds for image segmentation with Learning Automata. Mach. Vis. Appl. **22**, 805–818 (2011)
68. Ibrahim, R.A., Elaziz, M.A., Oliva, D., Cuevas, E., Lu, S.: An opposition-based social spider optimization for feature selection. Soft. Comput. **23**, 13547–13567 (2019)

69. Zaldivar, D., Morales, B., Rodríguez, A., Valdivia-G, A., Cuevas, E., Pérez-Cisneros, M.: A novel bio-inspired optimization model based on Yellow Saddle Goatfish behavior. Biosystems **174**, 1–21 (2018)
70. Ramírez-Ortegón, M.A., Tapia, E., Ramírez-Ramírez, L.L., Rojas, R., Cuevas, E.: Transition pixel: a concept for binarization based on edge detection and gray-intensity histograms. Pattern Recogn. **43**(4), 1233–1243 (2010)
71. Cuevas, E., Wario, F., Osuna-Enciso, V., Zaldivar, D., Pérez-Cisneros, M.: Fast algorithm for multiple-circle detection on images using learning automata. IET Image Proc. **6**(8), 1124–1135 (2012)

Chapter 3
Dynamic Multimodal Function Optimization: An Evolutionary-Mean Shift Approach

Numerous metaheuristic-based dynamic optimization algorithms have emerged recently. While these algorithms excel in locating a single global optimum, they fall short in identifying multiple optimal solutions. The primary objective of dynamic multimodal optimization is to identify several optimal solutions for a problem where the objective function changes over time. The discovery of multiple optimal solutions, both global and local, holds significant importance for numerous applications. The best implementable alternative may not always be the optimal solution due to practical limitations. Surprisingly, the literature has paid scant attention to the problem of dynamic multimodal optimization using evolutionary principles. Conversely, mean shift, a nonparametric iterative process, is capable of detecting local maxima in a density function represented by a sample set. Mean shift exhibits adaptive qualities that enable it to identify local maxima in dynamic environments. In this chapter, we present a mean shift approach to detect global and local optima in dynamic optimization problems. The presented method modifies the search strategy of mean shift by considering both density and fitness values of candidate solutions. To expedite the convergence process, a competitive memory and dynamic strategy are incorporated to leverage valuable information from previous environments. Consequently, the present approach effectively identifies a significant number of global and local optima in dynamic environments. To showcase its performance, we compare the present algorithm with well-known dynamic optimization methods using the CEC competition benchmark generator. Experimental results demonstrate the present scheme's high accuracy and robustness, positioning it as a highly competitive solution.

E. Cuevas et al., *New Metaheuristic Schemes: Mechanisms and Applications*, Intelligent Systems Reference Library 246, https://doi.org/10.1007/978-3-031-45561-2_3

3.1 Introduction

The majority of research in the field of metaheuristics focuses on static optimization, disregarding the fact that many optimization problems are inherently dynamic [1]. Dynamic optimization problems (DOPs) involve objective functions and constraints that change over time [2]. DOPs can take various forms, and the complexity of a DOP depends on the frequency and magnitude, known as severity, of the changes made to the objective function [3].

One way to approach dynamic optimization is from a static optimization perspective. In this approach, a static metaheuristic method can be employed to solve a DOP by treating each change in the objective function as a new optimization problem that must be solved from scratch [4]. However, this simplistic scheme proves to be ineffective since solving an optimization problem from scratch without utilizing information from past iterations is time-consuming [5]. Consequently, traditional static metaheuristic methods need to be modified to incorporate specialized mechanisms that can adapt to dynamic environments [6]. These modified approaches are designed to adjust themselves based on changes in the dynamic objective function by leveraging information obtained during the optimization process [7].

Static metaheuristic approaches lack specific mechanisms to adapt their search strategies in dynamic environments [8]. They employ operators that encourage convergence, allowing the algorithms to focus solely on the best solution within the stationary environment. Nevertheless, the lack of adaptability and high convergence are unfavorable in dynamic optimization scenarios [9]. Excessive convergence leads to a loss of solution diversity, preventing the algorithm from adequately responding to changes in the dynamic environment [10].

Consequently, several dynamic optimization algorithms based on metaheuristic methods have been proposed. These algorithms incorporate adaptability by combining static metaheuristic methods with specialized schemes that modify individual generations in response to changes in the objective function [11–14]. Based on these mechanisms, dynamic metaheuristic methods can be categorized into three classes [6, 8, 15]: those that enhance diversity, those that preserve diversity, and those that utilize memory elements.

Dynamic metaheuristic methods that aim to enhance diversity employ strategies to relocate existing elements to new positions when a significant difference in the objective function is detected [16]. Representative examples of this category include the quantum particle method [17] and Gaussian relocation [18]. In the quantum particle technique, hypothetical spheres are positioned at the center of each existing element. When a change in the environment is detected, existing elements are randomly relocated within the spheres, mimicking the movement of particles according to quantum mechanics. In Gaussian relocation, the algorithm generates new random elements by sampling from a normal distribution after identifying a change in the objective function. While these methods improve population diversity, they do not leverage the experience gained from previous searches.

Methods that uphold diversity prevent population convergence by eliminating similar individuals or reinitializing individuals that remain unchanged for a certain period of time. Li and Yang [19] introduced the random immigrant method as a dynamic scheme to automatically reduce redundant elements in the search space. The Self-Adaptive Differential Evolution (jDE) algorithm [20] incorporates an intriguing strategy where each individual includes an aging factor that keeps track of the number of iterations in which the element remains in the same position, allowing agents trapped in local optima to be relocated. The adaptive differential evolution algorithm (DDEBQ) [2] also includes a mechanism to eliminate individuals stagnating at local optima. In [21], a modified crossover operation is proposed to maintain diversity by generating new individuals far from their close parents, thereby enhancing the diversity of the entire population. Despite their competitive performance, diversity-maintaining schemes tend to slow down convergence towards an optimal solution, resulting in diminished accuracy due to inadequate exploitation.

Methods that aim to increase and preserve diversity do not rely on information from previous environments to guide their search strategies in the current environment. On the other hand, memory-based methods store valuable knowledge, such as promising positions obtained in past trials [22]. The goal is to utilize this information when detecting the current optimal positions in the presence of changes in the objective function. Several dynamic metaheuristic methods based on memory principles have been documented in the literature. For example, the clustering particle swarm optimization (CPSO) [23] records the best solution from each subpopulation just before a change is detected and adds these solutions to the newly reinitialized subpopulations. In [14], the concept of the inferior set is introduced, where a set of candidate solutions is collected during the optimization process. When a change in the environment is identified, the elements in the set are included in their respective subpopulations. Finally, the evolution algorithm is reinitialized to detect the changed optimal value in the new environment. In the dynamic differential evolution algorithm based on K-means (CDDE) [1], an external register is used to accumulate the optimal solutions detected in past environments. However, these memory-based algorithms only utilize the memory to store the best solutions obtained during the optimization process and do not leverage the information contained in the memory to adapt their operation when a change in the objective function is detected.

Most of the efforts in metaheuristic computation have been focused on finding a single optimal value, which represents the best possible solution for an optimization problem. However, in certain circumstances, using the optimal value can be excessively expensive or even infeasible to implement. Under such conditions, being limited to accessing only one solution can restrict the practical use of typical metaheuristic methods in real-world applications. In many practical scenarios, the desire is to obtain a sufficiently good solution with low implementation costs rather than a global optimum that offers only a marginal advantage at a high cost [24]. Therefore, it is desirable for a metaheuristic scheme to not only identify the best solution but also detect several solutions of good quality. Multimodal optimization refers to the process of identifying both the global optimum and multiple local optima for a specific optimization problem. Metaheuristic schemes deliver competitive results when used to locate the optimal solution of an optimization problem. However, they exhibit poor performance in detecting multiple solutions. The problem of identifying most of the optimal elements from an optimization problem in a single execution represents a higher level of complexity compared to global optimization. Metaheuristic methods were originally designed to identify only one optimal solution. Therefore, they need to incorporate additional mechanisms to detect and store multiple solutions in order to solve multimodal problems. Various schemes have been suggested in the literature to equip typical metaheuristic methods with multimodal capabilities.

These mechanisms encompass fitness sharing [25], crowding [26], speciation [27], and clearing [28]. Most of these mechanisms encounter the challenge of detecting multiple optima while avoiding the convergence of several solutions into a single point. Their ultimate goal is to maintain individual diversity within the population. Among these mechanisms, fitness sharing [25] is the most popular approach for introducing diversity in a population, particularly in multimodal scenarios. It achieves this by penalizing the concentration of solutions, degrading the fitness of an individual based on the number of neighboring solutions. Several multimodal schemes have been proposed in the literature based on fitness sharing, such as the Fitness Euclidean-distance ratio method [29], Fitness Sharing Differential Evolution (SDE) [30, 31], and Information Sharing Artificial Bee Colony algorithm [32].

Under the crowding mechanism [33, 34], each newly generated solution is compared with existing individuals in the population. The individuals that bear the closest resemblance to the new solution are deleted and replaced by the new solution. Numerous multimodal metaheuristic schemes have been put forth in the literature that employ the crowding mechanism, utilizing various metaheuristic algorithms such as Gravitational Search Algorithm (GSA) [35], Differential Evolution (DE) [30, 31], Genetic Algorithms (GA) [36], and Particle Swarm Optimization (PSO) [37].

Speciation-based mechanisms represent a recent approach to incorporating multimodal capabilities into metaheuristic methods. In these mechanisms, two solutions are considered to belong to the same species if their distance falls below a specified radius. As a result, the entire population is divided into distinct subpopulations, each associated with a specific concentration of individuals. Once the subpopulations are identified, only one representative individual is retained from each subpopulation, while the remaining solutions are removed. Examples of speciation-based techniques include the Differential Evolution with Self-adaptive strategy [30], Elitist-population strategy (AEGA) [38], and Distance-based PSO [39].

Clearing [28] is another intriguing mechanism used to preserve diversity in the population. During the clearing process, the best solutions in the population are first identified, and the remaining population members within a clearing radius around these solutions are deleted. Numerous multimodal algorithms have been proposed based on this mechanism, including the Topographical Clearing Differential Evolution [40] and Clearing Procedure (CP) [28].

Dynamic multi-modal optimization refers to the solution of optimization problems that involve dynamic objective functions capable of detecting multiple optima. The primary objective in dynamic multi-modal optimization is to identify several optimal solutions for a problem where the objective function changes over time

[41]. Finding multiple global and local optima in dynamic multi-modal optimization is particularly important for various applications, as the best solution may not always be feasible due to practical constraints. Real-world applications of dynamic multi-modal optimization include dynamic vehicle routing problems [42, 43], pricing techniques for transportation systems [44, 45], taxi sharing models [46], dynamic Knapsack problems [47, 48], and musculoskeletal models for understanding human movements [49]. Despite its significance, the application of evolutionary principles to dynamic multi-modal optimization has received limited attention in the literature. A few exceptions include the dynamic multimodal clonal selection algorithm (DMMCSA) [50], which incorporates a niching mechanism. In this approach, the hypermutation scale factor is adapted based on the distances among individuals in the population. Another interesting approach is the Clustering Particle Swarm Optimizer for dynamic optimization proposed in [51], which employs a hierarchical clustering scheme and a nearest neighbor search strategy to track multiple optima.

Conversely, mean shift [52] is an iterative and nonparametric technique employed for identifying local maxima within a density function represented by a collection of samples. Mean shift possesses adaptive characteristics that enable it to locate local maxima even in dynamic environments. These advantageous properties have led to its application in various domains, including image segmentation [53, 54], classification [55, 56], video processing [57], and object tracking [58]. The mean shift algorithm operates by iteratively shifting each sample point (a potential solution) towards the nearest local optimum in the direction of the gradient of the estimated density function. While mean shift's effectiveness in identifying optimal values within the density space is well-established [59, 60], its potential as an optimization tool has been largely overlooked.

In the presented approach, the functionality of the mean shift is altered to incorporate not only the density of solutions but also their corresponding fitness values. This modification enables the adapted mean shift to effectively approximate the objective function based on the positions of particles. Additionally, a competitive memory and dynamic strategy are introduced in the present framework to expedite the convergence process by leveraging valuable information from previous environments. The memory component stores all the optimal solutions observed thus far in the optimization process, employing a competition-based strategy to ensure the inclusion of diverse elements while avoiding redundancy. The dynamic strategy involves analyzing the memory to detect changes in the objective function and adapt to environmental modifications. To evaluate the effectiveness of the present algorithm, comprehensive comparisons are conducted with established dynamic optimization methods using the benchmark generator of the CEC competition for dynamic optimization. The experimental results demonstrate the competitive performance of the present scheme in terms of accuracy and robustness.

The remaining sections of the chapter are organized as follows: Sect. 3.2 describes the original mean shift method, while Sect. 3.3 introduces the present multimodal dynamic approach. In Sect. 3.4, the experimental results and a comparative study with state-of-the-art dynamic algorithms are presented. Finally, Sect. 3.5 offers concluding remarks and discussions.

3.2 The Method of Mean-Shift

Mean shift [27] is a nonparametric and iterative method designed to identify local maxima in a density function that is represented by a collection of samples. When performing mean shift, the process begins at a specific point **x**, and a new location is determined by moving in the direction of the greatest increase in the density function until a local maximum $\mathbf{x}_i^*$ is reached. The mean shift operation comprises two stages. First, the density function of the feature space is estimated using a kernel density estimation (KDE) technique. Next, an optimization step employing gradient ascent is performed to locate the local maxima (density attractors) within the density function [61].

3.2.1 Preliminary Concepts

The nonparametric kernel density estimator (KDE) is a valuable approach for generating a feature map of an unknown density. In the KDE method, the density at each point is determined by utilizing kernel functions that calculate the weighted sum of neighboring elements. A kernel function K is characterized as a symmetric, non-negative model that satisfies the following condition:

$$K(\mathbf{x}) \geq 0;\ K(-\mathbf{x}) = K(\mathbf{x});\ \int K(\mathbf{x})dx = 1 \tag{3.1}$$

Assuming a set of d elements $\{\mathbf{x}_1, \mathbf{x}_2, \ldots, \mathbf{x}_d\}$, the density function $\widehat{f}(\mathbf{x})$, can be estimated in terms of K as follows:

$$\widehat{f}(\mathbf{x}) = \frac{1}{dh}\sum_{i=1}^{d} K\left(\frac{\mathbf{x} - \mathbf{x}_i}{h}\right), \tag{3.2}$$

where h represents the influence distance of K.

While various kernel functions have been suggested for KDE techniques, the majority of KDE approaches typically employ the Gaussian function $K_G(x)$ as the kernel model. The Gaussian or normal kernel, defined in Eq. (3.3), is commonly used, and its visual representation is depicted in Fig. 3.1a.

$$K_G(x) = \left(\frac{1}{\sqrt{2\pi}}\right)e^{\left(\frac{-x^2}{2h}\right)} \tag{3.3}$$

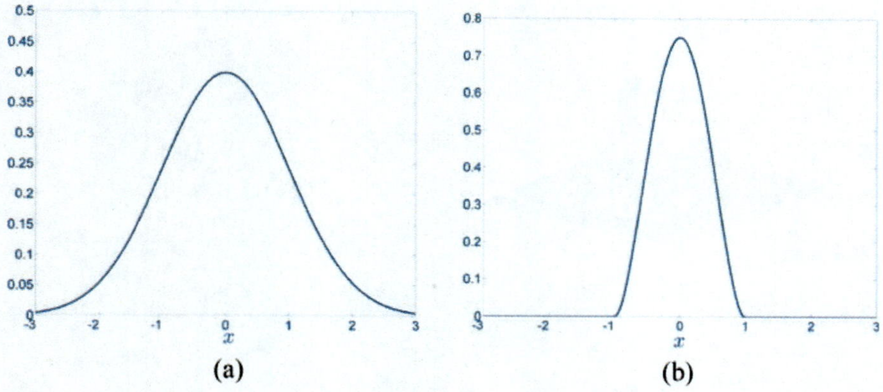

Fig. 3.1 Kernel elements **a** Gaussian $K_G(x)$ and **b** Epanechnikov $K_E(x)$

Another interesting alternative for KDE is the Epanechnikov function $K_E(x)$ (see Fig. 3.1b) [62], which is defined by the following model:

$$K_E(x) = \begin{cases} \left(\frac{3}{4}\right)(1 - x^2) & \text{if} |x| \leq 1 \\ 0 & \text{otherwise} \end{cases} \tag{3.4}$$

The Epanechnikov kernel $K_E(x)$ demonstrates superior accuracy in estimating the density of a feature map when the number of data points is severely limited. To illustrate the modeling capabilities of the Epanechnikov kernel $K_E(x)$, consider the following example. Let's assume a two-dimensional Gaussian mixture $M(x_1, x_2)$ represents the probability density function (PDF). The mixture $M(x_1, x_2)$ combines three Gaussian functions ($j = 1, 2, 3$) based on the model $N_j(\boldsymbol{\mu}_j, \Sigma_j)$, where μ represents the central point and Σ represents the covariance. Figure 3.2a depicts a Gaussian mixture defined as $(x_1, x_2) = \left(\frac{2}{3}\right)N_1 + \left(\frac{1}{6}\right)N_2 + \left(\frac{1}{6}\right)N_3$. From this PDF, a reduced set of 120 data points is sampled (refer to Fig. 3.2b). Using these feature points, the Gaussian kernel $K_G(x)$ and the Epanechnikov kernel $K_E(x)$ are employed to estimate the density map $\widehat{f}(x_1, x_2)$, resulting in Fig. 3.2c and d respectively. It is evident from Fig. 3.2c, d that the Epanechnikov kernel $K_E(x)$ provides a more accurate estimation compared to the Gaussian kernel $K_G(x)$. The Gaussian kernel generates multiple false local maxima in $\widehat{f}(x_1, x_2)$ due to its inability to appropriately model the density map when the available data set is significantly reduced.

3.2.2 Attractors

The fundamental concept behind estimating a density function $\widehat{f}(\mathbf{x})$ for a given dataset is to identify characteristic points $\mathbf{x}_i^*$ that serve as cluster centers, effectively representing the density function $\widehat{f}(\mathbf{x})$ modeled by the data points. These $\mathbf{x}_i^*$ points

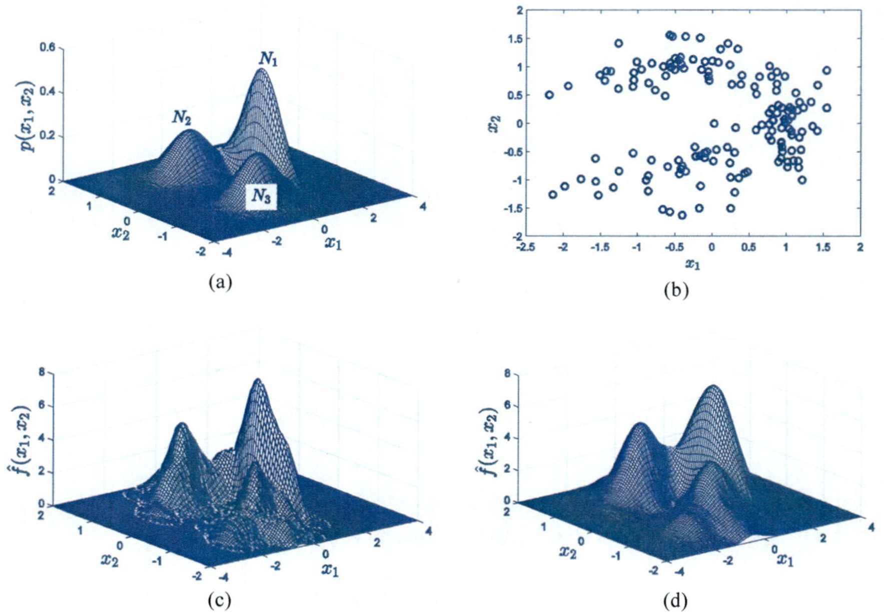

Fig. 3.2 The modeling abilities of the Gaussian kernel $K_G(x)$ and the Epanechnikov kernel $K_E(x)$ are examined when the dataset is limited. **a** The two-dimensional probability density function (PDF), **b** 120 features sampled from the PDF, **c** the estimated density map $\widehat{f}(x_1, x_2)$ using the Gaussian kernel $K_G(x)$, and **d** the density estimation with the Epanechnikov kernel $K_E(x)$

correspond to the local maxima found within the estimated density function $\widehat{f}(\mathbf{x})$. To locate such points, a gradient-based optimization method is employed. In this approach, considering a dataset $\mathbf{X}$ consisting of d elements ($\mathbf{X} = \{\mathbf{x}_1, \ldots, \mathbf{x}_d\}$), the process begins by randomly selecting a point $\mathbf{x}_j$ (where $j \in 1, \ldots, d$). Subsequently, a new location $\mathbf{x}_j^{t+1}$ is computed by moving in the direction of the density gradient $\nabla\widehat{f}(\mathbf{x}_j)$ until the local maximum $\mathbf{x}_i^*$ is reached. This updating rule can be formulated as follows:

$$\mathbf{x}_j^{t+1} = \mathbf{x}_j^t + \delta \cdot \nabla\widehat{f}\left(\mathbf{x}_j^t\right), \tag{3.5}$$

In the given expression, the variable t represents the current iteration number, and δ denotes the step size employed. In this formulation, the density gradient at a specific point $\mathbf{x}$ is determined by evaluating the derivative of the probability density function (PDF) using the following equation:

$$\nabla\widehat{f}(\mathbf{x}) = \frac{\partial}{\partial x}\widehat{f}(\mathbf{x}) = \frac{1}{dh}\sum_{i=1}^{d}\frac{\partial}{\partial x}K\left(\frac{\mathbf{x} - \mathbf{x}_i}{h}\right) \tag{3.6}$$

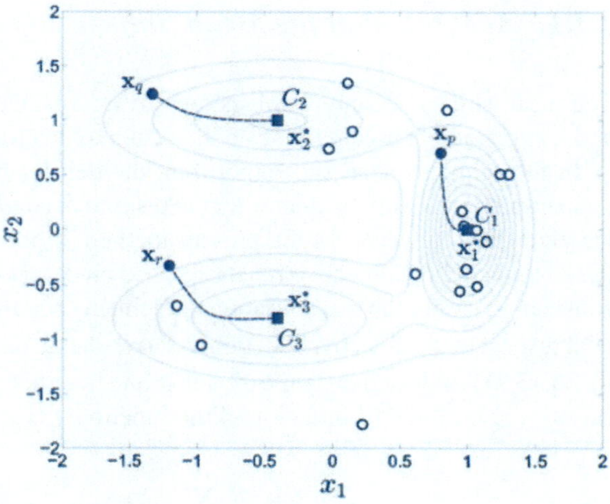

Fig. 3.3 Gradient method to obtain the local maxima

After applying the gradient method to all d elements in the dataset $\mathbf{X}$, multiple data points will converge towards the same local maximum $\mathbf{x}_i^*$. These points will then be considered as part of a cluster C_i. Consequently, the final outcome will be a collection of c density attractors $\{\mathbf{x}_1^*, \ldots, \mathbf{x}_c^*\}$, where each attractor $\mathbf{x}_i^*$ represents a local maximum that attracts a set of elements $\mathbf{X}_i = (\mathbf{x}_a, \ldots, \mathbf{x}_g)$ towards it (with $\mathbf{x}_a$ and $\mathbf{x}_g$ belonging to the cluster C_i).

Figure 3.3 provides a visual representation of the gradient process for identifying local maxima. The figure displays contour lines representing the density map $\widehat{f}(x_1, x_2)$. In this scenario, three starting points, denoted as $\mathbf{x}_p$, $\mathbf{x}_q$ and $\mathbf{x}_r$, are selected. From these points, new locations are determined by moving in the direction that corresponds to the steepest ascent of the density function $\widehat{f}(x_1, x_2)$. The process continues until reaching the respective local maxima, namely $\mathbf{x}_1^*$, $\mathbf{x}_2^*$ and $\mathbf{x}_3^*$.

3.3 Optimization Approach

The present approach enhances the mean shift operation by considering not only the solution density but also their fitness values. Additionally, a memory component is introduced to record the optimal solutions encountered during the optimization process using a competitive strategy. This memory mechanism selectively includes diverse elements while avoiding redundant solutions that represent the same optimal position.

This section introduces the dynamic optimization algorithm presented in this study. It is divided into four parts. Part A examines the modification of the mean shift scheme. Part B explores the memory mechanism employed in the approach. Part C explains the dynamic strategy implemented to adapt to changes in the objective function. Finally, Part D presents the complete dynamic multimodal algorithm, which combines all the aforementioned schemes.

3.3.1 Adaptation of the Mean Shift

The mean-shift procedure involves two main steps: estimating the density function and using gradient ascent to locate its local maxima (density attractors).

In the original mean shift algorithm, the density function is estimated using a nonparametric density estimator that takes into account the spatial concentration of the elements. However, in the present method, a new estimation process is introduced to model the objective function based on data points. This modified approach considers not only the concentration of elements but also their corresponding values with respect to the objective function. As a result, the density function $\widehat{f}(\mathbf{x})$ described by Eq. (3.2) is adjusted to define the substitute function $\widehat{f}(\mathbf{x})$. With a set of n elements $\{\mathbf{x}_1, \mathbf{x}_2, \ldots, \mathbf{x}_n\}$, the formulation of the function $\tilde{f}(\mathbf{x})$ is as follows:

$$\tilde{f}(\mathbf{x}) = \frac{1}{nh}\sum_{i=1}^{n} Q(\mathbf{x}_i)\cdot K\left(\frac{\mathbf{x}-\mathbf{x}_i}{h}\right) \tag{3.7}$$

The function $Q(\mathbf{x}_i)$ is defined to assign the evaluation of the objective function $J(\mathbf{x}_i)$ for each element $\mathbf{x}_i$, except for the element $\mathbf{x}_b$ (where $b \in 1, \ldots, n$) that yields the maximum value in terms of the objective function $J(\mathbf{x}_b)$. This particular element $\mathbf{x}_b$, which represents the best objective function value, is given more emphasis to improve the function value of $\tilde{f}(\mathbf{x})$. Consequently, this position has a higher likelihood of being selected in the subsequent application of the gradient method in the mean-shift process. Thus, the function $Q(\mathbf{x}_i)$ is formulated using the following model:

$$Q(\mathbf{x}_i) = \begin{cases} 2\cdot J(\mathbf{x}_i) & \text{if } \mathbf{x}_i = \mathbf{x}_b \\ J(\mathbf{x}_i) & \text{Otherwise} \end{cases} \tag{3.8}$$

The function $Q(\mathbf{x}_i)$ assigns the evaluation of the element $\mathbf{x}_i$ based on its objective function value $J(\mathbf{x}_i)$. In this approach, each element is assigned its corresponding objective function value $J(\mathbf{x}_i)$, while the best element $\mathbf{x}_b$ is assigned twice its objective function value $2\cdot J(\mathbf{x}_b)$.

The function $\tilde{f}(\mathbf{x})$ serves as an estimate of the objective function $J(\mathbf{x})$, possessing desirable qualities of smoothness and continuity. The accuracy of this approximation within a specific local area depends on the density of elements used to calculate $\tilde{\mathrm{f}}(\mathrm{x})$. When a greater number of elements are considered in a particular region, $\tilde{f}(\mathbf{x})$ becomes more similar to $J(\mathbf{x})$. To illustrate the modeling properties of the substitute function $\tilde{f}(\mathbf{x})$, we will consider an example. Let's assume a two-dimensional objective function defined as follows:

$$J(x_1, x_2) = 3(1-x_1)^2 e^{-(x_1^2 - x_2^2)} - 10\left(\frac{x_1}{5} - x_1^3 - x_2^5\right)e^{(-x_1^2 - x_2^2)} - 1/3e^{(-(x_1+1)^2 - x_2^2)} \tag{3.9}$$

Considering the domain Ω defined by the interval $-3 \leq x_1, x_2 \leq 3$, the objective function exhibits a global maximum and two local maxima. Figure 3.4a visually represents this function through a three-dimensional plot. Subsequently, a set of 150 randomly distributed points (n) is selected within Ω to form a point distribution, as depicted in Fig. 3.4b. Employing a value of $h = 0.8$, the model described by Eq. (3.7) is utilized to generate a substitute function $\tilde{f}(x_1, x_2)$ that serves as an approximation of the objective function $J(x_1, x_2)$. Figure 3.4c displays the resulting substitute function $\tilde{f}(x_1, x_2)$ obtained from this process. Notably, despite the limited number of points, the substitute function $\tilde{f}(x_1, x_2)$ demonstrates a reasonable approximation to the objective function $J(x_1, x_2)$ for optimization purposes. It is worth emphasizing the smoothness and continuity exhibited by the obtained function. In contrast, Fig. 3.4d showcases the density function $\widehat{f}(x_1, x_2)$ estimated by computing the model defined in Eq. (3.2), which solely considers the point concentration.

After obtaining the substitute function $\tilde{f}(\mathbf{x})$ for a given set of elements, the subsequent task is to identify the local maxima points $\mathbf{x}_i^*$. To accomplish this, a gradient-based optimization method is employed. Considering a population X comprising n agents ($\mathbf{X} = \{\mathbf{x}_1, \ldots, \mathbf{x}_n\}$), the procedure involves computing a new location $\mathbf{x}_j^{t+1}$

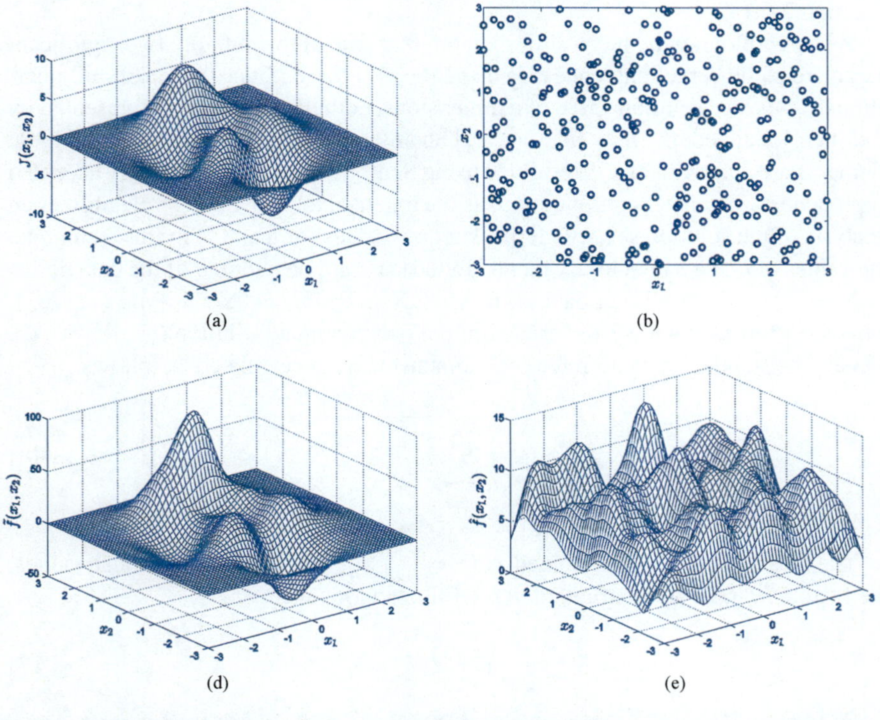

Fig. 3.4 The substitute function $\tilde{f}(\mathbf{x})$ and its modeling characteristics are illustrated in the following figures: **a** the original objective function $J(x_1, x_2)$, **b** the point distribution employed for the approximation, **c** the obtained substitute function $\tilde{f}(\mathbf{x})$ resulting from the modified mean shift scheme, and **d** the density function $\widehat{f}(x_1, x_2)$ estimated using the original mean shift method

for each agent $\mathbf{x}_j^t$ $(j \in 1, \ldots, n)$ in the direction of the gradient of the substitute function $\nabla \tilde{f}(\mathbf{x})$. The update rule can be expressed as follows:

$$\mathbf{x}_j^{t+1} = \mathbf{x}_j^t + \delta \cdot \nabla \tilde{f}\left(\mathbf{x}_j^t\right), \tag{3.10}$$

In the above equation, the variable t indicates the current iteration number, and δ represents the step size. The substitute function gradient at any given point $\mathbf{x}$ is determined by evaluating the derivative of each kernel result K, which is weighted by its corresponding function value $J(\mathbf{x}_i)$. The calculation can be expressed as:

$$\nabla \tilde{f}(\mathbf{x}) = \frac{\partial}{\partial x} \tilde{f}(\mathbf{x}) = \frac{1}{nh} \sum_{i=1}^{n} Q(\mathbf{x}_i) \frac{\partial}{\partial x} K\left(\frac{\mathbf{x} - \mathbf{x}_i}{h}\right) \tag{3.11}$$

After applying the gradient method to each element in the population $\mathbf{X}$ for multiple iterations, various elements will converge towards the same local maximum point $\mathbf{x}_i^*$, leading to the formation of clusters. These points will be associated with the respective local maxima. Ultimately, we obtain a collection of c attractors $\{\mathbf{x}_1^*, \ldots, \mathbf{x}_c^*\}$, where each attractor $\mathbf{x}_i^*$ represents a local maximum of the objective function $J(\mathbf{x})$.

When employing the mean shift method, it is crucial to configure two significant parameters: the kernel influence area h and the step size δ. Since the mean shift operation deals with a solution space of d dimensions, both parameters are represented by d-dimensional vectors: $\boldsymbol{h} = [h_1, \ldots, h_d]$ and $\boldsymbol{\delta} = [\delta_1, \ldots, \delta_d]$. The values of these elements are automatically determined using Scott's rule, which is a widely accepted approach for parameter configuration, as documented in [63]. Scott's rule utilizes an analysis of the solution set X to derive suitable values for h and δ. In this technique, the values of h and δ are closely related to the standard deviation σ of the data points in $\mathbf{X} = \{\mathbf{x}_1, \ldots, \mathbf{x}_n\}$. Each data point $\mathbf{x}_i \in \mathbf{X}$ is defined as $\mathbf{x}_i = \{x_{i,1}, \ldots, x_{i,d}\}$, where $x_{i,j}$ represents the j-th variable of the i-th candidate solution $(j \in 1, \ldots, d)$. Hence, the standard deviation for each dimension σ_j is calculated as follows:

$$\sigma_j = \sqrt{\frac{1}{n} \sum_{i=1}^{n} \left(x_{i,j} - \overline{x}^j\right)^2} \tag{3.12}$$

The mean value $\overline{x}^j$ is determined as $\overline{x}^j = \left(\frac{1}{n}\right) \sum_{i=1}^{n} x_{i,j}$. Using these mean values, the factors $\boldsymbol{h}$ and $\boldsymbol{\delta}$ are configured in the following manner:

$$h_j = 3.5 \cdot \sigma_j \cdot n^{-\left(\frac{1}{3}\right)}, \delta_j = \sqrt{0.5} \cdot \sigma_j, \tag{3.13}$$

To demonstrate the complete mean shift process for optimization, an illustrative example is conducted. The example focuses on identifying the local maxima of a two-dimensional objective function $J(x_1, x_2)$ defined in Eq. (3.8). A population $\mathbf{X}$ of n agents $(\mathbf{X} = \{\mathbf{x}_1, \ldots, \mathbf{x}_n\})$ is considered, where each agent $\{\mathbf{x}_i = (x_1^i, x_2^i)\}$

is initially assigned random values uniformly distributed within the range of $3 \leq x_1, x_2 \leq 3$. The modified mean shift algorithm is then applied. During the initial phase of the search strategy, the algorithm encourages the attraction of solutions towards local groups. This stage aims to locate the local maxima. As the iterations progress, the exploitation phase intensifies to refine the quality of the solutions. Figure 3.5 visualizes the behavior of the modified mean shift algorithm when solving the multimodal problem presented in Eq. (3.8). In this particular case, the algorithm uses 100 agents (n) and performs 100 iterations. Figure 3.5a illustrates the objective function $J(x_1, x_2)$ that needs to be optimized, along with its corresponding contour representation. It is evident from Fig. 3.5a that the objective function consists of one global maximum and two local maxima. Throughout the progression of the mean shift process, the operation is showcased using five selected points: (b) after 1 iteration, (c) after 40 iterations, (d) after 60 iterations, (e) after 80 iterations, and (f) after 100 iterations. Point (b) depicts an early stage of the algorithm, where the elements are predominantly located in their initial random positions. Figure 3.5c–e demonstrate how the mean shift gradually forms groups as the number of iterations increases. Finally, in Fig. 3.5f, all the elements converge to the global and local maxima, representing the culmination of the mean shift process.

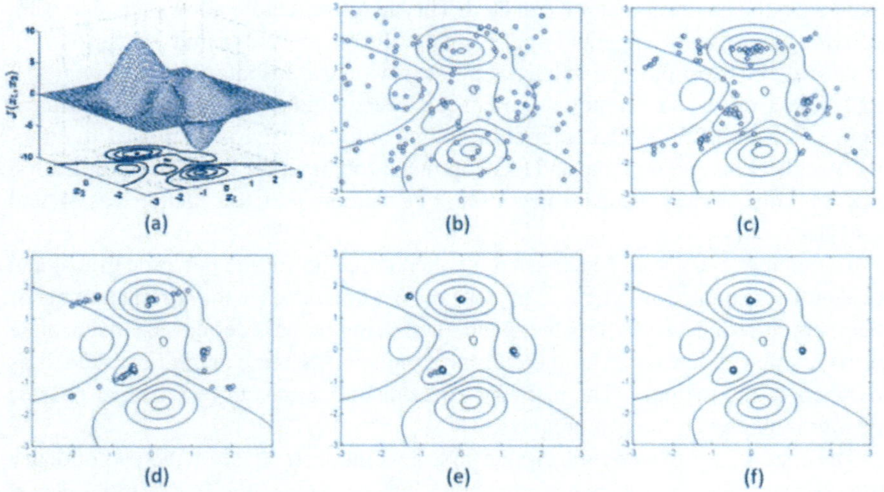

Fig. 3.5 The behavior of the modified mean shift algorithm is demonstrated over 100 iterations while solving the multimodal problem described in Eq. (3.8). The following figures illustrate this behavior: **a** the objective function accompanied by its contour representation, **b** the distribution of individuals after 1 iteration, **c** the distribution after 40 iterations, **d** the distribution after 60 iterations, **e** the distribution after 80 iterations, and **f** the distribution after 100 iterations

3.3.2 Process of Memory

In the present approach, a memory component has been introduced to store all the optimal solutions encountered during the optimization process, utilizing a competition strategy. This memory mechanism selectively includes diverse elements, ensuring that only distinct solutions are recorded, thus avoiding redundant representations of the same optimal position.

The incorporation of the memory serves two main purposes. Firstly, it allows for the registration of all optimal solutions seen thus far, providing the user with the ability to make informed decisions. This becomes especially important in dynamic multi-modal optimization problems where the best solution may not always be the most feasible option due to practical constraints. By maintaining a list of multiple optimal solutions, the user can evaluate and select an alternative based on various cost considerations.

Secondly, the memory plays a role in adapting the algorithm's behavior to changing environments in dynamic optimization problems. Different strategies are employed to continuously adjust the algorithm's performance when confronted with new circumstances. In the present scheme, a memory mechanism is utilized to retain previous optimal solutions, enabling their utilization in future iterations when changes in the environment are detected. This adaptive feature allows the algorithm to leverage past knowledge and improve its performance in dynamic scenarios.

Global and local optimal solutions possess two key attributes: (1) they exhibit a high fitness value and (2) they represent the best solutions within a particular local region. In light of these characteristics, the memory strategy is employed to effectively register potential global and local optima throughout the dynamic optimization process. This memory mechanism consists of a memory structure, denoted as $\mathbf{M}$, and a storing process.

The memory structure $\mathbf{M}$ serves as a collection point for the potential global and local optima, denoted as $\{\mathbf{m}_1, \ldots, \mathbf{m}_S\}$, where S corresponds to the current number of members stored in $\mathbf{M}$. On the other hand, the storing process defines the mechanism by which new solutions from the current population $\mathbf{X}$ are captured and added as members to the memory. The memory mechanism operates in two distinct phases: the initial phase and the capture phase.

The initial step is taken during the first generation ($t = 1$) of the evolutionary process and whenever a change in the environment is detected. In the initial phase, the memory structure $\mathbf{M}$ is formed by selecting the best element, based on the objective function, from the initial population $\mathbf{X}^1$ (or the current population in the event of an environmental change). The initial population, denoted as $\mathbf{X}^1 = \{\mathbf{x}_1^1, \ldots, \mathbf{x}_n^1\}$, consists of n elements, where each element's position $\mathbf{x}_i^1$ is represented by a d-dimensional vector ($\mathbf{x}_i^1 = \{x_{i,1}^1, \ldots, x_{i,d}^1\}$) containing the values of the design variables to be optimized. In this notation, $x_{i,j}^t$ represents the j-th parameter of the i-th element computed in the t-th iteration ($j \in 1, \ldots, d$). The values of $\mathbf{X}^1$ are generated randomly, following a uniform distribution within specified lower and upper bounds, denoted as a_j^{low} and a_j^{high}, respectively.

The capture phase occurs after the initialization and continues until a change in the objective function is detected. After applying the mean shift method to the population $\mathbf{X}^t$, a new population $\mathbf{X}^{t+1}$ is generated, consisting of n new positions. The next step involves updating the memory $\mathbf{M}$ through the capture process. The capture process utilizes a competitive strategy, where the p best elements from the new population $\mathbf{X}^{t+1}$ are selected to form the set $\mathbf{B} = \{\mathbf{b}_1, \ldots, \mathbf{b}_p\}$. Each element $\mathbf{b}_k$ from $\mathbf{B}$ is compared pairwise with every memory element $\mathbf{m}_h$ from $\mathbf{M}$. If the distance between $\mathbf{b}_k$ and $\mathbf{m}_h$ is less than a threshold value ρ, the element that performs better in terms of the fitness function is retained while the other one is removed. If the element $\mathbf{b}_k$ has a distance greater than ρ with all members of $\mathbf{M}$, it is added to the memory as a new member. The value of ρ has significant implications for the algorithm's behavior. Inappropriate values of ρ can lead to lower convergence rates, longer computational time, and convergence to local optima or unreliable solutions. To avoid such issues, the value of ρ is automatically computed using the following equation:

$$\rho = \frac{\prod_{j=1}^{d}(a_j^{high} - a_j^{low})}{10 \cdot d} \tag{3.14}$$

The value of ρ in this model is determined based on the lower bound a_j^{low}, the upper bound a_j^{high}, and the dimensions d of the optimization problem. This formulation, which is widely employed in the metaheuristic literature, has been extensively utilized to calculate a fixed neighbor distance between solutions.

During the mean shift scheme, multiple new optimal positions are identified. However, when a prominent objective function value is found, there is a possibility that local optima with lower objective function values may be overlooked. In such cases, the memory serves as a safeguard by preserving these positions. The memory retains vital information about the population's evolution, specifically focusing on elements that possess two key attributes: (1) a good fitness value and (2) representation of the best solutions within a specific local region. The memory excludes two types of information: (A) population elements that do not correspond to optima, and (B) elements that closely resemble existing memory members (i.e., duplicates). With this understanding, it can be asserted that although the population carries valuable information accumulated over multiple iterations, the most crucial information is already stored in the memory. Therefore, utilizing the memory elements alone is sufficient for reconstructing the primary characteristics of the objective function, rendering the remaining population solutions irrelevant.

To demonstrate the memory mechanism under the capture phase, Fig. 3.6 exhibits a simple maximization example that considers a two-dimensional objective function $J(x_1, x_2)$. In the example, it is assumed a set $\mathbf{B}$ of two $(m = 2)$ different elements $\{\mathbf{b}_1, \mathbf{b}_2\}$ and memory with two memory elements $\{\mathbf{m}_1, \mathbf{m}_2\}$ $(S = 2)$. According to Fig. 3.6, in the case of $\mathbf{b}_1$, the first step is to calculate the correspondent distances $D_{1,1}$ and $D_{1,2}$. $\mathbf{m}_1$ represents the nearest memory element to $\mathbf{b}_1$. As $D_{1,1} < \rho$, solutions $\mathbf{m}_1$ and $\mathbf{b}_1$ are quite related and could refer to the same local maximum. Under such conditions, both elements compete for the same slot. As $J(\mathbf{b}_1) > J(\mathbf{m}_1)$, the slot of

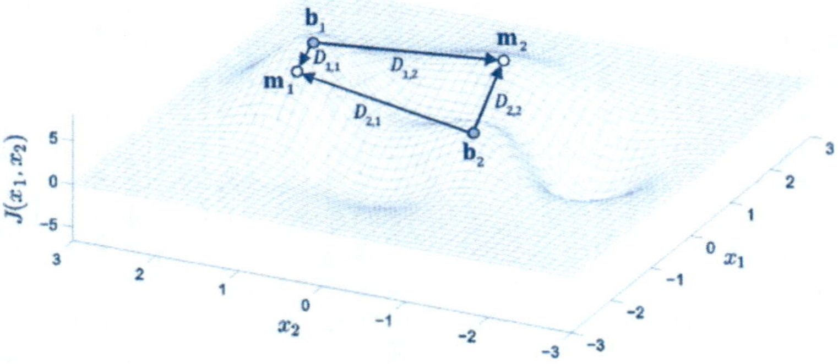

Fig. 3.6 Mechanism of memory during the capture phase

$\mathbf{m}_1$ is replaced by $\mathbf{b}_1$ in the memory **M**. On the other hand, since the element $\mathbf{b}_2$ is located far from both memory members $\mathbf{m}_1$ and $\mathbf{m}_2$, the element $\mathbf{b}_2$ is included as the member $\mathbf{m}_3$ in the memory **M**.

3.3.3 Dynamical Procedure

In order to tackle dynamic optimization problems, various strategies are employed to continuously adapt the algorithm's behavior to changing environments. In our present scheme, we utilize a memory mechanism to retain past optimal solutions for future use when a change in the environment is detected. A change in the environment is defined by a variation in the objective function $J(\mathbf{x})$ between two consecutive iterations, where $J^t(\mathbf{x}) \neq J^{t+1}(\mathbf{x})$.

The mean shift algorithm possesses inherent adaptability due to its reliance on the positions of local maxima in the objective function for the search strategy. As the mean shift method is applied over multiple iterations, diverse elements converge towards the same local maxima, forming clusters. While the single mean shift can track changes in the objective function, it does so with a relatively slow convergence rate. To expedite this process, a dynamic strategy incorporating memory is employed. This strategy involves utilizing the memory structure, denoted as **M**, to detect changes in the objective function and respond accordingly to modifications in the environment.

To identify changes in the objective function, a selective evaluation process is conducted during each iteration by reevaluating a small subset of memory elements. If any of these elements exhibit a change, it indicates a modification in $J(\mathbf{x})$. To achieve this, 10% of the top-performing elements from the memory **M**, based on their objective function values, are chosen for reevaluation. This subset, denoted as **R**, consists of representative positions $\{\mathbf{r}_1, \ldots, \mathbf{r}_e\}$ in the search space that encompass the distribution of objective function values. The size of **R** is determined by $e =$

$ceil(0.1 \cdot |\mathbf{M}|)$, where $|\mathbf{M}|$ represents the number of members in the memory $|\mathbf{M}|$ and $ceil(g)$ returns the smallest integer greater than or equal to g.

Once a change is identified in the objective function, a reaction mechanism is triggered with the aim of expediting the convergence process when the algorithm encounters a modified objective function. This mechanism involves replacing the elements in the current population $\mathbf{X}^t$ with members from the memory $\mathbf{M}$. The rationale behind this strategy is to leverage the knowledge gained from previous environments as a starting point for the subsequent search. Consequently, this approach accelerates the process of identifying local maxima in the altered objective function. In the present process, the replacement of elements from the current population $\mathbf{X}^t$ is determined by two scenarios: when $|\mathbf{M}| > n$ or when $|\mathbf{M}| < n$. If the number of members in $\mathbf{M}$ exceeds the elements in $\mathbf{X}^t$, the best n members from $\mathbf{M}$ are selected to replace the entire population of $\mathbf{X}^t$. On the other hand, if the number of members in $\mathbf{M}$ is fewer than the elements in $\mathbf{X}^t$, the complete number of members $|\mathbf{M}|$ from $\mathbf{M}$ are utilized to replace an equal number of elements from $\mathbf{X}^t$. Since a new environment has been detected ($J^t(\mathbf{x}) \neq J^{t+1}(\mathbf{x})$), the existing information stored in the memory is no longer valid for the new conditions. Consequently, the memory $\mathbf{M}$ is reinitialized using the initial phase procedure. Figure 3.7 provides an illustrative diagram of the dynamic search strategy in operation.

3.3.4 The Dynamical Optimization Method (EMDO)

In this subsection, we provide a comprehensive explanation of the Evolutionary-Mean shift for Dynamic Optimization (EMDO) approach, which encompasses all the schemes discussed earlier. The algorithm comprises four distinct processes: initialization, mean shift operation, memory updating mechanism, dynamic strategy, and termination test. These processes collectively form the EMDO algorithm, which we will now elaborate on.

Initialization

At the beginning of the algorithm, in the first iteration ($t = 1$), the present method generates a set $\mathbf{X}^t$ of n candidate solutions, denoted as $\{\mathbf{x}_1^t, \ldots, \mathbf{x}_n^t\}$. Each location $\mathbf{x}_i^1$ represents a d-dimensional element $\{x_{i,1}^t, \ldots, x_{i,d}^t\}$, which serves as a candidate solution for the dynamic optimization problem at hand. The values of each design variable $x_{i,j}^t$ (where j ranges from 1 to d) are initialized with random numbers following a uniform distribution between predefined lower limits (a_j^{low}) and upper limits (a_j^{high}). In this initial iteration, the memory $\mathbf{M}$ operates in the initial phase, where it is populated with the best element, in terms of the objective function, from the initial population $\mathbf{X}^t$.

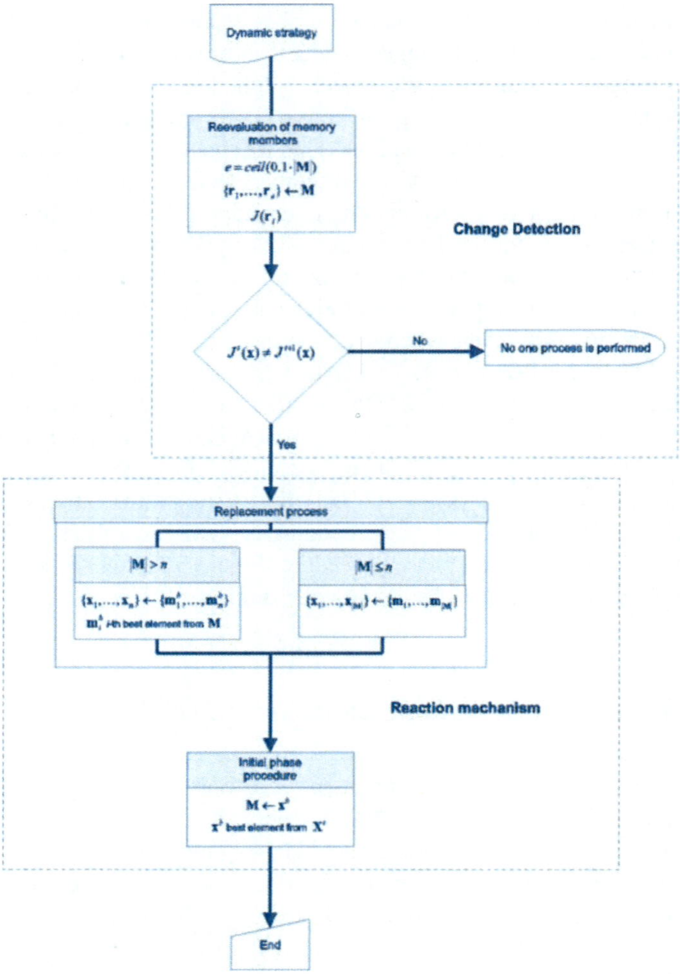

Fig. 3.7 Dynamical search procedure

Mean shift operation

The mean shift procedure is executed on the current population $\mathbf{X}^t$. Following its execution, a new population $\mathbf{X}^{t+1}$ is generated, consisting of n new positions. The complete process of mean shift is elaborated and explained extensively in Sect. 3.3.1.

Memory procedure

During this phase, the memory is updated by implementing the capture step discussed in Sect. 3.3.2. To update the memory $\mathbf{M}$, a competitive strategy is employed. In this strategy, the p best elements from the new population $\mathbf{X}^{t+1}$ are selected to form the set $\mathbf{B} = \{\mathbf{b}_1, \ldots, \mathbf{b}_m\}$. Each element $\mathbf{b}_k$ from $\mathbf{B}$ is compared to every memory element $\mathbf{m}_h$ from $\mathbf{M}$. If the distance between $\mathbf{b}_k$ and $\mathbf{m}_h$ is below a threshold value ρ, the element that exhibits superior performance in terms of the fitness function takes precedence, while the other element is removed. If the distance between $\mathbf{b}_k$ and all members of $\mathbf{M}$ exceeds ρ, $\mathbf{b}_k$ is added to the memory as a new member.

Dynamic mechanism

During this stage, the memory $\mathbf{M}$ undergoes a process to identify any changes in the objective function and respond accordingly to adapt to the modified environment. If a change in the environment is detected, a reaction mechanism is triggered. However, if there are no changes in the objective function, the dynamic strategy does not impact the present method.

The reaction mechanism involves two actions: replacing the elements of $\mathbf{X}^t$ with memory members from $\mathbf{M}$ and re-initializing the memory $\mathbf{M}$. The detailed explanation of the complete dynamic strategy can be found in Sect. 3.3.3.

Termination criterion

The mean shift operation, memory updating mechanism, and dynamic strategy steps are iteratively executed until a termination criterion is met, typically the maximum number of function evaluations.

3.4 Experiments and Results

Evolutionary algorithms are commonly proposed to tackle complex stationary optimization problems. However, real-world problems often exhibit dynamic changes caused by various factors such as randomness, state transitions, physical motion, and environmental shifts. To evaluate the performance of evolutionary methodologies in dynamic environments, a unified approach is necessary, which emulates certain random characteristics of dynamic problems. This section presents an experimental study of the Evolutionary-Mean shift for Dynamic Optimization (EMDO) using five problems generated by the Generalized Dynamic Benchmark Generator (GDBG) for CEC'2009 Competition on Dynamic Optimization [64]. The performance of the method is compared with four state-of-the-art dynamic multimodal algorithms: Clustering Particle Swarm Optimization (CPSO) [23], Simple Genetic Algorithm (SGA) [50], Simple Particle Swarm Optimization (SPSO) [51], and Self-Adaptive Differential Evolution (JDE) [65]. To ensure clarity, this section is divided into five parts: Sect. 3.4.1 describes the GDBG generator, Sect. 3.4.2 presents the performance criteria used in the experimental study, Sect. 3.4.3 provides the parameter settings of the method and its competitors, Sect. 3.4.4 presents the numerical results for the performance comparison, including a convergence analysis of the method. Finally, Sect. 3.4.5 discusses the computational effort of each dynamic multimodal approach.

3.4.1 Benchmark Generator (GDBG)

The Generalized Dynamic Benchmark Generator (GDBG) incorporates both rotation and shifting techniques to construct dynamic environments. In the GDBG framework, a Dynamic Optimization Problem (DOP) is defined as follows:

$$F = f(x, \phi, t) \tag{3.15}$$

In this context, the dynamic optimization problem (DOP) is denoted by F, with f representing the objective function, x representing the decision variable, t representing the time, and ϕ representing the system control parameter. The system control parameter ϕ determines the degree of deviation of the solution within the fitness landscape.

The dynamism generated by the GDBG system is achieved by modifying the control parameter ϕ such as:

$$\varphi(t + 1) = \varphi(t) + \Delta\varphi \tag{3.16}$$

The new environment at time $t + 1$ is mathematically described by incorporating a deviation term $\Delta\varphi$ into the current system control parameters. This deviation term modifies the environment as follows:

$$f(x, \varphi, t + 1) = f(x, \varphi(t) + \Delta\varphi, t) \tag{3.17}$$

In this research, we examine five distinct types of changes, each characterized by varying levels of complexity. The description of each change is provided below:

Small Step (SS):

$$\Delta\phi = \alpha \cdot \|\phi\| \cdot r \cdot \phi_{severity} \tag{3.18}$$

Large step (LS):

$$\Delta\phi = \|\phi\| \cdot (\alpha \cdot sign(r) + (\alpha_{max} - \alpha) \cdot r) \cdot \phi_{severity} \tag{3.19}$$

Random (RND):

$$\Delta\phi = N(0, 1) \cdot \phi_{severity} \tag{3.20}$$

Recurrent (REC):

$$\phi(t + 1) = \phi_{min} + \|\phi\| \left(\sin\left(\frac{2\pi}{P} t + \varphi \right) + 1 \right) / 2 \tag{3.21}$$

Recurrent with noise (RECN):

$$\phi\phi(t+1) = \phi_{min} + \|\phi\|\left(\sin\left(\frac{2\pi}{P}t + \varphi\right) + 1\right)/2 + N(0,1)\cdot noisy_{severity} \tag{3.22}$$

In this context, $\|\phi\|$ denotes the range of values that ϕ can take. ϕ_{severity} is a constant that determines the level of randomness associated with ϕ. ϕ_{min} represents the minimum value that ϕ can assume, while $noisy_{severity} \in (0, 1)$ indicates the severity of noise in recurrent changes.

By examining Eq. 3.15, it is evident that f represents the objective function that incorporates the dynamism influenced by the control parameter ϕ. In the GDBG system, accounting for the dynamic rotation of the peaks, the objective function f can be expressed as follows:

$$f(x,\phi,t) = \max_{i=1}^{m}\left(\vec{H}_i(t)\Big/\left(1 + \vec{W}_i(t)\cdot\sqrt{\sum_{j=1}^{n}\frac{\left(x_j - \vec{X}^i_j(t)\right)^2}{n}}\right)\right) \tag{3.23}$$

where $\phi = \left(\vec{H}, \vec{W}, \vec{X}\right)$, $\vec{H}$, $\vec{W}$ and $\vec{X}$ correspond to the peak height, width and position respectively, m represents the number of peaks and n corresponds to the number of dimensions.

To visually demonstrate each type of modification, Fig. 3.8 provides an example for each change type. Figure 3.8 SS (a) shows a 3D surface plot of an objective function generated by the GDBG system with a small step change. The corresponding contour plot is presented in Fig. 3.8 SS (b). Similarly, Fig. 3.8 SS (c) displays another objective function created with a small change, accompanied by its contour plot in Fig. 3.8 SS (d). For the large step change, Fig. 3.8 LS (a) exhibits an example of an objective function, and its contour plot is shown in Fig. 3.8 LS (b). Additionally, Fig. 3.8 LS (c) presents another instance of an objective function generated with the large step change, along with its corresponding contour plot in Fig. 3.8 LS (d). Regarding the random change (RND), Fig. 3.8 RND (a) and (c) display two examples of this type of change, while their contour plots can be seen in Fig. 3.8 RND (b) and (d). The recurrent change is illustrated in Fig. 3.8 REC (a) and (b), and their corresponding contour plots are depicted in Fig. 3.8 REC (b) and (d). Finally, the recurrent with noise (RECN) change type is presented graphically in Fig. 3.8 RECN (a) and (b), accompanied by their respective contour plots in Fig. 3.8 RECN (b) and (d).

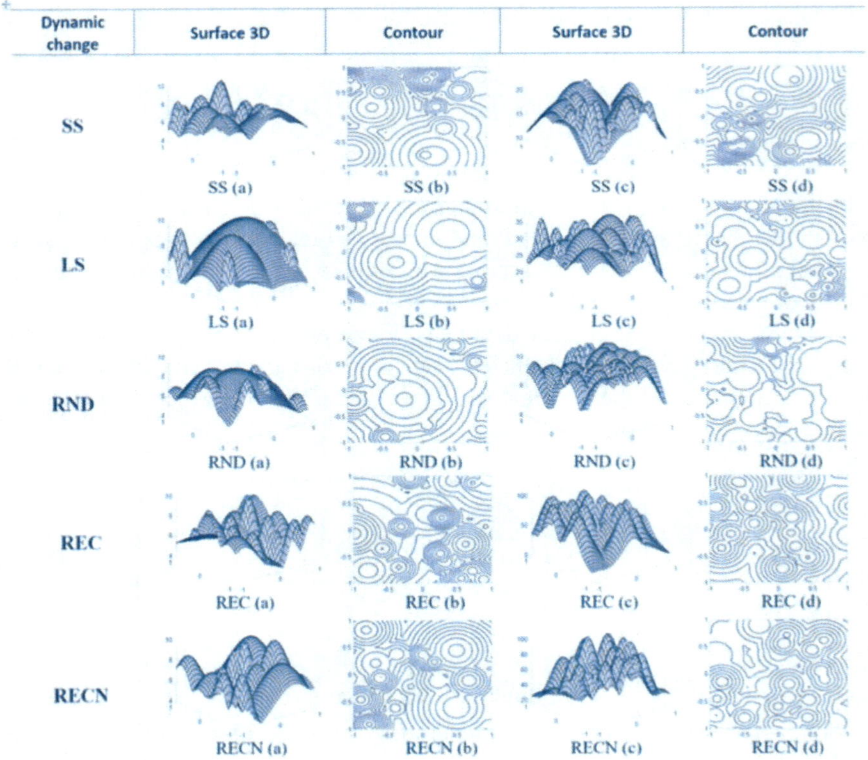

Fig. 3.8 Examples considering each kind of modification

3.4.2 Performance Metrics

In this chapter, the evaluation of dynamic multimodal approaches is conducted using the performance metrics proposed in the GDBG system. These metrics consist of four indices that assess the effectiveness of the approaches in accurately identifying the maximum number of peaks during environment changes. Among these indices is the Average Best ($AvgBest$) metric, which quantifies the minimum average value of the best solutions obtained by each evolutionary approach tested. The mathematical representation of $AvgBest$ is given by the following equation:

$$AvgBest = \left(\sum_{i=1}^{runs} Min_{j=1}^{num_{change}} E_{i,j}^{last}(t) \right) / runs \tag{3.24}$$

where num_{change} is the total number of changes, $runs$ corresponds to the number of single executions for each type of change and $E_{i,j}^{last}(t) = |f(x_{best}(t)) - f(x^*(t))|$ represents the error among the global optimum $x^*(t)$ and the local $x_{best}(t)$ optimum.

Contrary to the $AvgBest$, the Average Worst ($AvgWorst$) computes the maximum mean value as follows:

$$AvgWorst = \sum_{i=1}^{runs} Max_{j=1}^{num_{change}} E_{i,j}^{last}(t)/runs \tag{3.25}$$

Additionally, the Average Mean ($AvgMean$) computes the mean value according to:

$$AvgMean = \sum_{i=1}^{runs} \sum_{j=1}^{num\_change} E_{i,j}^{last}(t)/(runs * num\_change) \tag{3.26}$$

Lastly, the standard deviation metric quantifies the level of variation exhibited by each error value in relation to its corresponding mean value. It is calculated using the following equation:

$$STD = \sqrt{\frac{1}{runs * num_{change} - 1} \sum_{i=1}^{runs} \sum_{j=1}^{num_{change}} \left(E_{i,j}^{last}(t) - Avg\_mean\right)^2} \tag{3.27}$$

3.4.3 Configuration

In this section, we provide the details of the environmental configuration used in the dynamic GDBG generator and the parameter settings for each evolutionary scheme. To ensure reliable results, each type of change is executed in 20 independent runs. The population size for all evolutionary algorithms is set to 100 individuals, and the maximum number of iterations is set to 2000. A total of 60 changes are introduced by the GDBG system to generate dynamic environments. To maintain consistency with previously published works, the number of dimensions is set to 2.

For the comparative study, the EMDO algorithm is compared with CPSO, SGA, SPSO, and JDE methodologies. The parameter configuration for each evolutionary approach follows the reported guidelines that yield the best performance in multi-modal dynamic environments. Note that these parameter settings are carefully chosen to ensure a fair comparison among the different algorithms. The specific parameter configuration is as follows:

- For CPSO, we have implemented the algorithm based on the guidelines provided in [23]. The cognitive and social acceleration coefficients are set to $c1 = 1.7$ and $c2 = 1.7$, respectively. Additionally, a linearly decreasing weight factor is used, starting from 0.6 and gradually decreasing to 0.3.

- In the case of SGA, we have implemented a restart mechanism along with an elitism scheme, as described in [23]. The restart mechanism helps in exploring different regions of the search space, while the elitism scheme ensures that the best-found solutions are preserved throughout the optimization process.
- The implementation of SPSO follows the guidelines presented in [51]. These guidelines provide the necessary instructions and considerations for an effective implementation of SPSO.
- For JDE, we have adopted the search mechanism known as rand/1/bin, as suggested in [65]. This mechanism involves the random selection of three individuals, followed by a binomial crossover operation to generate a trial solution.
- Regarding EMDO, the kernel function employed is the Epanechnikov kernel, as specified in Eq. 3.4 of the method. This kernel function plays a crucial role in the mean shift operation within EMDO.

By adhering to these specific configurations and guidelines for each algorithm, we ensure that the comparative study is conducted in a consistent and fair manner.

To assess the robustness, scalability, and flexibility of the EMDO algorithm, we have performed an experimental study considering two scenarios: one with 10 peaks ($m = 10$) and another with 50 peaks ($m = 50$). This analysis allows us to evaluate the algorithm's performance across different problem complexities and examine its ability to handle varying numbers of peaks effectively.

3.4.4 Outcomes

In this section, we conduct a numerical performance comparison between EMDO and other state-of-the-art evolutionary approaches for dynamic multimodal optimization problems. The comparison study involves evaluating the algorithms on a set of 5 benchmark changes, which are designed to simulate real-world dynamic optimization scenarios. To ensure a fair comparison, we perform a total of 20 independent runs to mitigate the impact of randomness. Each run is terminated when the maximum number of iterations, $MaxIter = 2000$, is reached. Throughout the optimization process, a new dynamic environment is generated by the GDBG system every 33 iterations ($num_{change} = 60$). The purpose of generating new dynamic environments is to assess the ability of each evolutionary methodology to effectively identify all possible peaks within each change type. The experimental environment is implemented in MATLAB® 2020a, running on a Windows 10 Home ×64 operating system, with a CPU speed of 3.60 GHz (8× Intel Core i7-7700), 12.0 GB RAM, and a GPU with 8.0 GB + VRAM 2.0 GB (NVIDIA GeForce GT 730).

The numerical results of the method and its competitors are presented in Table 3.1, showcasing the outcomes obtained for the task of finding 10 peaks. The table includes several performance metrics: ***AvgBest*** represents the average best value, ***AvgWorst*** denotes the average worst value, ***AvgMean*** indicates the average mean value, and STD represents the standard deviation. Each column of the table corresponds to a specific type of change: SS represents Small Step (Eq. 3.18), LS denotes Large Step (Eq. 3.19), RND signifies Random change (Eq. 3.20), **REC** represents Recurrent (Eq. 3.21), and **RECN** stands for Recurrent with noise change (Eq. 3.22). The best entries in the table are highlighted in boldface to distinguish them from the rest.

Based on the data presented in Table 3.1, the performance of the EMDO approach surpasses that of the CPSO and SGA algorithms in terms of the ***AvgBest*** metric for detecting 10 peaks within a dynamic search space. The results obtained by the SPSO and JDE algorithms show similar performance to EMDO. However, considering the ***AvgMean*** and STD metrics, it becomes evident that EMDO consistently delivers more reliable outcomes compared to its competitors. These numerical findings suggest that EMDO effectively registers and maintains the optimal dynamic values for each type of change. This ability is achieved through the combination of the memory mechanism, which records potential candidate solutions, and the density attractor of the mean shift method, which identifies local maxima within the estimated density function. Consequently, the ***AvgMean*** performance index indicates that EMDO produces robust and stable solutions throughout the total number of changes for each change type. Moreover, the STD metric value clearly demonstrates that the method not only yields superior results but also exhibits greater consistency.

Table 3.1 Results with $m = 10$ peaks

		SS	LS	RND	REC	RECN
EMDO	AvgBest	0.00E+00	0.00E+00	0.00E+00	0.00E+00	0.00E+00
	AvgWorst	5.20E−02	1.35E−01	7.65E−02	7.62E+00	6.63E+00
	AvgMean	2.37E−03	6.64E−03	2.76E−03	2.89E−01	2.59E−01
	STD	1.92E−02	8.64E−02	2.82E−02	1.47E+00	1.53E+00
CPSO	AvgBest	1.05E−07	5.21E−08	4.31E−08	2.56E−07	4.33E−06
	AvgWorst	1.24E+00	2.71E+01	2.82E+01	2.17E+01	2.66E+01
	AvgMean	3.51E−02	2.72E+00	4.13E+00	1.87E+00	1.06E+00
	STD	4.26E−01	6.52E+00	8.99E+00	4.49E+00	4.81E+00
JDE	AvgBest	**0.00E+00**	**0.00E+00**	**0.00E+00**	**0.00E+00**	**0.00E+00**
	AvgWorst	9.10E−01	32.1E+00	31.7E+00	1.83E+01	32.7E+00
	AvgMean	2.88E−02	3.58E+00	2.99E+00	2.17E+00	1.14E+00
	STD	4.42E−01	7.83E+00	7.12E+00	4.32E+00	5.72E+00
SPSO	AvgBest	**0.00E+00**	**0.00E+00**	**0.00E+00**	**0.00E+00**	**0.00E+00**
	AvgWorst	31.0E+00	48.2E+00	43.2E+00	35.7E+00	78.9E+00
	AvgMean	5.66E+00	10.2E+00	11.7E+00	6.73E+00	32.0E+00
	STD	7.72E+00	12.6E+00	13.5E+00	8.71E+00	25.6E+00
SGA	AvgBest	4.01E−07	4.30E−05	5.54E−05	1.00E−04	2.89E−04
	AvgWorst	4.32E+01	5.21E+01	4.55E+01	4.02E+01	8.03E+01
	AvgMean	5.61E+00	1.01E+01	1.31E+01	7.90E+00	2.93E+01
	STD	9.35E+00	1.32E+01	1.39E+01	9.41E+00	2.57E+01

To assess the scalability of the method in comparison to its competitors, it is necessary to examine the performance of each evolutionary approach on functions with higher levels of multimodality. Accordingly, the comparison among these methods is conducted on a search space containing 50 peaks. Table 3.2 displays the numerical results obtained for the task of finding 50 peaks. The table includes the ***AvgBest*** metric, which represents the average best value, ***AvgWorst*** for the average worst value, ***AvgMean*** for the average mean value, and STD for the standard deviation. Each column corresponds to a specific type of change: SS denotes Small Step (Eq. 3.18), LS represents Large Step (Eq. 3.19), RND denotes Random change (Eq. 3.20), **REC** represents Recurrent change (Eq. 3.21), and **RECN** indicates Recurrent with noise change (Eq. 3.22). The table highlights the best entries in boldface.

The exceptional performance of the method is clearly evident from the results presented in Table 3.2. The AvgBest values indicate that the EMDO algorithm surpasses its competitors by successfully discovering all the optimal dynamic values within a search space containing a higher number of peaks, regardless of the type of change. There are a few exceptions, such as the performance of the SPSO method in the case of Small Step (SS) and Large Step (LS) changes, where the presented approach exhibits comparable performance. Moreover, the JDE method demonstrates

Table 3.2 Results with $m = 50$ peaks

		SS	LS	RND	REC	RECN
EMDO	AvgBest	0.00E+00	0.00E+00	0.00E+00	0.00E+00	0.00E+00
	AvgWorst	2.22E−02	3.54E−02	4.32E−02	5.71E−01	6.48E−01
	AvgMean	1.88E−03	1.61E−03	1.30E−03	6.11E−02	6.84E−02
	STD	5.77E−03	6.59E−03	8.30E−03	1.01E−01	1.16E−01
CPSO	AvgBest	2.45E−06	2.06E−07	9.89E−07	2.12E−06	9.03E−05
	AvgWorst	4.92E+00	2.21E+01	2.57E+01	9.61E+00	2.21E+01
	AvgMean	2.62E−01	3.28E+00	6.32E+00	8.48E−01	1.48E+00
	STD	9.36E−01	5.30E+00	7.44E+00	1.78E+00	4.39E+00
JDE	AvgBest	**0.00E+00**	**0.00E+00**	**0.00E+00**	**0.00E+00**	**0.00E+00**
	AvgWorst	3.92E+00	30.1E+00	27.6E+00	9.08E+00	33.1E+00
	AvgMean	1.72E−01	4.08E+00	4.29E+00	9.48E−01	1.76E+00
	STD	7.63E−01	6.45E+00	6.74E+00	1.76E+00	5.82E+00
SPSO	AvgBest	**0.00E+00**	**0.00E+00**	12.4E−05	98.9E−05	7.79E−13
	AvgWorst	33.3E+00	46.0E+00	45.3E+00	28.2E+00	78.3E+00
	AvgMean	7.95E+00	12.2E+00	14.8E+00	5.42E+00	36.2E+00
	STD	8.16E+00	11.5E+00	12.5E+00	6.34E+00	26.2E+00
SGA	AvgBest	1.43E−04	1.44E−04	2.53E−04	5.46E−01	1.03E−03
	AvgWorst	4.02E+01	4.48E+01	4.78E+01	2.80E+01	7.82E+01
	AvgMean	7.61E+00	1.13E+01	1.52E+01	5.29E+00	3.49E+01
	STD	9.75E+00	1.13E+01	1.30E+01	6.19E+00	2.65E+01

similar performance to EMDO across all entries. However, in terms of consistency, EMDO outperforms JDE in all conducted experiments. Therefore, EMDO proves to be highly capable of addressing dynamic optimization problems characterized by higher levels of multimodality. The AvgMean metric values suggest that EMDO consistently produces robust results across multiple independent runs. Furthermore, the ability of EMDO to identify all the peaks in each change type every 33 iterations is demonstrated by the low STD metric values, showcasing its capacity to deliver both optimal and consistent outcomes in comparison to its competitors.

To statistically evaluate the significance in the numerical results from Tables 3.1 and 3.2, a non-parametric test is conducted. In this study, the Wilcoxon rank-sum test [66] has been employed to validate the numerical outcomes from the performance analysis. The rank-sum test has been applied using the 0.05 significance value over 30 independent runs for each type of change for 10 and 50 peaks. Table 3.3 reports the p-values from the AvgMean of Table 3.1 (where $m = 10$). Additionally, Table 3.4 reports the p-values from the AvgMean of Table 3.2 (where $m = 50$). In Tables 3.3 and 3.4, a pair-wise comparison among EMDO and the rest of the dynamic approaches is presented. As a result, four groups are considered in the statistical comparison; EMDO versus CPSO, EMDO versus JDE, EMDO versus SPSO and EMDO versus SGA.

In the statistical test, the null hypothesis H0 indicates that there is no significant difference in a group [28, 67–70]. As a counterpart, the alternative hypothesis H1 indicates that there is a significant difference in a group. To make an understandable visualization of the significance results, Tables 3.3 and 3.4, adopt the symbols ▲ ▼, and ►. The symbol ▲, represents that the EMDO performs significantly better than a given competitor. The symbol ▼, represents that EMDO is able to produce worse results than its competitor, and the symbol ► represents the situation when the Wilcoxon test is not able to distinguish between the numerical results.

According to the p-values from Table 3.3 ($m = 10$), it is demonstrated that the EMDO obtains the best outcome for each type of change in a dynamic environment. Also, from Table 3.4 ($m = 50$), it is quite evident the scalability of the EMDO

Table 3.3 The *p*-values resulting from the Wilcoxon rank sum test for the comparisons between EMDO and CPSO, EMDO and JDE, EMDO and SPSO, and EMDO and SGA, based on the AvgMean values for each change type from Table 3.1

EMDO versus	CPSO	JDE	SPSO	SGA
SS	1.077E−07▲	2.558E−08▲	2.234E−07▲	5.765E−05▲
LS	8.932E−05▲	4.019E−11▲	3.343E−11▲	6.010E−07▲
RND	5.324E−10▲	2.034E−11▲	5.675E−06▲	7.016E−11▲
REC	6.513E−07▲	1.022E−11▲	1.677E−09▲	5.356E−11▲
RECN	3.021E−11▲	1.054E−11▲	3.456E−11▲	4.775E−11▲
▲	5	5	5	5
▼	0	0	0	0
►	0	0	0	0

Table 3.4 The *p*-values generated by the Wilcoxon rank-sum test for the comparisons between EMDO and CPSO, EMDO and JDE, EMDO and SPSO, and EMDO and SGA, based on the AvgMean values for each change type from Table 3.2

EMDO versus	CPSO	JDE	SPSO	SGA
SS	5.353E−07▲	3.356E−08▲	3.234E−07▲	6.765E−05▲
LS	7.454E−05▲	4.334E−11▲	4.463E−11▲	7.775E−07▲
RND	2.129E−10▲	3.197E−11▲	5.985E−06▲	4.732E−11▲
REC	1.953E−07▲	5.674E−11▲	5.687E−09▲	6.968E−11▲
RECN	1.345E−11▲	6.874E−11▲	6.465E−11▲	5.813E−11▲
▲	5	5	5	5
▼	0	0	0	0
►	0	0	0	0

by solving dynamic optimization problems containing a higher number of peaks, since the p-values from each group comparison suggest that EMDO produces the best entries for all the change types. In order to provide a more comprehensive assessment of the performance of dynamic multi-modal optimization algorithms, a convergence analysis of the five compared schemes was conducted in this section. The convergence graphs for the EMDO approach, considering each change type for 10 and 50 peaks, are presented in Fig. 3.9. These graphs illustrate the ability of EMDO to consistently converge to the global maximum every 33 iterations (num_{change} = 60). From the convergence graphs in Fig. 3.9, it is evident that EMDO maintains population diversity throughout the entire optimization process and successfully reaches the global maxima at every change for each change type. This indicates that EMDO effectively adapts to the dynamic environments and is capable of exploring and exploiting the search space to find the optimal solutions.

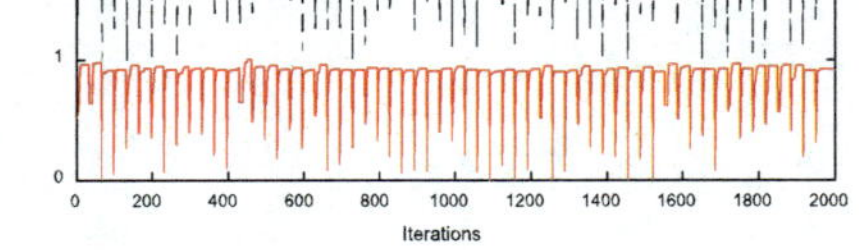

(a)

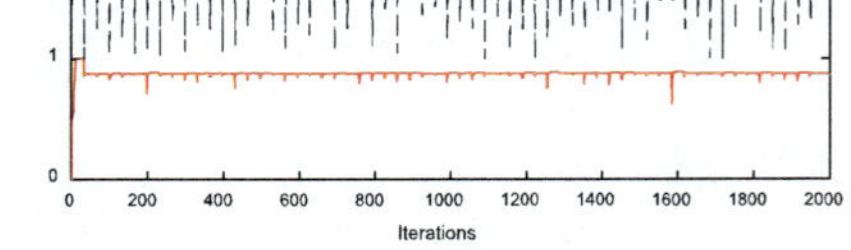

(b)

Fig. 3.9 The convergence graphs for each change type are depicted in this figure. In **a**, the convergence graph is shown for 10 peaks, while **b** represents the convergence graph for 50 peaks

In summary, the EMDO algorithm combines the competitive memory mechanism and the density attractor functionality, enabling it to effectively capture and retain all optimal values in a dynamic search space. The key features of EMDO include its ability to generate robust, consistent, and scalable results regardless of the type of change encountered. Additionally, the convergence graphs for both 10 and 50 peaks demonstrate that EMDO maintains population diversity for each type of change.

3.4.5 Computational Complexity

In this section, we analyze the computational effort required by each compared approach used in the experimental study. The computational effort is assessed in terms of computational complexity to ensure consistency with related literature [71–74]. For the SPSO algorithm, the original PSO method is executed multiple times to maintain potential solutions. The computational complexity of SPSO is expressed as $O(Np * d * runs)$, where Np represents the population size, d corresponds to the number of dimensions, and runs indicates the number of executions of the original PSO structure. The SGA method has a complexity order of $O(Np * d * Nc * runs)$, where Nc denotes the number of generated offspring solutions in each iteration.

Regarding the CPSO algorithm, the hierarchical structure of the clustering algorithm introduces a nonlinear order of $O(n^2)$. Therefore, the complete complexity of CPSO is $O(n^2 + Ns * (Np * d))$, where Ns represents the number of sub-swarms involved in the optimization process. The JDE algorithm consists of three main procedures: the aging procedure, the overlapping search, and the reinitialization mechanism. The total complexity of JDE is $O(3 \cdot S * (Ns * d))$, where S represents the number of sub-populations and Ns indicates the number of elements in each sub-population. Finally, the EMDO algorithm operates with a complexity of $O(Np * (d + (H * C) + H))$, where H and C correspond to the size of the historical and competitive memory, respectively. By analyzing the computational complexity of each approach, we can gain insights into their respective computational efforts and compare their efficiency.

As it can be shown in the complexity analysis from each algorithm, only the CPSO, the JDE and the EMDO are able to operate within dynamic environments without considering multiple runs, as in the case of the SPSO and SGA. Even when the complexity order from the SPSO, SGA, JDE and EMDO tend to be linear, the CPSO method suggests a quadratic order, which makes it more computationally expensive than the rest of the methods. The comparison among JDE and EMDO suggests that both algorithms produce a linear complexity; however, EMDO computes more steps due to the implementation of the competitive memory mechanism. Even when the EMDO computes more steps than JDE, EMDO produces more consistent solutions, and its complexity order presents an interesting alternative since it reflects a linear tendency of the computational cost [75–77].

3.5 Conclusions

Evolutionary optimization algorithms have traditionally been used to solve complex stationary optimization problems. However, many real-world applications involve dynamic scenarios, where the objective function changes over time. Traditional metaheuristic techniques are not suitable for efficiently locating multiple dynamic optimal solutions in such cases. Recently, several evolutionary approaches have been adapted to address this challenge by dynamically locating a single global optimum. However, these methods fail to effectively handle the presence of multiple dynamic optimal solutions.

To address this limitation, this chapter presents the use of the mean shift methodology for efficiently locating and tracking both local and global maxima in dynamic environments. The presented method modifies the operation of the mean shift by incorporating not only the solution density but also their fitness values. This modification enables the detection of most local optima and the adaptive response to changes in the objective function.

In addition to the mean shift modification, the presented scheme incorporates a competitive memory and a dynamic strategy to enhance the convergence process. The competitive memory is employed to register all the optimal solutions encountered during the optimization process using a competition strategy. This memory mechanism ensures that only distinct and relevant elements are included, avoiding the inclusion of solutions that represent the same optimal position.

The dynamic strategy involves a process over the memory to detect changes in the objective function and respond accordingly to these environmental modifications. By combining these components, the method aims to efficiently register and maintain multiple optima in dynamic multi-modal optimization problems.

To assess the effectiveness of the EMDO method, its performance is compared to other state-of-the-art metaheuristic techniques designed for dynamic multi-modal environments. The evaluation is conducted using dynamic test environments generated by the Generalized Dynamic Benchmark Generator (GDBG), which is a widely used benchmarking tool in the field of dynamic optimization. The GDBG allows for the reproduction of various types of changes in the objective functions, thereby creating realistic dynamic environments.

The experimental results obtained from the comparison indicate that the EMDO algorithm outperforms its competitors in terms of robustness, scalability, and accuracy. The EMDO method demonstrates superior performance and effectiveness in adapting to the changing dynamics of the objective functions, thereby achieving better optimization outcomes compared to other state-of-the-art metaheuristic techniques [75–77].

References

1. Halder, U., Das, S., Maity, D.: A cluster-based differential evolution algorithm with external archive for optimization in dynamic environments. IEEE Trans. Cybern. **43**(3), 881–897 (2013)
2. Das, S., Mandal, A., Mukherjee, R.: An adaptive differential evolution algorithm for global optimization in dynamic environments. IEEE Trans. Cybern. **44**(6), 966–978 (2014)
3. Mavrovouniotis, M., Müller, F.M., Yang, S.: Ant colony optimization with local search for dynamic traveling salesman problems. IEEE Trans. Cybern. **47**(7), 1743–1756 (2017)
4. Sadollah, A., Sayyaadi, H., Yadav, A.: A dynamic metaheuristic optimization model inspired by biological nervous systems: neural network algorithm. Appl. Soft Comput. **71**, 747–782 (2018)
5. Baykasoğlu, A., Ozsoydan, F.B.: Evolutionary and population-based methods versus constructive search strategies in dynamic combinatorial optimization. Inf. Sci. **420**, 159–183 (2017)
6. Nguyen, T.T., Yang, S., Branke, J.: Evolutionary dynamic optimization: a survey of the state of the art. Swarm Evolut. Comput. **6**, 1–24 (2012)
7. Tinós, R., Yang, S.: Analysis of fitness landscape modifications in evolutionary dynamic optimization. Inf. Sci. **282**, 214–236 (2014)
8. Mavrovouniotis, M., Li, C., Yang, S.: A survey of swarm intelligence for dynamic optimization: algorithms and applications. Swarm Evol. Comput. **33**, 1–17 (2017)
9. Topcuoglu, L.A.R.: Impact of sensor-based change detection schemes on the performance of evolutionary dynamic optimization techniques. Soft. Comput. **22**(14), 4741–4762 (2018)
10. Kazemi, J., Hossein, K., Firouzjaee, A., Meybodi, M.R.: An adaptive bi-flight cuckoo search with variable nests for continuous dynamic optimization problems. Appl. Intell. **48**(1), 97–117 (2018)
11. Lissovoi, A., Witt, C.: A runtime analysis of parallel evolutionary algorithms in dynamic optimization. Algorithmica **78**(2), 641–659 (2017)
12. Kazemi, J., Alireza, K., Mohammad, R., Meybodi, R.: New measures for comparing optimization algorithms on dynamic optimization problems. Nat. Comput. (2019) (in press)
13. Au, C.-K., Leung, H.-F.: Cooperative coevolutionary algorithms for dynamic optimization: an experimental study. Evol. Intell. **7**(4), 201–218 (2014)
14. Zuo, X., Xiao, L.: A DE and PSO based hybrid algorithm for dynamic optimization problems. Soft. Comput. **18**(7), 1405–1424 (2014)
15. Cao, L., Xu, L., Goodman, E.D.: A collaboration-based particle swarm optimizer with history-guided estimation for optimization in dynamic environments. Expert Syst. Appl. **120**, 1–13 (2019)
16. Liu, L., Yang, S., Wang, D.: Particle swarm optimization with composite particles in dynamic environments. IEEE Trans. Syst. Man Cybern. Part B (Cybern.) **40**(6), 1634–1648 (2010)
17. Blackwell, T., Branke, J.: Multiswarms, exclusion, and anti-convergence in dynamic environments. IEEE Trans. Evol. Comput. **10**(4), 459–472 (2006)
18. Mendes, R., Mohais, A.S.: DynDE: a differential evolution for dynamic optimization problems. In: Proceeding of the IEEE Congress on Evolutionary Computation, vol. 3, pp. 2808–2815 (2005)
19. Li, C., Yang, S.: A general framework of multipopulation methods with clustering in undetectable dynamic environments. IEEE Trans. Evol. Comput. **16**(4), 556–577 (2012)
20. Brest, J., Zamuda, A., Boskovic, B., Maucec, M.S., Zumer, V.: Dynamic optimization using self-adaptive differential evolution. In: Proceedings of the IEEE Congress on Evolutionary Computation, pp. 415–422 (2009)
21. Mukherjee, R., Debchoudhury, S., Das, S.: Modified differential evolution with locality induced genetic operators for dynamic optimization. Eur. J. Oper. Res. **253**(2), 337–355 (2016)
22. Bravo, Y., Luque, G., Alba, E.: Global memory schemes for dynamic optimization. Nat. Comput. **15**(2), 319–333 (2016)
23. Li, C., Yang, S.: A clustering particle swarm optimizer for dynamic optimization. In: Proceedings of the 2009 IEEE Congress on Evolutionary Computation, pp. 439–446 (2009)

24. Wong, K.-C., Chun-Ho, W., Mok, R.K.P., Peng, C.: Evolutionary multimodal optimization using the principle of locality. Inf. Sci. **194**, 138–170 (2012)
25. Goldberg, D.E., Richardson, J.: Genetic algorithms with sharing for multimodal function optimization. In: Proceedings of 2nd International Conference Genetic Algorithms, pp. 41–49 (1987)
26. De Jong, K.A.: An Analysis of the Behavior of a Class of Genetic Adaptive Systems. Ph.D. dissertation, University of Michigan, Ann Arbor (1975)
27. Li, J.-P., Balazs, M.E., Parks, G.T., Clarkson, P.J.: A species conserving genetic algorithm for multimodal function optimization. Evol. Comput. **10**(3), 207–234 (2002)
28. Ramírez-Ortegón, M.A., Tapia, E., Ramírez-Ramírez, L.L., Rojas, R., Cuevas, E.: Transition pixel: a concept for binarization based on edge detection and gray-intensity histograms. Pattern Recogn. **43**(4), 1233–1243 (2010)
29. Liang, J.J., Qu, B.Y., Mao, X.B., Niu, B., Wang, D.Y.: Differential evolution based on fitness Euclidean distance ratio for multimodal optimization. Neurocomputing **137**, 252–260 (2014)
30. Miller, B.L., Shaw, M.J.: Genetic algorithms with dynamic niche sharing for multimodal function optimization. In: Proceedings of the 3rd IEEE Conference on Evolutionary Computation, pp. 786–791 (1996)
31. Thomsen, R.: Multimodal optimization using crowding-based differential evolution. In: Evolutionary Computation, CEC2004. Congress (2004)
32. Biswas, S., Das, S., Kundu, S., Patra, G.R.: Utilizing time-linkage property in DOPs: an information sharing based articial bee colony algorithm for tracking multiple optima in uncertain environments. Soft. Comput. **18**, 1199–1212 (2014)
33. Das, S., Maity, S., Qu, B.Y., Suganthan, P.N.: Real-parameter evolutionary multimodal optimization—a survey of the state-of-the-art. Swarm Evol. Comput. **1**(2), 71–88 (2011)
34. Mengshoel, O.J., Galán, S.F., De Dios, A.: Adaptive generalized crowding for genetic algorithms. Inf. Sci. **258**, 140–159 (2014)
35. Yazdani, S., Nezamabadi-pour, H., Kamyab, S.: A gravitational search algorithm for multimodal optimization. Swarm Evol. Comput. **14**, 1–14 (2014)
36. Chen, C.-H., Liu, T.-K., Chou, J.-H.: A novel crowding genetic algorithm and its applications to manufacturing robots. IEEE Trans. Ind. Inf. **10**(3), 1705–1716 (2014)
37. Chang, W.-D.: A modified particle swarm optimization with multiple subpopulations for multimodal function optimization problems. Appl. Soft Comput. **33**, 170–182 (2015)
38. Lianga, Y., Kwong-Sak, L.: Genetic algorithm with adaptive elitist-population strategies for multimodal function optimization. Appl. Soft Comput. **11**, 2017–2034 (2011)
39. Qu, B.Y., Suganthan, P.N., Das, S.: A distance-based locally informed particle swarm model for multimodal optimization. IEEE Trans. Evol. Comput. **17**(3), 387–402 (2013)
40. Sacco, W.F., Henderson, N., Rios-Coelho, A.C.: Topographical clearing differential evolution: a new method to solve multimodal. Prog. Nucl. Energy **71**, 269–278 (2014)
41. Cheng, S., Lu, H., Guo, Y., Lei, X., Liang, J., Chen, J., Shi, Y.: Dynamic multimodal optimization: a preliminary study. In: 2019 IEEE Congress on Evolutionary Computation (CEC), Wellington, New Zealand (2019)
42. Kaiwartya, O., Kumar, S., Lobiyal, D.K., Tiwari, P.K., Abdullah, A.H., Hassan, A.N.: Multiobjective dynamic vehicle routing problem and time seed based solution using particle swarm optimization. J. Sensors **2015** (2015)
43. Okulewicz, M., Mańdziuk, J.: A metaheuristic approach to solve dynamic vehicle routing problem in continuous search space. Swarm Evol. Comput. **48**(March), 44–61 (2019)
44. Lei, Y., Jasin, S., Sinha, A.: Joint dynamic pricing and order fulfillment for E−commerce retailers. Manuf. Serv. Oper. Manag. **20**(2), 269–284 (2018)
45. Saharan, S., Bawa, S., Kumar, N.: Dynamic pricing techniques for intelligent transportation system in smart cities: a systematic review. Comput. Commun. **150**, 603–625 (2020)
46. Manjunath, A., Raychoudhury, V., Saha, S.: Ant-taxi to pie-passenger: optimizing routes and time for distributed taxi ride sharing. In: 2020 International Conference on COMmunication System & NETworkS, COMSNETS 2020, pp. 736–743 (2020)

47. Wu, H., Qian, S., Liu, Y., Wang, D., Guo, B.: An immune-based response particle swarm optimizer for knapsack problems in dynamic environments. Soft. Comput. **24**(20), 15409–15425 (2020)
48. Hao, X. et al.: Dynamic Knapsack Optimization Towards Efficient Multi-channel Sequential Advertising (2020)
49. Nguyen, V.Q., Johnson, R.T., Sup, F.C., Umberger, B.R.: Bilevel optimization for cost function determination in dynamic simulation of human gait. IEEE Trans. Neural Syst. Rehabil. Eng. **27**(7), 1426–1435 (2019)
50. Luo, W., Lin, X., Zhu, T., Xu, P.: A clonal selection algorithm for dynamic multimodal function optimization. Swarm Evol. Comput. (2018) (in press)
51. Li, C., Yang, S.: A clustering particle swarm optimizer for dynamic optimization. In: 2009 IEEE Congress on Evolutionary Computation, Trondheim, Norway, pp. 439–446 (2009)
52. Fukunaga, K., Hostetler, L.D.: The estimation of the gradient of a density function, with applications in pattern recognition. IEEE Trans. Inf. Theory **21**, 32–40 (1975)
53. Tao, W., Jin, H., Zhang, Y.: Color image segmentation based on mean shift and normalized cuts. IEEE Trans. Syst. Man Cybern. Part B (Cybern.) **37**(5), 1382–1389 (2007)
54. Comaniciu, D., Meer, P.: Mean shift: a robust approach towards feature space analysis. IEEE Trans. Pattern Anal. Mach. Intell. **24**, 603–619 (2002)
55. Comaniciu, D.: An algorithm for data-driven bandwidth selection. IEEE Trans. Pattern Anal. Mach. Intell. **25**, 281–288 (2003)
56. Anand, S., Mittal, S., Tuzel, O., Meer, P.: Semi-supervised kernel mean shift clustering. EEE Trans. Pattern Anal. Mach. Intell. **36**(6), 1201–1215 (2014)
57. DeMenthon, D., Doermann, D.: Video retrieval using spatio-temporal descriptors pages. In: Proceedings of the Eleventh ACM International Conference on Multimedia, pp. 508–517 (2003)
58. Zivkovic, Z., Krose, B.: An EM-like algorithm for color-histogram-based object tracking. In: Proceedings of the International Conference on Computer Vision and Pattern Recognition, vol. 1, pp. 798–803 (2004)
59. Cheng, Y.: Mean shift, mode seeking, and clustering. IEEE Trans. Pattern Anal. Mach. Intell. **17**(8), 790–799 (1995)
60. Fashing, M., Tomasi, C.: Mean shift is a bound optimization. IEEE Trans. Pattern Anal. Mach. Intell. **27**(3), 471–474 (2005)
61. Horová, I., Koláček, J., Zelinka, J.: Kernel Smoothing in Matlab. World Scientific (2012)
62. Gramacki, A.: Nonparametric Kernel Density Estimation and Its Computational Aspects. Springer (2018)
63. Scott, D.W.: Scott's rule. Wires Comput. Statist. **2**(4), 497–502 (2010)
64. Li, S., Yang S., Nguyen T., Yu E., Yao X., Jin Y., Beyer H.G., Suganthan P.N.: Benchmark generator for CEC 2009 competition on dynamic optimization. Technical Report 2008, Department of Computer Science, University of Leicester, U.K., October 2009. http://hdl.handle.net/2086/11893
65. Brest, J., Zamuda, A., Bǒskovíc, B., Maŭcec, M.S., Žumer, V.: Dynamic optimization using self-adaptive differential evolution. In: 2009 IEEE Congress on Evolutionary Computation, CEC 2009, pp. 415–422 (2009)
66. Wilcoxon, F.: Individual comparisons by ranking methods. Biometrics 80–83 (1945)
67. Cuevas, E., González, A., Fausto, F., Zaldívar, D., Pérez-Cisneros, M.: Multithreshold segmentation by using an algorithm based on the behavior of locust swarms. Math. Probl. Eng. (2015)
68. Zaldivar, D., Morales, B., Rodríguez, A., Valdivia-G, A., Cuevas, E., Pérez-Cisneros, M.: A novel bio-inspired optimization model based on Yellow Saddle Goatfish behavior. Biosystems **174**, 1–21 (2018)
69. Cuevas, E., Gálvez, J., Hinojosa, S., Avalos, O., Zaldívar, D., Pérez-Cisneros, M.: A comparison of evolutionary computation techniques for IIR model identification. J. Appl. Math. (2014)
70. Bandyopadhyay, R., Basu, A., Cuevas, E., Sarkar, R.: Harris Hawks optimisation with simulated annealing as a deep feature selection method for screening of COVID-19 CT-scans. Appl. Soft Comput. **111**, 107698 (2021)

71. Cuevas, E., Wario, F., Osuna-Enciso, V., Zaldivar, D., Pérez-Cisneros, M.: Fast algorithm for multiple-circle detection on images using learning automata. IET Image Proc. **6**(8), 1124–1135 (2012)
72. Hinojosa, S., Dhal, K.G., Abd Elaziz, M., Oliva, D., Cuevas, E.: Entropy-based imagery segmentation for breast histology using the stochastic fractal search. Neurocomputing **321**, 201–215 (2018)
73. Cuevas, E., Zaldivar, D., Pérez-Cisneros, M.: Seeking multi-thresholds for image segmentation with learning automata. Mach. Vis. Appl. **22**, 805–818 (2011)
74. Ibrahim, R.A., Elaziz, M.A., Oliva, D., Cuevas, E., Lu, S.: An opposition-based social spider optimization for feature selection. Soft. Comput. **23**, 13547–13567 (2019)
75. Xu, Q., Lei, W., Si, J.: Predication based immune network for multimodal function optimization. Eng. Appl. Artif. Intell. **23**, 495–504 (2010)
76. Cuevas, E., González, M., Zaldivar, D., Pérez-Cisneros, M., García, G.: An Algorithm for Global Optimization Inspired by Collective Animal Behavior. Discrete Dynamics in Nature and Society, vol 2012, Article ID 638275
77. Cuevas, E., Gonzalez, M.: An optimization algorithm for multimodal functions inspired by collective animal behavior. Soft. Comput. **17**, 489–502 (2013)

Chapter 4
Trajectory-Driven Metaheuristic Approach Using a Second-Order Model

To obtain the optimal solution in an optimization problem where multiple local solutions exist, one must navigate the search space using an objective function and effective search patterns. However, this task becomes challenging due to ruggedness and the presence of multiple local optima, which can lead to stagnation or premature convergence. Stagnation refers to settling for a suboptimal solution as the best solution. Recently, various metaheuristic schemes have emerged in the literature, each employing different metaphors but using similar search patterns to explore the search space. Meanwhile, second-order systems exhibit diverse temporal behaviors based on their parameter values, which can be considered as search patterns with multiple behaviors and simple configurations. In this study, a novel set of search patterns inspired by the response of second-order systems is introduced. These search patterns are integrated into a comprehensive search strategy called Second-Order Algorithm (SOA) to efficiently explore the search space and obtain the global solution for complex optimization problems. To evaluate the performance of the approach, it is compared against representative optimization problems encompassing multimodal, unimodal, and hybrid benchmark formulations. The competitive results obtained highlight the promising outcomes achieved by employing the presented search patterns.

4.1 Introduction

Metaheuristic algorithms encompass versatile optimization strategies that replicate the behavior of various natural or social processes. These approaches employ a group of search agents to execute the optimization process. Each agent holds a potential solution to the optimization problem, initially generated through a random feasible solution. An objective function evaluates the quality of each agent's solution. Utilizing the objective function values, the search agents' positions are adjusted at

E. Cuevas et al., *New Metaheuristic Schemes: Mechanisms and Applications*, Intelligent Systems Reference Library 246, https://doi.org/10.1007/978-3-031-45561-2_4

each iteration, guided by a collection of search patterns that govern their movements within the search space. These search patterns serve as conceptual models inspired by natural or social processes [1]. This iterative process continues until a predefined termination criterion is met. The effectiveness of metaheuristic schemes has been demonstrated across numerous real-world applications, particularly in scenarios where traditional methods are impractical [1].

In essence, there is no definitive classification of metaheuristic methods. Nonetheless, several categories have been proposed based on criteria such as the source of inspiration, type of operators, or agent cooperation. With respect to inspiration, metaheuristic algorithms can be categorized into three groups: evolution-based, swarm-based, and physics-based. Evolution-based approaches are well-established search strategies that utilize evolutionary operators to generate search patterns. These operations, including reproduction, mutation, recombination, and selection, play a crucial role in the production of search patterns. Examples of evolution-based techniques include Evolutionary Strategies (ES) [2–4], Genetic Algorithms (GA) [5], Differential Evolution (DE) [6], and Self-Adaptive Differential Evolution (JADE) [7].

Swarm-inspired techniques draw on behavioral patterns observed in the collaborative interactions of various animals or insect species to formulate search strategies. Numerous swarm-based approaches have been published recently. Popular examples of swarm-inspired methods include the Crow Search Algorithm (CSA) [8], Artificial Bee Colony (ABC) [9], Particle Swarm Optimization (PSO) algorithm [10–12], Firefly Algorithm (FA) [13–15], Cuckoo Search (CS) [16], Bat Algorithm (BA) [17], Gray Wolf Optimizer (GWO) [18], and Moth-Flame Optimization Algorithm (MFO) [19].

Metaheuristic algorithms that adopt a physics-based scheme employ simplified physical models to generate search patterns for their agents. Noteworthy examples of physics-based techniques include the States of Matter Search (SMS) [20, 21], Simulated Annealing (SA) algorithm [22–24], Gravitational Search Algorithm (GSA) [25], Water Cycle Algorithm (WCA) [26], Big Bang-Big Crunch (BB-BC) [27], and Electromagnetism-like Mechanism (EM) [28]. Although these approaches draw inspiration from diverse phenomena, the search patterns used to explore the search space exhibit striking similarities. Many metaheuristic methods aim to configure and adapt successful, recycled search patterns from previous approaches to generate novel optimization schemes.

On the contrary, the differential equation's order pertains to the highest derivative's degree considered in the model. Consequently, a model formulated as a second-order differential equation, indicating a second-order system, is of significance [29]. The notable characteristic of a second-order model lies in its ability to manifest diverse behaviors based on the parameter configuration. These behaviors encompass oscillatory, underdamped, or overdamped responses, resulting in distinct temporal characteristics [30]. From the perspective of metaheuristic schemes, these behaviors can be interpreted as search trajectories. Therefore, second-order systems offer the possibility of generating oscillatory movements within specific regions or constructing intricate search patterns targeting different points or segments of the search space.

This study introduces a novel collection of search patterns designed to effectively explore the search space by emulating the response of a second-order system. These search patterns are integrated into a comprehensive search strategy called Second-Order Algorithm (SOA) to achieve the global solution for complex optimization problems. To evaluate the performance of the approach, it is compared against a diverse set of representative optimization problems, encompassing multimodal, unimodal, and hybrid benchmark formulations. The obtained results exhibit competitive outcomes, demonstrating the promising efficacy of the introduced search patterns.

The subsequent sections of this chapter are organized as follows: Sect. 4.2 provides a concise introduction to second-order systems; Sect. 4.3 discusses the key search patterns utilized in metaheuristic methods; Sect. 4.4 presents the definitions of the search patterns; Sect. 4.5 describes the measurement of exploration–exploitation; Sect. 4.6 introduces the metaheuristic scheme; Sect. 4.7 presents the numerical results; and, lastly, Sect. 4.8 draws the conclusions.

4.2 Description of Second-Order Models

A second-order system is defined as a model with a closed-loop transfer function where the input $R(s)$ and output $C(s)$ are connected. This type of model is significant due to its capacity to exhibit diverse behaviors based on the parameter configuration. A general representation of a second-order model can be expressed as follows [29]:

$$\frac{C(s)}{R(s)} = \frac{\omega_n^2}{s^2 + 2\zeta\omega_n s + \omega_n^2}, \tag{4.1}$$

In the given expression, the symbols ζ and ω_n represent the damping ratio and natural frequency, respectively. The variable s denotes the Laplace domain.

The evaluation of a system's dynamic behavior is based on the temporal response observed when a unitary step signal is applied as the input $R(s)$. The dynamic behavior refers to how the system responds over time, aiming to reach a value of one. For a second-order system, the dynamic behavior is characterized by the parameters ζ and ω_n [30]. With these parameters considered, the second-order system exhibits three distinct behaviors: underdamped ($0 < \zeta < 1$), critically damped ($\zeta = 1$), and overdamped ($\zeta > 1$).

4.2.1 Underdamped Behavior ($0 < \zeta < 1$)

In this scenario, the poles (which are the roots of the denominator) of Eq. (4.1) are complex conjugates and positioned in the left-half of the s plane. Under these circumstances, the underdamped response of the system, denoted as $C_U(s)$, in the

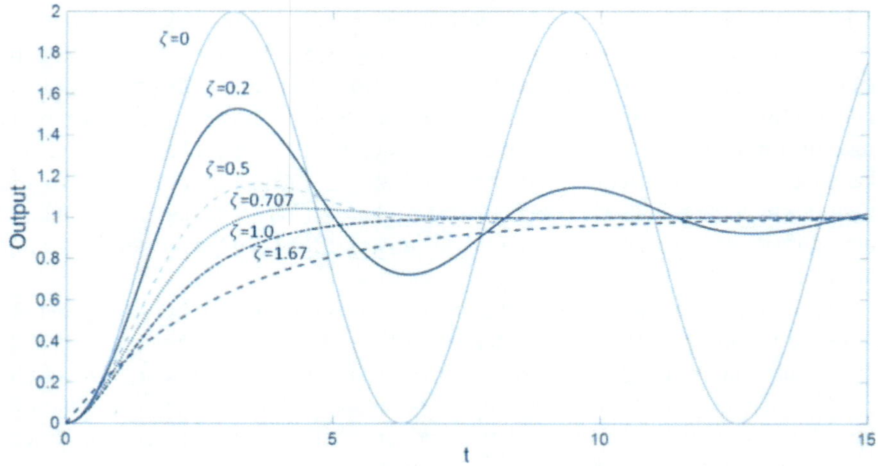

Fig. 4.1 The temporal responses of a second-order system exhibit distinct behaviors: Underdamped ($0 < \zeta < 1$), critically damped ($\zeta = 1$), and overdamped ($\zeta > 1$). These behaviors manifest as different patterns in the system's temporal response

Laplace domain can be described in the following manner:

$$C_U(s) = \frac{\omega_n^2}{s(s + \zeta\omega_n)^2 + \omega_n^2(1 - \zeta^2)}, \tag{4.2}$$

By employing partial fraction decomposition and the inverse Laplace transform, we derive the temporal response that characterizes the underdamped behavior, denoted as $c_U(t)$, as indicated in Eq. (4.3).

$$c_U(t) = 1 - \frac{e^{-\zeta\omega_n t}}{\sqrt{1 - \zeta^2}} \sin\left(\omega_n\sqrt{1 - \zeta^2} + \tan^{-1}\left(\frac{\sqrt{1 - \zeta^2}}{\zeta}\right)\right) \tag{4.3}$$

When $\zeta = 0$, a unique scenario arises where the temporal response of the system becomes oscillatory. Figure 4.1 provides a visualization of the output for various cases: $\zeta = 0$, $\zeta = 0.2$, $\zeta = 0.5$ and $\zeta = 0.707$. In the underdamped behavior, the system response initiates with significant acceleration, resulting in an overshoot that exceeds the value of one. The magnitude of the overshoot is inversely proportional to the value of ζ.

4.2.2 Critically Damped Behavior ($\zeta = 1$)

In this particular behavior, the transfer function of Eq. (4.1) exhibits two real poles with identical values. As a result, the response of the system in the critically damped behavior, denoted as $C_C(s)$, in the Laplace domain can be expressed in the following manner:

$$C_C(s) = \frac{\omega_n^2}{s(s+\omega_n)^2}, \tag{4.4}$$

By applying the inverse Laplace transform to Eq. (4.4), we can determine the temporal response of the critically damped behavior, denoted as $c_C(t)$, according to the following model:

$$c_C(t) = 1 - e^{-\omega_n t}(1+\omega_n t) \tag{4.5}$$

In the case of critically damped behavior, the system response exhibits a temporal pattern resembling that of a first-order system. It achieves the desired value of one without exhibiting an overshoot. Figure 4.1 illustrates the output of the critically damped behavior.

4.2.3 Overdamped Behavior ($\zeta > 1$)

In the scenario of overdamped behavior, the transfer function of Eq. (4.1) features two real poles with distinct values. The response in the Laplace domain, denoted as $C_O(s)$, is modeled using the following formulation:

$$C_O(s) = \frac{\omega_n^2}{s(s+\zeta\omega_n+\omega_n\sqrt{\zeta^2-1})(s+\zeta\omega_n-\omega_n\sqrt{\zeta^2-1})}, \tag{4.6}$$

By performing the inverse Laplace transform, we obtain the temporal response of the overdamped behavior, denoted as $c_O(t)$, defined as follows:

$$c_O(t) = 1 + e^{-(\zeta-\sqrt{\zeta^2-1})\omega_n t} \tag{4.7}$$

In the case of the Overdamped behavior, the system gradually responds and eventually reaches the value of one. The deceleration of the response is determined by the value of ζ. A higher value of ζ corresponds to a slower response. Figure 4.1 provides a visualization of the output for the Overdamped behavior, specifically for the case when $\zeta = 1.67$.

4.3 Movement Patterns as Search Strategies

Finding effective search patterns to navigate through a fitness landscape and explore it properly can be challenging, especially when dealing with ruggedness and multiple local optima. In recent literature, various new metaheuristic schemes have been introduced. Despite their different conceptual inspirations, these approaches share similar search patterns to explore the search space. A search pattern refers to a set of movements or rules designed to explore promising solutions within the search space.

Two critical characteristics of a search pattern are exploration and exploitation. Exploration refers to the ability of a search pattern to explore a wide range of solutions distributed across different regions of the search space. On the other hand, exploitation involves the capacity of a search pattern to refine existing solutions by focusing on local examination. The combination of both exploration and exploitation mechanisms within a search pattern is essential for achieving successful outcomes when solving specific optimization problems.

From a metaheuristic perspective, the optimization problem is solved by evolving a population of N candidate solutions (individuals), denoted as $\mathbf{P}^k\left(\{\mathbf{x}_1^k, \ldots, \mathbf{x}_N^k\}\right)$, over a series of generations from an initial point ($k = 1$) to a maximum number of generations ($k = Maxgen$). Each individual $\mathbf{x}_i^k$, where i ranges from 1 to N, represents a d-dimensional element $\{x_{i,1}^k, \ldots, x_{i,d}^k\}$ that represents the decision variables of the optimization problem. At each generation, search patterns are applied to the individuals in the population $\mathbf{P}^k$ to generate a new population $\mathbf{P}^{k+1}$. The quality of each individual $\mathbf{x}_i^k$ is assessed based on its solution with respect to the objective function $J\left(\mathbf{x}_i^k\right)$, which provides the fitness value of $\mathbf{x}_i^k$. As the metaheuristic method progresses, the best current individual $\mathbf{b}\ \{b_1, \ldots, b_d\}$ is retained as it represents the best solution found thus far.

In general, a search pattern is applied to each individual $\mathbf{x}_i^k$, taking the best element $\mathbf{b}$ as a reference. Based on a specific model, a set of movements is generated to modify the position of $\mathbf{x}_i^k$ until it reaches the location of $\mathbf{b}$. The purpose of this mechanism is to explore solutions along the trajectory from $\mathbf{x}_i^k$ to $\mathbf{b}$, aiming to discover a better solution than the current $\mathbf{b}$. Search patterns vary in terms of the model used to generate the trajectories $\mathbf{x}_i^k$ towards $\mathbf{b}$.

Two commonly used search models are the attraction and spiral trajectories. The attraction model generates movements that attract individuals from $\mathbf{x}_i^k$ towards $\mathbf{b}$. This model is widely utilized in various metaheuristic methods such as PSO [10–12], FA [13–15], CS [16], BA [17], GSA [25], EM [33], and DE [6]. On the other hand, the spiral model generates a spiral trajectory that revolves around the best element $\mathbf{b}$. This model is employed in recently published schemes like WOA and GWO. Figures 4.2a and b depict the trajectories produced by the attraction and spiral models, respectively.

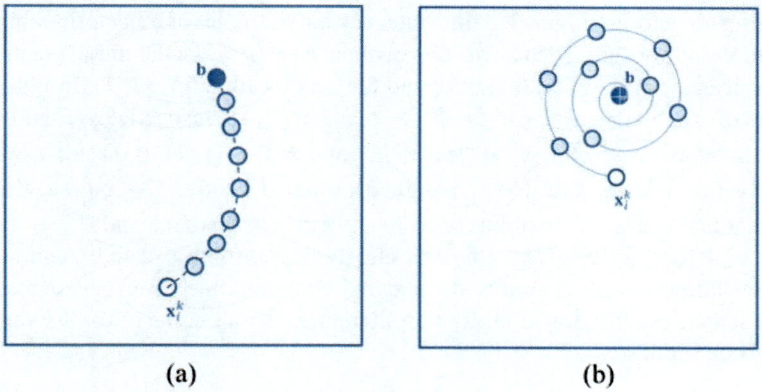

Fig. 4.2 Patterns generated by the **a** attraction and **b** spiral models

4.4 Movement Pattern Through Second-Order Models

This chapter presents a novel collection of search patterns that are designed to effectively explore the search space. These patterns mimic the behavior of a second-order system. The introduced set of search patterns is incorporated into a comprehensive search strategy aimed at achieving the global solution for challenging optimization problems.

Our approach utilizes the time-based behavior of a second-order system to generate a path from the initial position $\mathbf{x}_i^k = \{x_{i,1}^k, \ldots, x_{i,d}^k\}$ to the target location $\mathbf{b} = \{b_1, \ldots, b_d\}$. By employing such models, we can create more intricate paths that offer a more comprehensive exploration of the search space. In this context, we consider three different responses of a second-order system to generate three distinct search patterns: underdamped, critically damped, and overdamped. These patterns are represented by Eqs. (4.8, 4.9), and (4.10) respectively.

$$x_{i,j}^k = \left(1 - \frac{e^{-\zeta\omega_n k}}{\sqrt{1-\zeta^2}} \sin\left(\omega_n \sqrt{1-\zeta^2} + \tan^{-1}\left(\frac{\sqrt{1-\zeta^2}}{\zeta}\right)\right)\right)$$
$$\left(b_j - x_{i,j}^k\right); 0 \le \zeta < 1 \tag{4.8}$$

$$x_{i,j}^k = (1 - e^{-\omega_n k}(1 + \omega_n k))\left(b_j - x_{i,j}^k\right); (\zeta = 1) \tag{4.9}$$

$$x_{i,j}^k = \left(1 + e^{-\left(\zeta - \sqrt{\zeta^2 - 1}\right)\omega_n k}\right)\left(b_j - x_{i,j}^k\right); (\zeta > 1), \tag{4.10}$$

In the given context, i ($\in [1, N]$) represents the search agent, while j ($\in [1, d]$) symbolizes the decision variable or dimension. The behavior of each search pattern is influenced by the value of ζ, enabling the combination of elements to generate interesting trajectories. Figure 4.3 illustrates several examples of trajectories obtained using different ζ values. In this two-dimensional case ($d = 2$), the initial position of the search agent $\mathbf{x}_i^k$ is set at (0.5, 0.5), and the best location $\mathbf{b}$ is (1, 1). In Fig. 4.3a, ζ values of 0 and 1 are assigned to $x_{i,1}^k \leftarrow \zeta$ and $x_{i,2}^k \leftarrow \zeta$ respectively. Figure 4.3b demonstrates the case of $x_{i,1}^k \leftarrow \zeta$ set to 0.1 and $x_{i,2}^k \leftarrow \zeta$ set to 0.5 for ζ values. Similarly, Fig. 4.3c shows $x_{i,1}^k \leftarrow \zeta$ assigned a value of 1 and $x_{i,2}^k \leftarrow \zeta$ set to 1.67 for ζ values. Finally, Fig. 4.3d exhibits of $x_{i,1}^k \leftarrow \zeta$ with a value of 0.5 and $x_{i,2}^k \leftarrow \zeta$ with a value of 1 for ζ values. These figures clearly demonstrate that the second-order responses allow the generation of diverse and complex trajectories, encompassing various search patterns found in existing literature. For all cases (a-d), the value of ω_n has been set to 1.

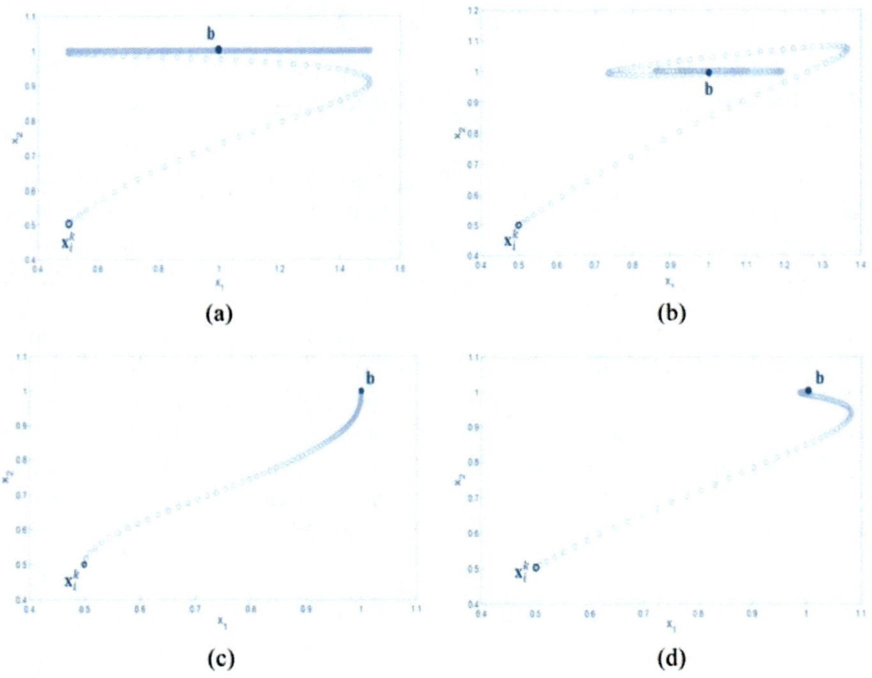

Fig. 4.3 Some examples of trajectories produced by using different values for ζ. **a** $x_{i,1}^k \leftarrow \zeta = 0$ and $x_{i,2}^k \leftarrow \zeta = 1$, **b** $x_{i,1}^k \leftarrow \zeta = 0.1$ and $x_{i,2}^k \leftarrow \zeta = 0.5$, **c** $x_{i,1}^k \leftarrow \zeta = 1$ and $x_{i,2}^k \leftarrow \zeta = 1.67$ and **d** $x_{i,1}^k \leftarrow \zeta = 0.5$ and $x_{i,2}^k \leftarrow \zeta = 1$

4.5 Analysis of Exploration and Exploitation

Metaheuristic methods utilize a group of search agents to explore the search space and find a satisfactory solution for an optimization problem. Within these methods, search agents with better fitness values exert influence on the search process, attracting other agents towards them. This phenomenon leads to a reduction in the distance between individuals and emphasizes the exploitation aspect of the optimization process. Conversely, when the distance between individuals increases, the exploration aspect becomes more prominent.

To quantify the relative distance between individuals in terms of increase and decrease, a diversity indicator called the dimension-wise diversity index [31] is employed. This diversity index is defined as follows:

$$Div_j = \frac{1}{N}\sum_{i=1}^{N}\left|median(x^j) - x_{i,j}\right|$$

$$Div = \frac{1}{d}\sum_{j=1}^{d} Div_j \tag{4.11}$$

The expression $median(x^j)$ represents the median value of dimension j across all search agents. The variable $x_{i,j}$ denotes the j-th decision variable of individual i. N represents the population size of $\mathbf{P}^k$, and d corresponds to the number of dimensions in the optimization problem. The diversity measure Div_j (specific to the j-th dimension) assesses the relative distance between the variable j of each individual and its corresponding median value. The overall diversity measure, Div, captures the average diversity across all dimensions. Both Div_j and Div are computed at each iteration.

Once the diversity values have been assessed, the exploration and exploitation levels can be determined by calculating the percentage of time a search strategy dedicates to exploration or exploitation based on its diversity values. These percentages are computed at each iteration using the following models:

$$XPL\% = \left(\frac{Div}{Div_{max}}\right) \times 100$$

$$XPT\% = \left(\frac{|Div - Div_{max}|}{Div_{max}}\right) \times 100 \tag{4.12}$$

The maximum diversity value achieved during the optimization process is denoted as Div_{max}. The exploration percentage ($XPL\%$) represents the extent of exploration and is calculated as the ratio of Div to Div_{max}. Conversely, the exploitation percentage ($XPT\%$) indicates the degree of exploitation. $XPT\%$ is derived as the complementary percentage to $XPL\%$ since the difference between Div_{max} and Div arises due to the concentration of individuals.

4.6 Movement Pattern

The complete description of the Second-Order Algorithm (SOA), which integrates the set of search patterns based on second-order systems, is presented in this section. The SOA method is designed to obtain global solutions for complex optimization problems. It consists of four stages: (A) Initialization, (B) trajectory generation, (C) reset of bad elements, and (D) avoidance of premature convergence mechanism. Steps (B–D) are executed sequentially until a stop criterion is met. The flowchart of the complete metaheuristic method is illustrated in Fig. 4.4.

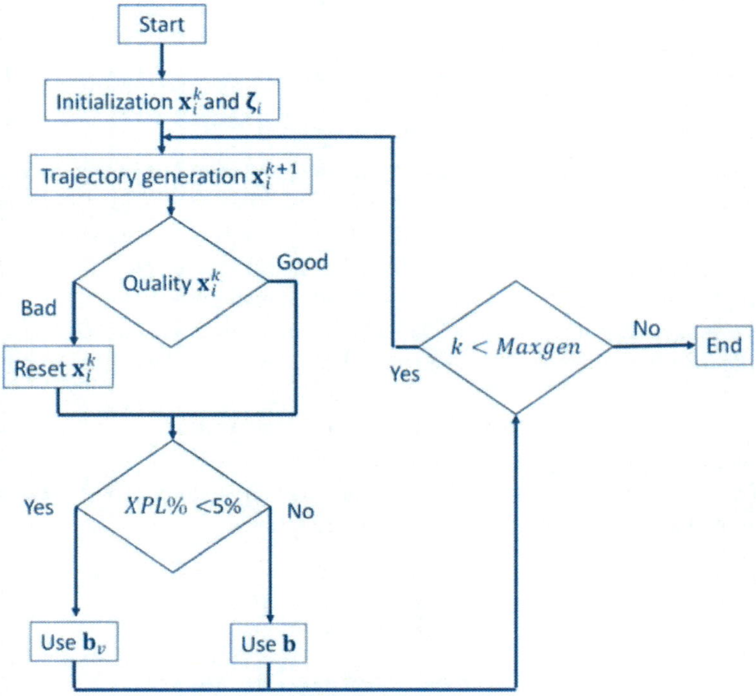

Fig. 4.4 Flowchart illustrating the metaheuristic method, which utilizes the response of second-order systems

4.6.1 Initialization Stage

In the initial iteration ($k = 0$), a population $\mathbf{P}^0$ consisting of N agents $\{\mathbf{x}_1^0, \ldots, \mathbf{x}_N^0\}$ is generated randomly using the following equation:

$$x_{i,j}^0 = rand \cdot \left(b_j^{high} - b_j^{low}\right) + b_j^{low}$$

$$i = 1, 2, \ldots, N; \; j = 1, \ldots, d \tag{4.13}$$

The j-th decision variable is bounded by the upper limit b_j^{high} and the lower limit b_j^{low}. The value of rand is a random number uniformly distributed between 0 and 1.

For each individual $\mathbf{x}_i$ in the population, a vector $\boldsymbol{\zeta}_i = \{\zeta_{i,1}, \ldots, \zeta_{i,d}\}$ is assigned. The elements $\zeta_{i,j}$ of this vector determine the trajectory behavior of the j-th dimension. Initially, each $\zeta_{i,j}$ is assigned a random value within the range of [0,2]. This range allows for all possible second-order behaviors: underdamped ($0 < \zeta < 1$), critically damped ($\zeta = 1$), and overdamped ($\zeta > 1$).

4.6.2 Generation of the Trajectory

After initializing the population, the best element $\mathbf{b}$ is determined. Next, the new position $\mathbf{x}_i^{k+1}$ of each agent $\mathbf{x}_i^k$ is calculated based on a trajectory generated by a second-order system. Once all the new positions in the population $\mathbf{P}^k$ have been computed, the best element $\mathbf{b}$ is redefined.

4.6.3 Reset of Agents with Bad Quality

Each agent $\mathbf{x}_i^k$ is given the freedom to follow its trajectory for ten iterations. If, after ten iterations, an agent $\mathbf{x}_i^k$ exhibits the poorest performance according to the fitness function, it is reinitialized in terms of both position and its vector $\boldsymbol{\zeta}_i$. This means that the agent will be placed in a different position and will adopt a new trajectory behavior.

4.6.4 Mechanism for Avoiding Premature Convergence

If the percentage of exploration $XPL\%$ is below 5%, the best value $\mathbf{b}$ is substituted with a virtual best value $\mathbf{b}_v$. The virtual best value $\mathbf{b}_v$ is obtained by calculating the average value of the best five individuals in the population. This approach aims to discover a new position that generates different trajectories and prevents the search process from becoming stuck in a local optimum.

4.7 Results

To assess the performance of the SOA algorithm, a series of experiments were conducted. The SOA algorithm was tested on a collection of 12 unimodal functions ($f_1(\mathbf{x}) - f_{12}(\mathbf{x})$), four unimodal functions ($f_{13}(\mathbf{x})$ and $f_{16}(\mathbf{x})$), and four hybrid functions ($f_{17}(\mathbf{x})$ and $f_{20}(\mathbf{x})$) listed in Tables A1, A2, and A3 of Appendix 1. In addition, the performance of the SOA algorithm was compared to well-known algorithms in the literature, including Artificial Bee Colony (ABC) [9], Covariance matrix adaptation evolution strategy (CMAES) [4], Crow Search Algorithm (CSA) [8], Differential Evolution (DE) [6], Moth-flame optimization algorithm (MFO) [19], and Particle Swarm Optimization (PSO) [10]. These algorithms are widely recognized as the most popular metaheuristic schemes in numerous optimization studies [32].

For the experiments, a population size of 50 search agents (N) was used, and the maximum iteration number ($Maxgen$) for all functions was set to 1000. This stop criterion was selected to maintain compatibility with similar works published in the literature [33, 34].

- ABC: Onlooker Bees = 50, acceleration coefficient = 1 [9].
- DE: crossover probability = 0.2, Betha = 1 [6].
- CMAES: Lambda = 50, father number = 25, sigma = 60, csigma = 0.32586, dsigma = 1.32586 [4].
- CSA: Flock = 50, awareness probability = 0.1, flight length = 2 [8].
- MFO: search agents = 50, "a" linearly decreases from 2 to 0 [19].
- SOA: The experimental results indicate that the best performance of the algorithm is achieved with the following parameter set: par1 = 0.7, par2 = 0.3, and par3 = 0.05.

In the comparisons, three different metrics are taken into account: the Average Best-so-far (AB) solution, the Median Best-so-far (MB) solution, and the Standard Deviation (SD) of the best-so-far solutions. For each optimization problem, all algorithms are applied 30 times, resulting in 30 sets of solutions. The Average Best-so-far (AB) solution is obtained by calculating the mean value of the best solutions found in each of the 30 runs. The Median Best-so-far (MB) solution is determined by computing the median of the 30 sets of solutions. The Standard Deviation (SD) is calculated as a measure of the dispersion of the best-so-far solutions across the 30 runs, providing an indication of the algorithm's robustness. The AB and MB metrics reflect the accuracy of the solutions, while the SD metric represents their dispersion.

4.7.1 Multimodal Experiments

Multimodal functions are characterized by having multiple local optima, making them more challenging to solve. In this analysis, the performance of the SOA method is compared with ABC, CMAES, CSA, DE, MFO, and PSO on multimodal functions. The multimodal objective functions considered in this study range from $f_1(\mathbf{x})$ to $f_{12}(\mathbf{x})$, as listed in Table 4.7 in the Appendix. These functions exhibit an increasing number of local minima as the dimensionality of the function increases. Thus, this study aims to evaluate the ability of each metaheuristic scheme to identify the global optimum in the presence of multiple local optima. The experiments conducted in this analysis assume objective functions operating in 30 dimensions ($n = 30$). The results, including the averaged best (AB) values obtained from 30 independent runs, are presented in Table 4.1. The table also displays the median values (MD) and standard deviations (SD) of the solutions. Based on Table 4.1, the SOA scheme outperforms ABC, CMAES, CSA, DE, MFO, and PSO in functions $f_1(\mathbf{x})$, $f_4(\mathbf{x})$, $f_5(\mathbf{x})$, $f_6(\mathbf{x})$, $f_8(\mathbf{x})$, $f_9(\mathbf{x})$, $f_{10}(\mathbf{x})$, $f_{11}(\mathbf{x})$ and $f_{12}(\mathbf{x})$. However, the results of SOA are comparable to those obtained by DE, CMAES, and MFO in functions $f_2(\mathbf{x})$, $f_3(\mathbf{x})$ and $f_7(\mathbf{x})$.

To validate the conclusions drawn from Table 4.1 in a statistical manner, a non-parametric study using the Wilcoxon rank-sum analysis [35] is conducted. This statistical test aims to assess the significance of performance differences between two methods. Therefore, pairwise comparisons are made, including SOA vs. ABC, SOA vs. CMAES, SOA vs. CSA, SOA vs. DE, SOA vs. MFO, and SOA vs. PSO. In the Wilcoxon analysis, the null hypothesis (H_0) assumes that there is no significant difference between the results, while the alternative hypothesis (H_1) assumes that the results have a similar structure. A significance value of 0.05 is adopted for the Wilcoxon analysis, considering 30 independent executions for each test function. Table 4.2 presents the p-values obtained from the Wilcoxon study, based on the results from Table 4.1 (where $n = 30$). For ease of interpretation, the symbols ▲, ▼, and ▶ are used in the table. The ▲ symbol indicates that the SOA algorithm produces significantly better solutions compared to its competitor, while the ▼ symbol indicates that SOA achieves worse results. The ▶ symbol indicates that both compared methods produce similar solutions. Upon examining Table 4.2, it is evident that for functions f_1, f_4, f_5,f_6, f_8, f_9, f_{10},f_{11} and f_{12}, the SOA scheme consistently achieves better solutions than the other methods. Conversely, for functions f_2, f_2 and f_7, it is clear that the groups SOA versus ABC, SOA versus CMAES, SOA versus DE, SOA versus MFO, and EA-HC versus SCA exhibit similar solution qualities.

4.7.2 Unimodal Experiments

In this subsection, we compare the performance of SOA with ABC, DE, CMAES, CSA, and MFO on functions that have a single optimum [36–40]. These functions are

Table 4.1 Results of minimization for multimodal functions

		ABC	DE	CMAES	CSA	PSO	MFO	SOA
	AB	8.9132622	0.7932535	2.8976E−19	55.918504	0.2012803	27.983271	0.1119774
$f_1\mathbf{x}$	**MD**	8.4392750	0.7993166	2.4779E−19	57.038263	4.1459E−23	25.365199	1.0714E−10
	SD	2.6748059	0.1378538	1.5343E−19	6.2032578	1.1022968	12.283254	0.2272439
	AB	2	2	2	1897783.3	27.4	2	2
$f_2\mathbf{x}$	**MD**	2	2	2	35691.155	2	2	2
	SD	9.9512E−12	0	0	9636632.1	113.43495	0	0
	AB	2	2	2	3620834.4	34.723128	2	2
$f_3\mathbf{x}$	**MD**	2	2	2	676981.46	9	2	2
	SD	2.3308E−11	0	0	7265352.7	113.36672	0	0
	AB	0.1371551	0.002	1.7942E−06	0.0862919	2.2285E−08	5.5194E−10	1.164E−11
$f_4\mathbf{x}$	**MD**	0.1349076	0.01	0	0.0892256	0	0	7.7118E−12
	SD	0.0399861	0.123	5.4833E−06	0.0213332	1.2206E−07	3.0231E−09	1.0995E−11
	AB	13781291	1331987.7	22307.195	44274761	82.539625	85.756149	71.964984
$f_5\mathbf{x}$	**MD**	13876263	1365502.1	72.377516	46153728	81.698488	85.665615	72.362277
	SD	3237147.7	306385.34	50923.637	10118180	7.1916726	3.2857088	0.9929591
	AB	1.152E+85	5.850E+81	1.812E+83	6.429E+83	1.397E+81	3.0883E+81	1.0051E+81
$f_6\mathbf{x}$	**MD**	4.622E+84	3.405E+81	7.928E+82	2.939E+83	5.977E+80	7.9607E+80	4.901E+80
	SD	1.685E+85	7.33E+81	3.047E+83	7.551E+83	2.197E+81	5.6651E+81	1.9457E+81
	AB	30.033333	30	30	58.766666	30	33.633333	30
$f_7\mathbf{x}$	**MD**	30	30	30	59	30	30	30
	SD	0.18257419	0	0	1.77498583	0	5.4550409	0
	AB	9.2	8.0666666	1.0666666	19543.266	0.0333333	2000.0333	0

(continued)

Table 4.1 (continued)

		ABC	DE	CMAES	CSA	PSO	MFO	SOA
$f_8\mathbf{x}$	**MD**	9	8	0	19797	0	0	0
	SD	2.5784250	1.9464084	3.1941037	2077.7459	0.1825741	4842.3277	0
	AB	−745.05202	−1125.4815	−1127.8626	−725.09353	−1071.7869	−1031.2617	−1146.3478
$f_9\mathbf{x}$	**MD**	−743.4462	−1174.9722	−1132.5748	−719.10777	−1068.9596	−1033.6178	−1145.2467
	SD	25.137593	78.706	25.809999	25.815652	34.055847	34.244188	10.928362
	AB	110282.54	665278.86	−4930	1170939.0	45556.260	222833.73	−501.79356
$f_{10}\mathbf{x}$	**MD**	96461.061	673449.49	−4930	1126234.2	5076.8152	71582.051	−332.82466
	SD	44933.417	129147.27	3.7318E−09	159175.64	75990.880	305159.37	663.92006
	AB	−18.26109	−26.056561	−29.6576	−16.504756	−28.367666	−28.863589	−30
$f_{11}\mathbf{x}$	**MD**	−18.131984	−26.092349	−29.9286	−16.183447	−28.14029	−29.070145	−30
	SD	1.6366873	0.5428906	0.466	1.16917142	1.5408536	1.2837415	0
	AB	1502.3129	369.60375	786.36819	519.17242	196.95838	261.52332	11.905761
$f_{12}\mathbf{x}$	**MD**	1457.6865	368.82916	778.72369	465.93238	213.00730	252.76747	0.3841574
	SD	420.70611	35.389136	215.93893	228.69860	86.542862	106.52353	29.544101

Table 4.2 Statistical analysis for multimodal functions

	SOA	SOA	SOA	SOA	SOA	SOA
Function	Versus	Versus	Versus	Versus	Versus	Versus
	ABC	CMAES	CSA	DE	MFO	PSO
$f_1(x)$	7.13E-09▲	2.61E-08▲	2.40E-11▲	9.77E-07▲	3.97E-11▲	6.87E-08▲
$f_2(x)$	1.21E-12►	1►	1.21E-12▲	1►	1►	4.13E-09▲
$f_3(x)$	1.21E-12►	1►	1.21E-12▲	1►	1►	8.33E-07▲
$f_4(x)$	6.48E-12▲	2.43E-08▲	6.48E-12▲	1.10E-07▲	1.18E-07▲	5.79E-08▲
$f_5(x)$	3.02E-11▲	2.18E-06▲	3.02E-11▲	3.02E-11▲	8.30E-01▲	9.94E-08▲
$f_6(x)$	3.69E-11▲	1.29E-09▲	2.87E-10▲	6.73E-08▲	1.75E-05▲	9.52E-04▲
$f_7(x)$	3.34E-01►	1►	1.57E-12▲	3.34E-01►	2.23E-05▲	3.34E-07▲
$f_8(x)$	3.96E-06▲	1.10E-07▲	7.87E-12▲	3.28E-06▲	2.45E-01▲	5.58E-07▲
$f_9(x)$	2.97E-11▲	5.75E-08▲	2.97E-11▲	7.72E-08▲	4.20E-04▲	1.83E-05▲
$f_{10}(x)$	3.02E-11▲	2.85E-11▲	3.02E-11▲	3.02E-11▲	2.57E-07▲	1.86E-03▲
$f_{11}(x)$	2.80E-11▲	1.12E-07►	2.80E-11▲	3.00E-11▲	8.88E-01►	8.86E-06▲
$f_{12}(x)$	3.02E-11▲	3.02E-11▲	3.02E-11▲	3.02E-11▲	1.78E-10▲	9.76E-10▲

represented by functions $f_{13}(\mathbf{x})$ to $f_{16}(\mathbf{x})$ in Table 4.7. All functions are evaluated in 30 dimensions ($d = 30$), and the experimental results are obtained from 30 independent executions, as shown in Table 4.3. The results in terms of AB, MB, and SD are reported in the table. According to Table 4.3, the SOA approach outperforms ABC, DE, CMAES, CSA, and MFO for all the functions considered. The study reveals significant performance differences among the metaheuristic schemes, which can be attributed to the better balance between exploration and exploitation achieved by the trajectories of the SOA scheme. Table 4.4 presents the results of the Wilcoxon analysis based on the information from Table 4.3. The p-values in Table 4.4 indicate that the SOA method consistently demonstrates superior performance compared to each of the metaheuristic algorithms considered in the experimental study.

4.7.3 Hybrid Experiments

This study focuses on the evaluation of optimization solutions using hybrid functions within the context of the SOA scheme. Hybrid functions are formed by combining multiple multimodal functions to create multimodal optimization problems. The specific formulations of these hybrid functions are provided in Table 4.4. A in the Appendix. For detailed information on the implementation of hybrid functions, please refer to references [61, 62]. In our experiments, we compare the performance of the SOA approach with other metaheuristic schemes on these hybrid functions [41–44].

The results of the simulations are presented in Table 4.5, which shows the performance of each algorithm based on the metrics AB, MB, and SD. It is evident from

Table 4.3 Minimization results of unimodal benchmark functions

		ABC	DE	CMAES	CSA	PSO	MFO	SOA
	AB	25.648891	23.234006	0.0382940	117864.48	2433.8148	16893.538	4.416E−16
$f_{13}(\mathbf{x})$	**MD**	26.054364	22.218747	1.503E−23	120885.37	5.9145E−10	10737.418	4.1862E−16
	SD	8.3305663	5.5158397	0.1428856	14742.855	4322.0341	19501.276	2.4373E−16
	AB	0.0136600	0.0145526	1.2398E−05	51.476735	4.0467E−13	7.8643202	1.5E−20
$f_{14}(\mathbf{x})$	**MD**	0.0144199	0.0144921	1.3237E−20	51.798031	7.5065E−14	2.8398E−07	1.3059E−20
	SD	0.0041468	0.0037530	3.908E−05	5.9378289	7.8227E−13	14.024260	9.6811E−21
	AB	0.5442777	0.569612	19.789197	2466.2017	20	403.33365	1.2053E−18
$f_{15}(\mathbf{x})$	**MD**	0.4952416	0.5733913	0.2257190	2478.7246	8.9126E−12	200.00000	1.025E−18
	SD	0.1986349	0.1376955	39.959487	344.56578	66.436383	520.92974	6.4946E−19
	AB	0.0006647	1.8659E−10	6.9743E−10	0.0068342	5.6588E−24	9.5555E−19	0
$f_{16}(\mathbf{x})$	**MD**	0.0004306	1.2946E−10	7.1112E−10	0.0066858	3.6687E−29	2.8339E−22	0
	SD	0.0006187	1.8814E−10	4.3714E−10	0.0032037	3.0973E−23	3.3608E−18	0

Table 4.4 Wilcoxon analysis for unimodal benchmark functions

Function	SOA Versus ABC	SOA Versus CMAES	SOA Versus CSA	SOA Versus DE	SOA Versus MFO	SOA Versus PSO
$f_{13}(\mathbf{x})$	2.80E-11▲	3.86E-01▲	2.80E-11▲	2.80E-11▲	1.75E-09▲	1.22E-04▲
$f_{14}(\mathbf{x})$	5.51E-09▲	2.11E-01▲	2.72E-11▲	3.22E-09▲	1.07E-04▲	1.74E-02▼
$f_{15}(\mathbf{x})$	1.58E-01▲	3.02E-11▲	3.02E-11▲	1.81E-01▲	5.26E-04▲	3.11E-01▲
$f_{16}(\mathbf{x})$	1.21E-12▲	1.21E-12▲	1.21E-12▲	1.21E-12▲	1.21E-12▲	1.21E-12▲

Table 4.5 that the SOA method outperforms the other techniques across all functions. The Wilcoxon analysis results are reported in Table 4.6, considering the AB values from Table 4.5. The presence of the ▲ symbol for all elements confirms that the SOA method consistently produces superior results compared to the other methods. The excellent performance of the SOA scheme on hybrid functions can be attributed to its ability to strike a better balance between exploration and exploitation, facilitated by the characteristics of the second system trajectories. This indicates that the SOA approach generates a suitable number of promising search agents, enabling effective exploration of the search space. Simultaneously, it also produces a balanced number of candidate solutions, leading to improvements in the quality of the identified solutions in terms of the objective function.

4.7.4 *Study of Convergence*

The assessment of an optimization algorithm's abilities cannot be solely based on the accuracy of the final solution. The convergence of a metaheuristic scheme plays a crucial role in evaluating its performance. In this subsection, a convergence analysis has been conducted, focusing on the speed at which a metaheuristic scheme reaches the optimal solution. To save space, the performance of the top four metaheuristic schemes is compared using a representative set of six functions (two multimodal, two unimodal, and two hybrid) operating in 30 dimensions. Convergence graphs are generated by processing the raw simulation data obtained from different experiments. Since each simulation is run 30 times for each metaheuristic method, the convergence data corresponding to the median result is selected. Figures 4.5, 4.6, and 4.7 depict the convergence graphs for the four best-performing metaheuristic methods. A careful examination of Fig. 4.5 reveals that the SOA scheme exhibits superior convergence compared to the other algorithms for all functions.

Table 4.5 Results for hybrid functions

		ABC	DE	CMAES	CSA	PSO	MFO	SOA
	AB	396.75458	7.7178366	3.1526E-09	20330.245	334.63082	23758.790	0.8147792
$f_{17}(\mathbf{x})$	**MD**	210.31173	7.7772734	2.7959E-09	19814.613	6.9181E-07	20077.849	4.3564E-11
	SD	497.02847	1.3190384	1.1535E-09	2074.4742	1832.8485	18166.603	2.5336865
	AB	212.40266	75.917033	105.99474	731.38151	65.728942	161.36081	30.785661
$f_{18}(\mathbf{x})$	**MD**	212.09269	75.575979	31.783896	741.08639	65.904649	116.86683	28.998449
	SD	25.805145	10.604761	84.809720	68.759154	14.502755	107.40820	3.5251022
	AB	221724.73	1264.0862	57.570417	70494770	87.951413	80.887783	31.999808
$f_{19}(\mathbf{x})$	**MD**	200128.19	1278.1579	32.661016	65110859	84.267959	78.040599	31.999808
	SD	114341.27	268.82145	54.602147	23890494	25.256576	23.169503	6.4582E-10
	AB	319.57592	49.741282	97.542580	867.75158	122.92508	802.21444	30.307556
$f_{20}(\mathbf{x})$	**MD**	299.50312	49.708103	29.002196	879.05543	65.254961	685.34608	29
	SD	64.709010	4.4720540	93.393879	103.61875	127.74698	500.43174	4.0252793

Table 4.6 Statistical analysis for hybrid functions

Function	SOA Versus ABC	SOA Versus CMAES	SOA Versus CSA	SOA Versus DE	SOA Versus MFO	SOA Versus PSO
$f_{17}(\mathbf{x})$	4.35E-11▲	6.63E-05▲	2.92E-11▲	8.16E-08▲	5.40E-10▲	6.55E-02▲
$f_{18}(\mathbf{x})$	1.16E-07▲	7.28E-04▲	3.02E-11▲	6.63E-07▲	2.01E-01▲	5.20E-06▲
$f_{19}(\mathbf{x})$	3.02E-11▲	3.02E-11▲	3.02E-11▲	3.02E-11▲	1.17E-04▲	4.94E-05▲
$f_{20}(\mathbf{x})$	4.91E-11▲	2.71E-05▲	2.98E-11▲	6.73E-05▲	5.43E-11▲	8.11E-05▲

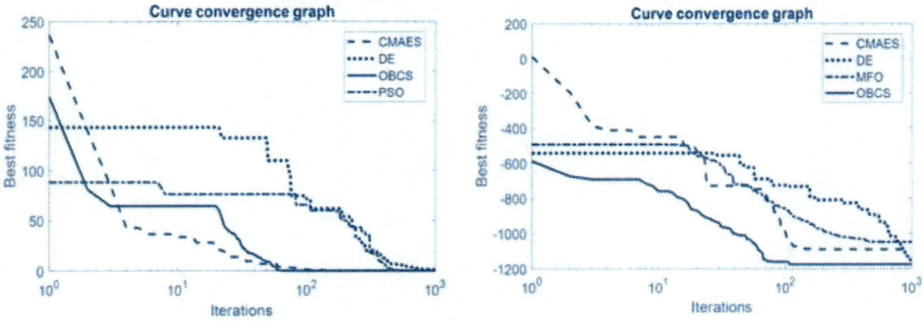

(a) f_{17} +x, (b) f_{20} +x,

Fig. 4.5 Convergence graphs in two representative multimodal functions

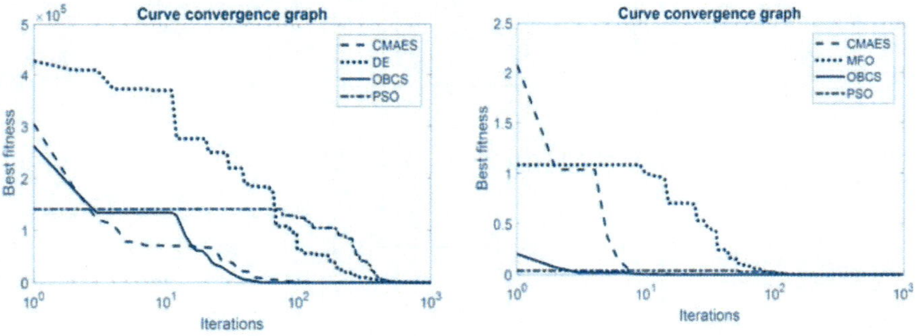

(a) f_{17} +x, (b) f_{20} +x,

Fig. 4.6 Convergence graphs in two representative unimodal functions

4.8 Conclusions

This chapter introduces a set of new search patterns that mimic the behavior of a second-order system, aiming to efficiently explore the search space. These search patterns, namely underdamped, critically damped, and overdamped, are employed

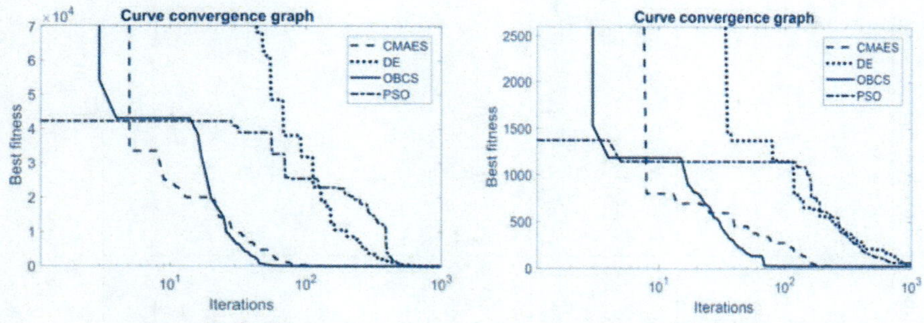

(a) f_{17} +x, (b) f_{20} +x,

Fig. 4.7 Convergence graphs in two representative hybrid functions

as part of a comprehensive search strategy called Second-Order Algorithm (SOA) to solve complex optimization problems and obtain global solutions.

The performance of the SOA is evaluated by testing it on a diverse set of 20 benchmark functions, including multimodal, unimodal, and hybrid functions. The results obtained from multimodal functions highlight the algorithm's strong exploration capabilities, while the performance on unimodal test functions demonstrates its ability to exploit the search space effectively. Moreover, the results on hybrid functions demonstrate the efficacy of the search patterns in handling more challenging problem formulations. To validate the results, the Wilcoxon analysis is employed.

Overall, the introduction of these new search patterns and the integration of the SOA provide a promising approach for efficient and effective optimization across various problem domains.

Appendix 1 List of Benchmark Functions

See Table 4.7.

	Name	Function	S	Dim	Minimum
$f_1(\mathbf{x})$	Levy	$sin^2(\pi\omega_1) + \sum_{i=1}^{d-1}(\omega_i - 1)^2[1 + 10sin^2(\pi\omega_i + 1) + (\omega_d - 1)^2[1 + sin^2(2\pi\omega_d)]]$	$[-10, 10]^n$	30	$f(\mathbf{x}^*) = 0$; $\mathbf{x}^* = (1, \ldots, 1)$
$f_2(\mathbf{x})$	Mishra 1	$(1 + x_n)^{x_n}; x_n = n - \sum_{i=1}^{n-1} x_i$	$[0, 1]^n$	30	$f(\mathbf{x}^*) = 2$; $\mathbf{x}^* = (1, \ldots, 1)$
$f_3(\mathbf{x})$	Mishra 2	$(1 + x_n)^{x_n}; x_n = n - \sum_{i=1}^{n-1} \frac{(x_i + x_{i+1})}{2}$	$[0, 1]^n$	30	$f(\mathbf{x}^*) = 2$; $\mathbf{x}^* = (1, \ldots, 1)$
$f_4(\mathbf{x})$	Mishra 11	$\left[\frac{1}{n}\sum_{i=1}^{n}\|x_i\| - \left(\prod_{i=1}^{n}\|x_i\|\right)^{\frac{1}{n}}\right]^2$	$[-10, 10]^n$	30	$f(\mathbf{x}^*) = 0$; $\mathbf{x}^* = (0, \ldots, 0)$
$f_5(\mathbf{x})$	Penalty 1	$\frac{\pi}{30}\left\{\begin{array}{c} 10\sin^2(\pi y_1) \\ +\sum_{i=1}^{n-1}(y_i - 1)^2\left[1 + 10\sin^2(\pi y_i + 1)\right] \\ +(y_n - 1)^2 \end{array}\right\} + \sum_{i=1}^{n} u(x_i, 10, 100, 4)$; $y_i = 1 + \frac{x_i + 1}{4}$; $u(x_i, a, k, m) = \begin{cases} k(x_i - a)^m, x_i > a \\ 0, -a \le x_i \le a \\ k(-x_i - a)^m, x_i < -a \end{cases}$	$[-50, 50]^n$	30	$f(\mathbf{x}^*) = 0$; $\mathbf{x}^* = (-1, \ldots, -1)$

(continued)

(continued)

	Name	Function	S	Dim	Minimum
$f_6(\mathbf{x})$	Perm1	$\sum_{k=1}^{n}\left[\sum_{i=1}^{n}(i^k+50)\{(x_i/i)^k-1\}\right]^2$	$[-n, n]^n$	30	$f(\mathbf{x}^*)=0$; $\mathbf{x}^*=(1,2,\ldots,n)$
$f_7(\mathbf{x})$	Plateau	$30+\sum_{i=1}^{n}\lvert x_i\rvert$	$[-5.12, 5.12]^n$	30	$f(\mathbf{x}^*)=30$; $\mathbf{x}^*=(0,\ldots,0)$
$f_8(\mathbf{x})$	Step	$\sum_{i=1}^{n}(\lfloor x_i+0.5\rfloor)^2$	$[-100, 100]^n$	30	$f(\mathbf{x}^*)=0$; $\mathbf{x}^*=(0,\ldots,0)$
$f_9(\mathbf{x})$	Styblinski tang	$\frac{1}{2}\sum_{i=1}^{n}\left(x_i^4-16x_i^2+5x_i\right)$	$[-5, 5]^n$	30	$f(\mathbf{x}^*)=-39.1659n$; $\mathbf{x}^*=(-2.90,\ldots,2.90)$
$f_{10}(\mathbf{x})$	Trid	$\sum_{i=1}^{n}(x_i-1)^2-\sum_{i=1}^{n}x_ix_i-1$	$[-n^2, n^2]^n$	30	$f(\mathbf{x}^*)=$ $-n(n+4)(n-1)/6$; $\mathbf{x}^*=[i(n+1-i)]$ for$i=1,\ldots,n$
$f_{11}(\mathbf{x})$	Vincent	$-\sum_{i=1}^{n}\sin(10\log x_i)$	$[0.25, 10]^n$	30	$f(\mathbf{x}^*)=-n$; $\mathbf{x}^*=(7.70,\ldots,7.70)$
$f_{12}(\mathbf{x})$	Zakharov	$\sum_{i=1}^{n}x_i^2+\left(\sum_{i=1}^{n}0.5ix_i\right)^2+\left(\sum_{i=1}^{n}0.5ix_i\right)^4$	$[-5, 10]^n$	30	$f(\mathbf{x}^*)=0$; $\mathbf{x}^*=(0,\ldots,0)$
$f_{13}(\mathbf{x})$	Rothyp	$\sum_{i=1}^{d}\sum_{j=1}^{i}x_j^2$	$[-65.536, 65.536]^n$	30	$f(\mathbf{x}^*)=0$; $\mathbf{x}^*=(0,\ldots,0)$
$f_{14}(\mathbf{x})$	Schwefel 2	$\sum_{i=1}^{n}\left(\sum_{j=1}^{i}x_i\right)^2$	$[-100, 100]^n$	30	$f(\mathbf{x}^*)=0$; $\mathbf{x}^*=(0,\ldots,0)$
$f_{15}(\mathbf{x})$	Sum2	$\sum_{i=1}^{d}\sum_{j=1}^{i}x_j^2$	$[-10, 10]^n$	30	$f(\mathbf{x}^*)=0$; $\mathbf{x}^*=(0,\ldots,0)$

(continued)

(continued)

	Name	Function	S	Dim	Minimum
$\boldsymbol{F}_{16}(\mathbf{x})$	Sum of different powers	$\sum_{i=1}^{d} \lvert x_i \rvert^{i+1}$	$[-1, 1]^n$	30	$f(\mathbf{x}^*) = 0$; $\mathbf{x}^* = (0, \ldots, 0)$
$\boldsymbol{f}_{17}(\mathbf{x})$	Rastringin + Schwefel22 + Sphere	$10n + \sum_{i=1}^{n}\left[{x_i}^2 - 10\cos(2\pi x_i)\right] + \left(\sum_{i=1}^{n}\lvert x_i\rvert + \prod_{i=1}^{n}\lvert x_i\rvert\right) + \left(\sum_{i=1}^{n} x_i^2\right)$	$[-100, 100]^n$	30	$f(\mathbf{x}^*) = 0$; $\mathbf{x}^* = (0, \ldots, 0)$
$\boldsymbol{f}_{18}(\mathbf{x})$	Griewank + Rastringin + Rosenbrock	$\frac{1}{4000}\sum_{i=1}^{n} x_i^2 - \prod_{i=1}^{n}\cos\left(\frac{x_i}{\sqrt{i}}\right) + 1 + 10n + \sum_{i=1}^{n}\left[x_i^2 - 10\cos(2\pi x_i)\right] + \sum_{i=1}^{n-1}\left[100(x_{i+1} - x_i^2)^2 + (x_i - 1)^2\right]$	$[-100, 100]^n$	30	$f(\mathbf{x}^*) = n - 1$; $\mathbf{x}^* = (0, \ldots, 0)$

(continued)

(continued)

	Name	Function	S	Dim	Minimum
$f_{19}(\mathbf{x})$	Ackley + Penalty2 + Rosenbrock + Schwefel2	$\left(-20exp\left(-0.2\sqrt{\frac{1}{n}\sum_{i=1}^{n}x_i^2}\right) - exp\left(\frac{1}{n}\sum_{i=1}^{n}cos(2\pi x_i)\right)\right.$ $\left.+20+exp\right)$ $+(0.1\{\sin(3\pi x_1)$ $+\sum_{i=1}^{n}(x_i-1)^2[1$ $\left.+sin^2(3\pi x_i+1)\right]$ $+\left[(x_n-1)^2\left[1+sin^2(2\pi x_n)\right]\right]\}$ $\left.+\sum_{i=1}^{n}u(x_i, 5.100, 4)\right)$ $+\left(\sum_{i=1}^{n-1}\left[100\left(x_{i+1}-x_i^2\right)^2\right.\right.$ $\left.\left.+(x_i-1)^2\right]\right)$ $+\left(\sum_{i=1}^{n}\lvert x_i\rvert+\prod_{i=1}^{n}\lvert x_i\rvert\right)$	$[-100, 100]^n$	30	$f(\mathbf{x}^*) = (1.1n) - 1$; $\mathbf{x}^* = (0, \ldots, 0)$

(continued)

(continued)

	Name	Function	S	Dim	Minimum
$f_{20}(\mathbf{x})$	Ackley + Griewnk + Rastringin + Rosenbrock + Schwefel22	$-20e^{-0.2\sqrt{\frac{1}{n}\sum_{i=1}^{n}x_i^2}} - e^{\frac{1}{n}\sum_{i=1}^{n}\cos(2\pi x_i)} + 20 + e$ $+\frac{1}{4000}\sum_{i=1}^{n}x_i^2$ $-\prod_{i=1}^{n}\cos\left(\frac{x_i}{\sqrt{i}}\right) + 1 + 10n$ $+\sum_{i=1}^{n}\left[x_i^2 - 10\cos(2\pi x_i)\right]$ $+\sum_{i=1}^{n-1}\left[100\left(x_{i+1} - x_i^2\right)^2 + (x_i - 1)^2\right]$ $+\sum_{i=1}^{n}\lvert x_i\rvert + \prod_{i=1}^{n}\lvert x_i\rvert$	$[-100, 100]^n$	30	$f(\mathbf{x}^*) = n - 1$; $\mathbf{x}^* = (0, \ldots, 0)$

References

1. Cuevas, E., Gálvez, J., Avila, K., Toski, M., Rafe, V.: A new metaheuristic approach based on agent systems principles. J Comput Sci **47**, 1–20 (2020)
2. Beyer, H.-G., Beyer, H.-G., Schwefel, H.-P., Schwefel, H.-P.: Evolution strategies—A comprehensive introduction. Nat. Comput. **1**, 3–52. https://doi.org/10.1023/A:1015059928466(2002)
3. Bäck, T., Hoffmeister, F., Schwefel, H.-P.: A survey of evolution strategies, in: Proc. Fourth Int. Conf. Genet. Algorithms, p. 8. 10.1.1.42.3375 (1991)
4. Hansen, N.: The CMA evolution strategy: A Tutorial **102**, 75–102 (2016)
5. Tang, K.S., Man, K.F., Kwong, S., He, Q.: Genetic algorithms and their applications. IEEE Signal Process. Mag. **13**, 22–37. https://doi.org/10.1109/79.543973 (1996)
6. Storn, R., Price, K.: Differential Evolution—A Simple and efficient heuristic for global optimization over continuous spaces. J. Glob. Optim. **11**, 341–359 (1997)
7. Zhang, J., Sanderson, A.C.: JADE: Self-adaptive differential evolution with fast and reliable convergence performance, 2007 IEEE Congr. Evol. Comput. CEC (2007)
8. Askarzadeh, A.: A novel metaheuristic method for solving constrained engineering optimization problems: crow search algorithm. Comput. Struct. **169**, 1–12 (2016)
9. Karaboga, D., Gorkemli, B., Ozturk, C., Karaboga, N.: A comprehensive survey: artificial bee colony (ABC) algorithm and applications. Artif. Intell. Rev. **42**(1), 21–57. https://doi.org/10.1007/s10462-012-9328-0 (2014)
10. Kennedy, J., Eberhart, R.: Particle swarm optimization, Neural Networks, 1995. Proceedings., IEEE Int. Conf. 4. **4**, 1942–1948 (1995)
11. Poli, R., Kennedy, J., Blackwell, T.: Particle swarm optimization. Swarm Intell. **1**, 33–57 (2007)
12. Marini, F., Walczak, B.: Particle swarm optimization (PSO). A tutorial, Chemom. Intell. Lab. Syst. **149**, 153–165 (2015)
13. Yang, X.-S., Firefly algorithms for multimodal optimization, In: Lect. Notes Comput. Sci. (Including Subser. Lect. Notes Artif. Intell. Lect. Notes Bioinformatics), pp. 169–178. https://doi.org/10.1007/978-3-642-04944-6_14 (2009)
14. Yang, X.-S.: Firefly algorithm, Lévy flights and global optimization, In: Res. Dev. Intell. Syst. XXVI, Springer London, London, pp. 209–218. https://doi.org/10.1007/978-1-84882-983-1_15 (2010)
15. Yang, X.S.: Firefly algorithms for multimodal optimization, Lect. Notes Comput. Sci. (Including Subser. Lect. Notes Artif. Intell. Lect. Notes Bioinformatics). 5792 LNCS 169–178 (2009)
16. Yang, X.S. Deb, S.: Cuckoo search via Lévy flights, in: 2009 World Congr. Nat. Biol. Inspired Comput. NABIC 2009—Proc., pp. 210–214 (2009)
17. Yang, X.-S.: A new metaheuristic Bat-inspired Algorithm. Stud. Comput. Intell. **284**, 65–74 (2010)
18. Mirjalili, S., Mirjalili, S.M., Lewis, A.: Grey wolf optimizer. Adv. Eng. Softw. **69**, 46–61 (2014)
19. Mirjalili, S.: Moth-flame optimization algorithm: A novel nature-inspired heuristic paradigm. Knowl.-Based Syst. **89**, 228–249 (2015)
20. Cuevas, E., Echavarría, A., Ramírez-Ortegón, M.A.: An optimization algorithm inspired by the States of Matter that improves the balance between exploration and exploitation. Appl. Intell. **40**, 256–272. https://doi.org/10.1007/s10489-013-0458-0 (2013)
21. Valdivia-Gonzalez, A., Zaldívar, D., Fausto, F., Camarena, O., Cuevas, E., Perez-Cisneros, M.: A states of matter Search-Based approach for solving the problem of intelligent power allocation in Plug-in hybrid electric vehicles. Energies **10**, 92 (2017). https://doi.org/10.3390/en10010092
22. Kirkpatrick, S., Gelatt, C.D., Vecchi, M.P.: Optimization by simulated annealing, Science (80-.). 220 (1983) 671–680. https://doi.org/10.1126/science.220.4598.671
23. Rutenbar, R.A.: Simulated annealing algorithms: An Overview. IEEE Circuits Devices Mag. **5**, 19–26 (1989)
24. Siddique, N., Adeli, H.: Simulated annealing, its variants and engineering applications. Int. J. Artif. Intell. Tools **25**, 1630001 (2016)

25. Rashedi, E., Nezamabadi-pour, H., Saryazdi, S.: GSA: A Gravitational Search Algorithm. Inf. Sci. (Ny) **179**, 2232–2248 (2009)
26. Eskandar, H., Sadollah, A., Bahreininejad, A., Hamdi, M.: Water cycle algorithm—A novel metaheuristic optimization method for solving constrained engineering optimization problems. Comput. Struct. **110–111**, 151–166 (2012)
27. Osman, K.: Erol Ib, rahim Eksin, A new optimization method: Big Bang-Big Crunch. Adv. Eng. Softw. **37**(2), 106–111 (2006)
28. Birbil, ŞI., Fang, S.C.: An electromagnetism-like mechanism for global optimization. J. Glob. Optim. **25**, 263–282 (2003)
29. Dennis G. Zill.: A first course in differential equations with modeling applications. Cengage Learning. ISBN 978-1-285-40110-2. (15 March 2012)
30. Mark A. Haidekker: Linear feedback controls. Elsevier, (2013)
31. Morales-Castañeda, B., Zaldívar, D., Cuevas, E., Fausto, F., Rodríguez, A.: A better balance in metaheuristic algorithms: Does it exist? Swarm Evol. Comput. **54**, 100671 (2020)
32. Boussaïda, I., Lepagnot, J., Siarry, P.: A survey on optimization metaheuristics. Inf. Sci. **237**, 82–117 (2013)
33. Han, M., Liu, C., Xing, J.: An evolutionary membrane algorithm for global numerical optimization problems. Inf. Sci. **276**, 219–241 (2014)
34. Meng, Z., Pan, J.-S.: Monkey King Evolution: A new memetic evolutionary algorithm and its application in vehicle fuel consumption optimization. Knowl.-Based Syst. **97**, 144–157 (2016)
35. Wilcoxon, F.: Individual comparisons by ranking methods. Biom. pp. 80–83, (1945)
36. Cuevas, E., Gálvez, J., Hinojosa, S., Avalos, O., Zaldívar, D., Pérez-Cisneros, M.: A comparison of evolutionary computation techniques for IIR model identification. J. Appl. Math. (2014)
37. Hinojosa, S., Dhal, K.G., Abd Elaziz, M., Oliva, D., Cuevas, E.: Entropy-based imagery segmentation for breast histology using the stochastic fractal search. Neurocomputing **321**, 201–215 (2018)
38. Ibrahim, R.A., Elaziz, M.A., Oliva, D., Cuevas, E., Lu, S.: An opposition-based social spider optimization for feature selection. Soft. Comput. **23**, 13547–13567 (2019)
39. Cuevas, E., Wario, F., Osuna-Enciso, V., Zaldivar, D., Pérez-Cisneros, M.: Fast algorithm for multiple-circle detection on images using learning automata. IET Image Proc. **6**(8), 1124–1135 (2012)
40. Ramírez-Ortegón, M.A., Tapia, E., Ramírez-Ramírez, L.L., Rojas, R., Cuevas, E.: Transition pixel: A concept for binarization based on edge detection and gray-intensity histograms. Pattern Recogn. **43**(4), 1233–1243 (2010)
41. Bandyopadhyay, R., Basu, A., Cuevas, E., Sarkar, R.: Harris Hawks optimisation with Simulated Annealing as a deep feature selection method for screening of COVID-19 CT-scans. Appl. Soft Comput. **111**, 107698 (2021)
42. Cuevas, E., González, A., Fausto, F., Zaldívar, D., & Pérez-Cisneros, M.: Multithreshold segmentation by using an algorithm based on the behavior of locust swarms. Math. Probl. Eng. (2015)
43. Zaldivar, D., Morales, B., Rodríguez, A., Valdivia-G, A., Cuevas, E., Pérez-Cisneros, M.: A novel bio-inspired optimization model based on Yellow Saddle Goatfish behavior. Biosystems **174**, 1–21 (2018)
44. Cuevas, E., Zaldivar, D., Pérez-Cisneros, M.: Seeking multi-thresholds for image segmentation with Learning Automata. Mach. Vis. Appl. **22**, 805–818 (2011)

Chapter 5
Collaborative Hybrid Grey Wolf Optimizer: Uniting Synchrony and Asynchrony

The Grey Wolf Optimizer (GWO) is a relatively new metaheuristic approach that has shown promising results in solving continuous optimization problems. The Grey Wolf Optimizer (GWO) is a relatively new metaheuristic approach that has shown promising results in solving continuous optimization problems. However, it faces challenges when applied to various optimization formulations that require different exploration–exploitation rates. This chapter introduces a modified version of GWO called the Group-based Synchronous-Asynchronous Grey Wolf Optimizer (GSAGWO). GSAGWO divides the population into three groups, each containing an equal number of candidate solutions. The optimization process employs a control strategy that allows for adjusting the exploration–exploitation balance. Additionally, the algorithm incorporates a mechanism to enhance population diversity by exchanging elements among the groups. The performance of GSAGWO is evaluated using the CEC2017 benchmark and several real-world engineering problems. Comparative analyses with other state-of-the-art metaheuristic algorithms demonstrate the effectiveness and accuracy of the discussed technique.

5.1 Introduction

In various domains, including economics, biology, physics, engineering, and industry, individuals encounter intricate optimization problems on a daily basis [1]. The objective of these problems is to identify the best solution by employing a search strategy [2]. Over the past few years, numerous metaheuristic algorithms have been introduced as effective optimization techniques [3]. These algorithms have demonstrated successful application in various fields, including clustering [4], image processing [5], economic dispatch [6], electric load forecasting [7, 8], filter design [9], heart disease prediction [10], among others [11–13].

E. Cuevas et al., *New Metaheuristic Schemes: Mechanisms and Applications*, Intelligent Systems Reference Library 246, https://doi.org/10.1007/978-3-031-45561-2_5

The majority of optimization algorithms have been developed by drawing inspiration from the natural world [14]. Animals, in particular, have exhibited clever strategies for hunting and survival, often collaborating to solve problems with a higher likelihood of success. The collective behavior observed in animals has served as a driving force behind the creation of numerous optimization algorithms. Examples of such algorithms include the Combined Evolutionary Game Particle Swarm Optimization (C-EGPSO) [15], the Selfish Herd Optimizer (SHO) [16], the Intelligent Learning strategy and a Turbulent operator to Artificial Bee Colony (ILTD-ABC) [17], the Grey Wolf Optimizer (GWO) [18], the Improved Crow Search Algorithm (ICSA) [19], the Whale Optimization Algorithm (WOA) [20], the Multi Measure improved Firefly Algorithm (MMFA) [21], the Enhanced Moth-Flame Optimization (EMFO) [22], the Yellow Saddle Goatfish Algorithm (YSGA) [23], the Ensemble Particle Swarm Optimizer (EPSO) [24], the Heterogeneous Comprehensive Learning Particle Swarm Optimization (HCLPSO) [25], among others [14].

Recently, additional bio-inspired approaches have been put forth in the field of optimization. One such method is the Bird Swarm Algorithm (BSA), which was introduced in 2016 [26] and draws inspiration from the social behavior observed in bird flocks. In BSA, the population navigates the search space using operators designed based on rules derived from bird social behavior. Another novel metaheuristic algorithm, the Spotted Hyena Optimizer (SHO), was proposed in 2017 [27]. This method emulates the social behavior of spotted hyenas and mathematically models the identification, encircling, and attacking processes of prey within a search strategy.

Introduced in 2015, the Earthworm optimization algorithm (EWA) takes inspiration from the reproductive process of earthworms [28]. It incorporates various crossover operators and Cauchy mutations in its search strategy. In 2018, the Moth Search (MS) algorithm was presented as a means to address global optimization problems [29]. It leverages the phototaxis process observed in moths and employs the Lévy flight technique in its modeling.

The Emperor Penguin Optimizer (EPO), proposed in 2018 [30], mimics the huddling behavior of emperor penguins. Mathematical models are used in EPO to simulate the generation of the huddle boundary, the temperature surrounding the huddle, the distance, and the effective mover. Another swarm-based algorithm introduced in 2019 is the Artificial Coronary Circulation System (ACCS) [31]. It emulates the growth of coronary arteries in the human heart and incorporates both local and global search mechanisms, where particles evolve based on laws inspired by coronary artery behavior.

Lastly, there is the L-SHADE method, an enhanced version of the Success-History based Adaptive DE (SHADE) algorithm [32], along with notable improvements such as LSHADE-cnEpSin [33], LSHADE-SPACMA [34], and DbL-SHADE [35]. These algorithms represent recent and highly accurate techniques in the field of optimization. All of these algorithms are classified as synchronous approaches as they update candidate solutions once the entire population has been evaluated. In this scheme, crucial elements utilized in the search process, such as the best agent, are selected based on the entire population. Typically, most population-based algorithms are designed under the synchronous mechanism, forming the foundation of their structures.

However, the asynchronous scheme has also been incorporated into metaheuristics to shift the paradigm and enhance their performance. In the asynchronous mechanism, each particle is updated immediately after its evaluation. In this scheme, specific individuals (e.g., the best agent) utilized by the metaheuristic operators are considered based on the currently available information, without considering the entire population. The asynchronous strategy has also been employed in the development of parallel metaheuristic approaches. In parallel computing, the terms synchronous and asynchronous generally refer to communications and how information is shared among processes.

The asynchronous strategy is employed as an update mechanism in several algorithms. For instance, the Hybrid chaos-based Particle Swarm Optimization-Ant Colony Optimization algorithm incorporates an asynchronous pheromone updating process [36]. This approach reduces search time by controlling the number of pheromones updated using an asynchronous strategy, thereby decreasing the number of cycles in the Ant Colony Optimization algorithm. Another notable scheme is the Asynchronous Simulated Kalman Filter Optimization algorithm (ASKF) [37], which improves upon the original synchronous Simulated Kalman Filter Optimization algorithm (SKF). ASKF updates particles asynchronously instead of synchronously, offering enhanced performance. Similarly, an asynchronous version of Particle Swarm Optimization has been documented [38], where the velocity and position of each particle are updated immediately after their evaluation. This approach enables faster convergence and exploration in the optimization process. Additionally, an asynchronous variant of the Gravitational Search Algorithm (GSA) called the Asynchronous Gravitational Search Algorithm (A-GSA) has been introduced [39]. This asynchronous version enhances the GSA method by updating individuals asynchronously, leading to potential performance improvements.

Both synchronous and asynchronous schemes dictate the sequence in which particles are evaluated and updated. The advantage of the asynchronous scheme lies in its utilization of partial information from the population, which fosters exploration and enhances diversity. In contrast, the synchronous mechanism facilitates exploitation of the search space by sharing complete information among all agents. To combine the strengths of both mechanisms and mitigate their shortcomings, hybrid synchronous-asynchronous schemes have been devised. One such approach is the Synchronous-Asynchronous Particle Swarm Optimization (SA-PSO) [40]. SA-PSO divides particles into sets of agents and updates elements within each set asynchronously or synchronously based on a complex decision system that evaluates the current state of the population. This strategy aims to increase swarm diversity, prevent stagnation in local optima, and enhance solution quality. Another synchronous-asynchronous approach is outlined in [41] with the name Synchronous-asynchronous switching Simulated Kalman Filter. This method incorporates both schemes in the update process but not simultaneously. The algorithm dynamically determines the active update mechanism at each iteration based on the success of solution improvement. Initially, the synchronous strategy is employed, but if the solution does not improve, the mechanism switches to asynchronous mode. This strategy promotes diversity, preventing the algorithm from getting trapped in local optima. The same update mechanism has been integrated into the Adaptive Switching Synchronous-Asynchronous Gravitational Search Algorithm (ASw-GSA) [42] and the Adaptive Switch Particle Swarm Optimization (Switch-PSO) [43]. These approaches dynamically adapt between synchronous and asynchronous modes based on the algorithm's progress, promoting diversity and avoiding stagnation in local optima.

Recently, to achieve higher accuracy and faster convergence, several variants [52–54] of the GWO algorithm have been proposed [49]. These modified versions of GWO focus on introducing new operators, alternative updating mechanisms, different encoding schemes for individuals, and diverse population structures and hierarchies. One of these improvements is the Modified Grey Wolf Optimizer (mGWO) [55], which enhances the exploration–exploitation capabilities of GWO by employing a different updating mechanism for the parameter 'a'. The mGWO algorithm nonlinearly decreases 'a' using an exponential function instead of a linear model. Another variant, the Proportional-based Grey Wolf Optimizer (PGWO) [56], suggests using a different selection method to choose the best solutions. The authors of PGWO employ the proportional selection method for selecting the top elements. Similarly, the Tournament-based Grey Wolf Optimizer (TGWO) [56] incorporates the tournament selection method to determine the best elements in the population.

In the Weighted Distance Grey Wolf Optimizer (wdGWO) [57], a novel weighted average scheme is utilized, where each element has a distinct influence on the generation of new individuals. The Complex-valued encoding Grey Wolf Optimizer (CGWO) [58] introduces a new encoding system based on complex numbers, as opposed to the real-valued versions. Lastly, the Evolutionary Population Dynamics Grey Wolf Optimizer (EPD-GWO) [59] addresses the drawbacks of the original GWO by eliminating the worst individuals in the population and replacing them with new particles generated around the best elements. For more detailed information about these variants, please refer to Sect. 5.3. These modified versions have demonstrated their superiority over the original GWO. However, they still possess certain limitations as they are mere simple modifications of the GWO algorithm.

This chapter presents a modified version of the Grey Wolf Optimizer (GWO) called as the Synchronous-Asynchronous Grey Wolf Optimizer (GSAGWO). The GSAGWO method introduces a set of nonlinear functions to alter the generation of candidate solutions, which are updated using a synchronous-asynchronous strategy. To enhance the balance between exploration and exploitation, three new operators have been presented. Furthermore, a novel mechanism is incorporated to promote population diversity through element interchange among the groups. In the GSAGWO process, the population is divided into three groups, each initially containing an equal number of candidate solutions. These groups are individually processed in a predetermined order based on a simple control strategy, which adjusts the exploration–exploitation balance during optimization. Unlike other synchronous-asynchronous approaches, our scheme utilizes a straightforward mechanism to determine the number of elements and the order of the strategy (synchronous or asynchronous) for processing. This mechanism does not employ complex methods to select agents or determine the strategy that best fits the current state of the population. Consequently, the presented method is a self-adaptive technique that autonomously configures the parameters controlling the operational process throughout iterations.

The chapter is structured as follows: Sect. 5.2 provides an introduction to the GWO algorithm, serving as background information. Section 5.3 offers a comprehensive review of the various modified versions of the GWO. The discussed variant of the GWO is presented in Sect. 5.4. Section 5.5 focuses on the discussion of experimental results and statistical analysis. Finally, Sect. 5.6 concludes the chapter by summarizing the findings and outlining potential directions for future work.

5.2 The Original Grey Wolf Optimizer Method

This section delves into the intricacies of the Grey Wolf Optimizer (GWO), a technique introduced by Mirjalili et al. in 2014 [18]. The GWO draws inspiration from the social behavior of grey wolves, specifically their pack dynamics. These packs exhibit remarkable characteristics related to leadership [60] and hunting strategies [61], enabling collective behavior that enhances their survival chances.

Within a pack of grey wolves, leadership plays a crucial role in fostering collaborative behavior. The pack operates on a hierarchical structure that determines social authority and dominance. Wolves are categorized into different ranks, namely alpha, beta, delta, and omega, based on their specific traits. Each wolf's role within the pack is determined by its position in the hierarchy. The alpha wolf holds the highest rank as the dominant leader, responsible for making decisions. The beta wolf follows as the second in command, supporting the alpha in guiding and controlling the lower-ranking wolves. The delta wolf holds the third position in the hierarchy, while the omegas are subordinate wolves. Figure 5.1 provides a visual representation of the grey wolf hierarchy.

The hunting strategy employed by grey wolves encompasses several key actions, including tracking, chasing, approaching, pursuing, encircling, and attacking their prey. The GWO algorithm incorporates some of these elements as operators to emulate the hunting behavior of grey wolves. The subsequent section provides a detailed explanation of the mathematical model employed in the GWO algorithm.

Fig. 5.1 Hierarchy imposed in grey wolf packs

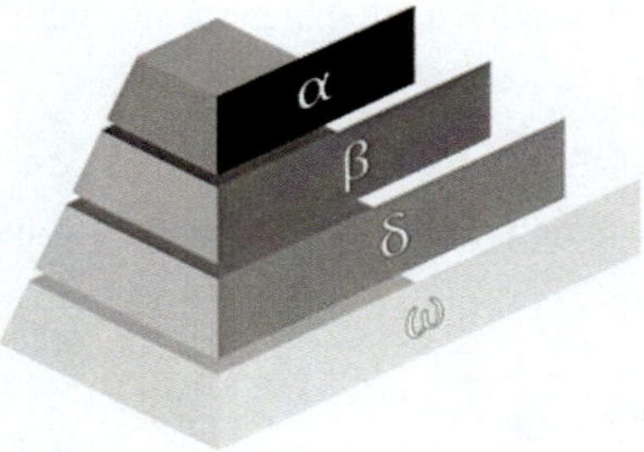

5.2.1 Computational Algorithm

The concept of leadership in the GWO algorithm is represented by the designation of alpha α, beta β, and delta δ particles based on their fitness values. The alpha particle denotes the best solution, the beta particle represents the second-best solution, and the delta particle corresponds to the third-best solution. These particles play a crucial role in guiding the search process. The remaining members of the population are referred to as omega ω elements.

In the GWO methodology, a pack of m grey wolves is represented as $\mathbf{X}^t = \{\mathbf{x}_1^t, \mathbf{x}_2^t, \ldots, \mathbf{x}_m^t\}$, where t denotes the current generation, and $\mathbf{x}_i^t$ represents a grey wolf whose elements correspond to n decision variables, such as $\mathbf{x}_i^t = \{x_i^1, x_i^2, \ldots, x_i^n\}$. The GWO algorithm incorporates the key hunting strategies of encircling and attacking prey. The encircling behavior is mathematically modeled as follows:

$$\mathbf{x}_i^{t+1} = \mathbf{x}_p^t - \mathbf{A} \cdot \mathbf{D} \tag{5.1}$$

Here, $\mathbf{x}_i^{t+1}$ denotes the subsequent position of individual i, while $\mathbf{x}_p^t$ represents the current location of the prey. The coefficient vector $\mathbf{A}$ and the distance $\mathbf{D}$ between the wolf and the prey play important roles in the calculation. The coefficient $\mathbf{A}$ can be determined using the following formula:

$$\mathbf{A} = 2a\mathbf{r}_1 - a \tag{5.2}$$

Let $\mathbf{r}_1$ represent a vector containing random values ranging from 0 to 1, and a be a variable that progressively decreases from 2 to 0 during the optimization procedure. The value of a is determined using the following equation:

$$a = 2\left(1 - \frac{t}{t_{\max}}\right) \tag{5.3}$$

here, $t_{\max}$ is the maximum number of iterations. The distance $\mathbf{D}$ is calculated as follows:

$$\mathbf{D} = \left|\mathbf{C} \cdot \mathbf{x}_p^t - \mathbf{x}_i^t\right| \tag{5.4}$$

where $\mathbf{C}$ is a coefficient vector with values among [0, 2] and is calculated as:

$$\mathbf{C} = 2\mathbf{r}_2 \tag{5.5}$$

where, $\mathbf{r}_2$ is a vector of random values among [0, 1].

Given that α, β, and δ wolves represent the most optimal solutions discovered thus far, it is assumed that they possess superior knowledge regarding the whereabouts of the prey $\mathbf{x}_p^t$. Based on this assumption, the prey's location can be assumed to be in close proximity to α, β, or δ. As a result, the search process is directed by these entities. Consequently, Eq. 2.1 can be reformulated to update the position of each potential solution, factoring in the influence of α, β, and δ:

$$\mathbf{x}_i^{t+1} = \frac{z_\alpha + z_\beta + z_\delta}{3} \tag{5.6}$$

Here,z_α, z_β and z_δ represent the movement contributions attributed to, β, and δ respectively.

$$z_\alpha = \mathbf{x}_\alpha^t - \mathbf{A}_1 \cdot \mathbf{D}_\alpha \tag{5.7}$$

$$z_\beta = \mathbf{x}_\beta^t - \mathbf{A}_2 \cdot \mathbf{D}_\beta \tag{5.8}$$

$$z_\delta = \mathbf{x}_\delta^t - \mathbf{A}_3 \cdot \mathbf{D}_\delta \tag{5.9}$$

$$\mathbf{D}_\alpha = \left|\mathbf{C}_1 \cdot \mathbf{x}_\alpha^t - \mathbf{x}_i^t\right| \tag{5.10}$$

$$\mathbf{D}_\beta = \left|\mathbf{C}_2 \cdot \mathbf{x}_\beta^t - \mathbf{x}_i^t\right| \tag{5.11}$$

$$\mathbf{D}_\delta = \left|\mathbf{C}_3 \cdot \mathbf{x}_\delta^t - \mathbf{x}_i^t\right| \tag{5.12}$$

The predatory behavior is simulated by reducing the value of a from 2 to 0 during the iterations. As a decreases, the variable $\mathbf{A}$ also decreases and takes on random values within the range $[-2a, 2a]$. This mechanism aims to exploit the solution space, causing the wolves to progressively approach the prey throughout the iterations when the magnitude of $\mathbf{A}$ is less than 1. Conversely, when the magnitude of $\mathbf{A}$ exceeds 1, the wolves move away from the prey in search of better solutions. This approach facilitates the exploration process and prevents the algorithm from becoming stuck at a local minimum.

Algorithm 5.1 The pseudocode of GWO method

Initialize population of m wolves $\mathbf{X}^0 = \{\mathbf{x}_1^0, \mathbf{x}_2^0, \ldots, \mathbf{x}_m^0\}$
Initialize a, $\mathbf{A}_1, \mathbf{A}_2, \mathbf{A}_3, \mathbf{C}_1, \mathbf{C}_2$, and $\mathbf{C}_3$
Evaluate population in the objective function $f(\mathbf{X}^0)$
Set α, β, and δ wolves as:
  $\mathbf{x}_\alpha^0 \leftarrow \text{best}(f(\mathbf{X}^0))$
  $\mathbf{x}_\beta^0 \leftarrow \text{second\_best}(f(\mathbf{X}^0))$
  $\mathbf{x}_\delta^0 \leftarrow \text{third\_best}(f(\mathbf{X}^0))$
while $(t < t_{max})$
| **for** each wolf
| | update the current position $\mathbf{x}_i^{t+1} = \frac{\mathbf{z}_\alpha + \mathbf{z}_\beta + \mathbf{z}_\delta}{3}$
| **end for**
| Update a, $\mathbf{A}_1, \mathbf{A}_2, \mathbf{A}_3, \mathbf{C}_1, \mathbf{C}_2$, and $\mathbf{C}_3$
| Evaluate population in the objective function $f(\mathbf{X}^t)$
| Update $\mathbf{x}_\alpha^t$, $\mathbf{x}_\beta^t$, and $\mathbf{x}_\delta^t$ wolves
| $t \leftarrow t + 1$
end while
Best solution $\mathbf{x}_\alpha^{t_{max}}$

The GWO algorithm is an iterative technique that begins by initializing a population of m grey wolves denoted as $\mathbf{X}^t = \{\mathbf{x}_1^t, \mathbf{x}_2^t, \ldots, \mathbf{x}_m^t\}$ within the search space. Subsequently, the population is evaluated based on the objective function. Following this, the three best individuals are chosen from the population to represent the alpha, beta, and delta elements, while the remaining candidate solutions are designated as omega elements. The position of the wolves is then updated by applying Eq. 5.6. This process continues until the maximum number of iterations is reached. The complete procedure of the GWO algorithm is outlined in Algorithm 5.1.

5.3 Improved Versions of the GWO

To enhance the performance of the GWO algorithm, numerous variations have been proposed. These modified versions primarily introduce novel operators, update mechanisms, encoding schemes for individuals, or alternative population structures and hierarchies. This section provides a detailed description of these algorithms.

5.3.1 The Modified Grey Wolf Optimizer (mGWO)

To enhance the exploration–exploitation capabilities of the GWO algorithm, a modified approach has been introduced that utilizes a distinct updating mechanism for the parameter a. The mGWO algorithm, proposed by Mittal et al. in [55], incorporates a nonlinear decrease of a by employing the exponential function described in Eq. 5.13, as opposed to the linear model depicted in Eq. 5.3. This method aims to improve the performance of GWO by adapting the parameter update scheme.

$$a = 2\left(1 - \frac{t^2}{{t_{\max}}^2}\right) \tag{5.13}$$

By employing this mechanism, the algorithm demonstrates improved exploration capabilities throughout the iterations, effectively preventing premature convergence. In the original GWO, the balance between exploration and exploitation shifts proportionally as the parameter a decreases. In the original GWO scheme, the exploration and exploitation processes occur at a 50 and 50% rate respectively. In contrast, the mGWO algorithm adjusts this proportion to 70% for exploration and 30% for exploitation. This modification significantly alters the exploration–exploitation trade-off, constituting the primary contribution of this enhancement.

5.3.2 The Proportional-Based Grey Wolf Optimizer (PGWO)

In metaheuristic algorithms, individual selection plays a vital role [62] and has a significant impact on the optimization process [63]. Various selection methods exist, including proportional selection [64] and tournament selection [65].

The PGWO [56] introduces an alternative selection method, deviating from the traditional approach where the α, β, and δ elements are chosen based on their rank in the population (i.e., the best solutions). Instead, the PGWO employs the proportional selection method to select these elements. This novel scheme proposed in PGWO promotes a more balanced exploration–exploitation process by facilitating the sharing of global information among all members of the population. This approach aims to enhance the algorithm's overall performance by diversifying the selection process and encouraging better information exchange among individuals.

5.3.3 The Tournament-Based Grey Wolf Optimizer (TGWO)

TGWO [56] introduces the utilization of the tournament selection method to determine the α, β, and δ elements. In TGWO, a subset of τ elements, where $\tau \geq 3$, is randomly chosen from the population. Subsequently, based on their fitness values, the three most promising candidate solutions are selected from this subset to represent the α, β, and δ elements. This approach allows for a competitive selection process that considers the performance of multiple individuals and ensures that the most favorable solutions are chosen as the guiding elements for the algorithm.

5.3.4 The Weighted Distance Grey Wolf Optimizer (wdGWO)

The original GWO algorithm directs the search strategy by utilizing the average combination of the α, β, and δ elements (as indicated in Eq. 5.6). In this approach, each element holds an equal influence on the final outcome. However, in the wdGWO method [57], a novel weighted average scheme is introduced, where each element carries a distinct level of influence on the final position. Consequently, a revised mechanism is established to update a specific candidate solution, defined as follows:

$$\mathbf{x}_i^{t+1} = \frac{w_1 z_\alpha + w_2 z_\beta + w_3 z_\delta}{3} \tag{5.14}$$

The weights w_1, w_2, and w_3 correspond to the α, β, and δ agents, respectively. These weights are determined through the following calculations:

$$\begin{aligned} w_1 &= \mathbf{A}_1 \cdot \mathbf{C}_1 \\ w_2 &= \mathbf{A}_2 \cdot \mathbf{C}_2 \\ w_3 &= \mathbf{A}_3 \cdot \mathbf{C}_3 \end{aligned} \tag{5.15}$$

5.3.5 Complex-Valued Encoding Grey Wolf Optimizer (CGWO)

CGWO [58] introduces a novel encoding system that utilizes complex numbers instead of real-valued units to represent the information of individuals. Consequently, each individual is represented by both a real part and an imaginary part. According to the authors in [58], this complex-valued encoding employed in CGWO enhances population diversity, leading to improved exploration capabilities within the algorithm.

5.3.6 *Evolutionary Population Dynamics Grey Wolf Optimizer (EPD-GWO)*

EDP-GWO [59] incorporates a mechanism where the poorest individuals within the population are eliminated and replaced with new particles generated in the vicinity of the best elements α, β, and δ. This approach aims to improve the overall quality of the population and facilitate the exploration process, as stated by the authors. By removing weaker individuals and introducing new ones around the best solutions, EDP-GWO enhances the algorithm's ability to explore the solution space and potentially discover better solutions.

5.4 The Synchronous-Asynchronous GWO

To address the limitations of the original GWO algorithm, the GSAGWO algorithm has been developed. The discussed approach introduces a Synchronous-Asynchronous mechanism to regulate the order of particle updates. In our method, a series of nonlinear functions are employed to generate new candidate solutions. Additionally, the search strategy incorporates varying levels of exploration and exploitation through a self-adaptive technique that automatically configures the parameters governing the operational process throughout the iterations. In the following section, the GSAGWO algorithm is introduced and comprehensively described, highlighting its key features and mechanisms.

5.4.1 *The Synchronous-Asynchronous Mechanism*

In the discussed method, we combine the synchronous and asynchronous strategies to achieve a balanced approach that considers both schemes.

Within the Synchronous-Asynchronous framework, the complete set of candidate solutions is divided into a predetermined number m of subpopulations. Each subpopulation contains an equal number of candidate solutions (N/m), randomly selected from the entire population of N individuals. The subpopulations are processed in a specific order, starting with the first subpopulation and ending with subpopulation m. When generating candidate solutions within each subpopulation, only the information from that particular subpopulation and the preceding subpopulations is taken into account. Consequently, if the best solution is found within subpopulation three, the candidate solutions generated in subpopulations one and two do not incorporate this information in their operators. By integrating both synchronous and asynchronous strategies, a better balance is achieved between exploration and exploitation of the search space, leading to improved optimization performance [40].

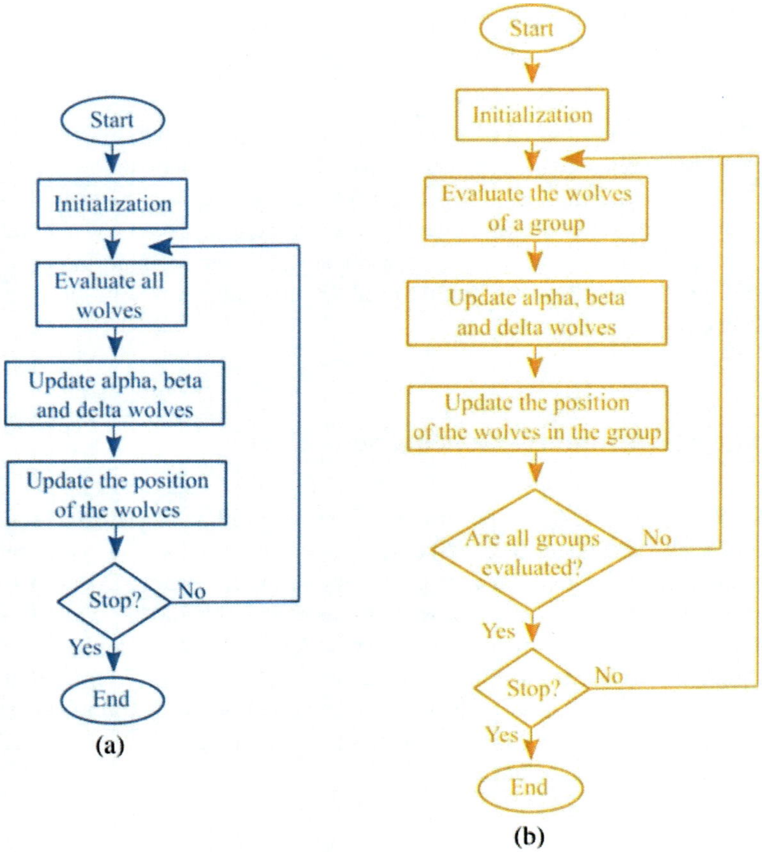

Fig. 5.2 Comparison between **a** the original GWO and **b** the group-based synchronous-asynchronous scheme, which is incorporated into the discussed approach

The discussed method combines the synchronous-asynchronous mechanism with the original GWO algorithm. Figure 5.2 provides a comparative representation of the original GWO (Fig. 5.2a) and the synchronous-asynchronous scheme (Fig. 5.2b), which is incorporated into the discussed algorithm. In this new scheme, the population structure is modified by dividing all its elements into three groups: $\mathbf{g}_1^t$, $\mathbf{g}_2^t$, and $\mathbf{g}_3^t$. An example of group configuration is illustrated in Fig. 5.3, considering a population of $N = 9$ individuals.

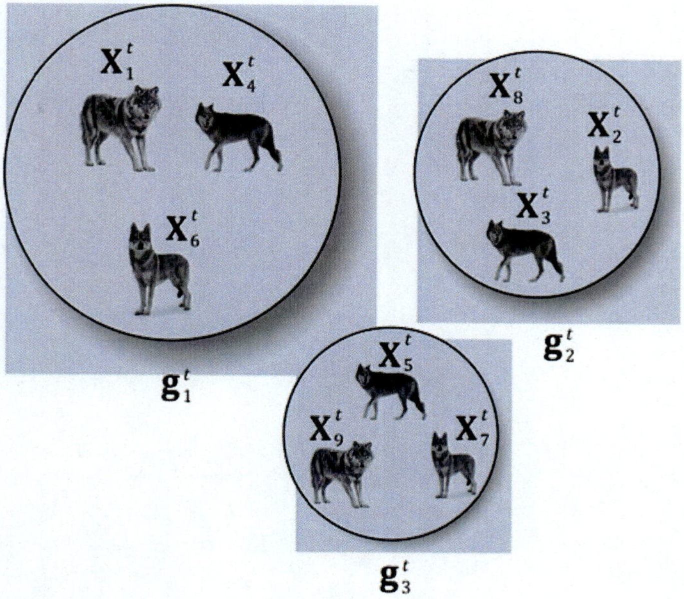

Fig. 5.3 Here is an illustration of a group configuration, considering a population of $N = 9$ elements

5.4.2 The Impact and Effect of the Parameter a

The original GWO algorithm employs a linear decrease of the parameter a to simulate the attacking prey, as described in Eq. 5.3. However, it has been demonstrated in previous studies [55] that this mechanism may not be the optimal choice. Nonlinear alternatives have been proposed in [55] and [66] that exhibit superior performance in reducing the parameter a. In this work, we also introduce a set of nonlinear functions to update this parameter.

Given that our discussed method is based on the synchronous-asynchronous scheme, where the population is divided into three groups, we suggest using a different nonlinear function to decrease the parameter a. The aim is to establish distinct levels of exploration and exploitation within each group. Thus, the update of a will be determined by employing a specific decreasing function, as defined below:

$$a = f(\mathbf{g}_i^t) = \begin{cases} e^{-4\left(\frac{t}{t_{\max}}\right)}, & i = 1 \\ 2\left(1 - \frac{t^2}{t_{\max}^2}\right), & i = 2 \\ 3e^{-\left(\frac{t}{t_{\max}}\right)^2}, & i = 3 \end{cases} \tag{5.16}$$

Here, $\mathbf{g}_i^t$ refers to group i in the current iteration t, and $t_{\max}$ denotes the maximum number of iterations.

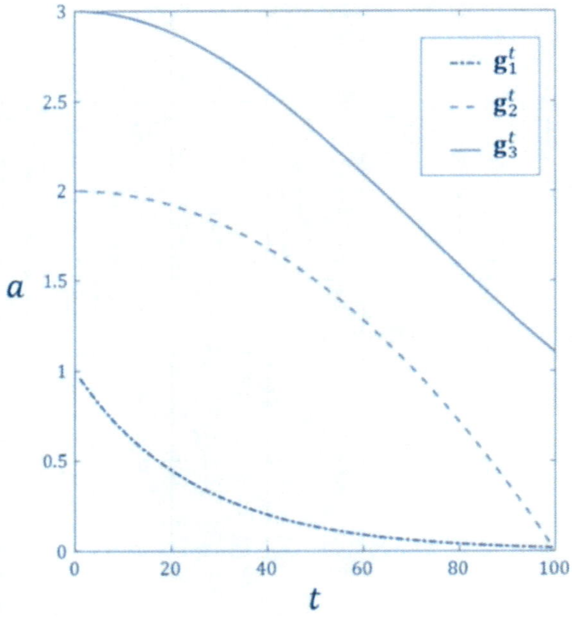

Fig. 5.4 Evolution of the parameter a throughout iterations for groups $\mathbf{g}_1^t$,$\mathbf{g}_2^t$, and $\mathbf{g}_3^t$

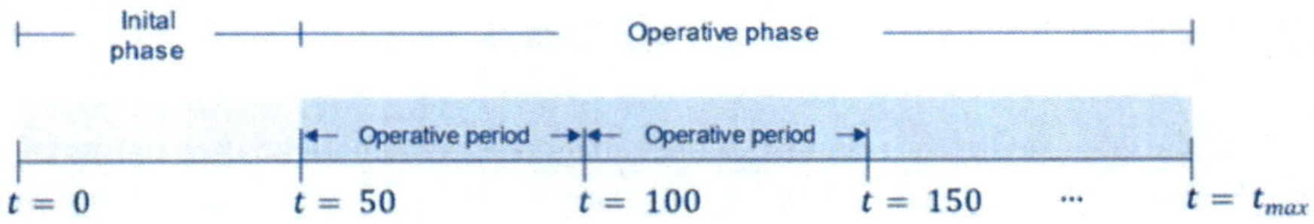

Fig. 5.5 Control procedure

With these functions, group $\mathbf{g}_1^t$ exhibits a distinct exploitation behavior, focusing primarily on exploiting the search space. Conversely, group $\mathbf{g}_3^t$ engages in a more exploratory process, emphasizing the exploration of the solution space. Meanwhile, group $\mathbf{g}_2^t$ strikes a balance between exploitation and exploration. To illustrate these behaviors, Fig. 5.4 depicts the progression of the parameter a across generations for each group.

5.4.3 Control Procedure

In the discussed method, each group exhibits a distinct behavior to guide the search process, either focusing on exploration or exploitation. The processing order of the groups plays a crucial role in determining the search strategy.

In this scenario, the assignment of the group order $\mathbf{g}_1^t$,$\mathbf{g}_2^t$, and $\mathbf{g}_3^t$ dynamically changes throughout the optimization process. The determination of the group order is based on an order control strategy, which is a self-adaptive technique commonly used in metaheuristic algorithms to automatically configure their parameters during iterations based on the operation process [67]. The strategy involves collecting information during an initial "operative period" and subsequently analyzing and utilizing this information for decision-making. The GSAGWO algorithm employs an order control strategy to determine the execution order of the groups. This strategy divides the optimization process into two phases: (a) the Initial phase and (b) the Operative phase. The behavior of the order control strategy varies depending on the current phase. The Initial phase spans the first 50 iterations, while the Operative phase encompasses the remaining evolution process (refer to Fig. 5.5 for a visual representation).

5.4.4 *Initialization*

The Initial phase is relatively shorter compared to the Operative phase. It encompasses only the first 50 iterations, ranging from $t = 1$ to $t = 50$. During these initial 50 iterations, the order of the groups is determined randomly. This random assignment enhances the exploration capability of the discussed search strategy. Additionally, in this stage, the occurrences v_i of the α element in each group i are recorded $(i \in 1, 2, 3)$. Based on this information, a probability p_i is assigned to each group, calculated as $p_i = v_i/50$. This probability is utilized to determine the order of the groups in the subsequent iterations.

5.4.5 *Processing Stage*

The processing stage commences after the initial 50 iterations, starting from $t = 51$ and continuing until $t = t_{\max}$. Within this phase, there are consecutive episodes of 50 iterations each, known as operative periods, during which performance information for each group is collected to support the order control strategy. The group that contains the α element most frequently during an operative period is considered the best group, while the group with the lowest occurrence frequency of the α element is regarded as the worst group. The intention is to provide the best group with more opportunities for exploiting the search space. Consequently, the best group is prioritized and processed first among the groups. Thus, there is a higher probability of selecting the best group as the first group $\mathbf{g}_1^t$. Conversely, the worst group should have more opportunities for exploring the search space. As a result, the worst group has a reduced probability of being chosen as the first group and a higher probability of being assigned as the third group $\mathbf{g}_3^t$ compared to the other groups.

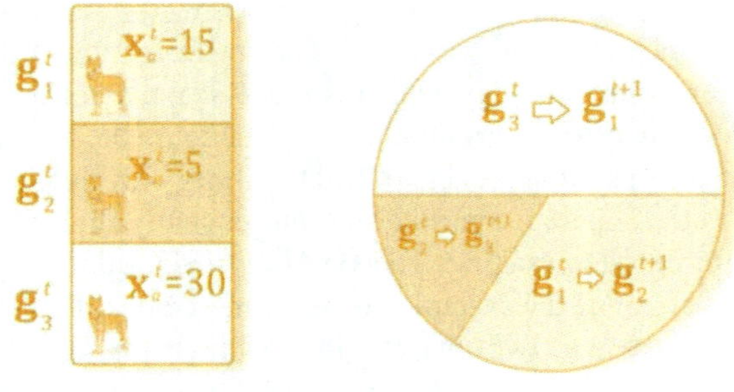

Fig. 5.6 One representative example of the order control strategy in the operative phase

During the operative phase, the order of the groups is no longer selected randomly. Instead, it is determined using the proportional selection method based on the information gathered in each operative period. To accomplish this, a roulette wheel is constructed based on the probability p_i assigned to each group i, which is derived from the number of times υ_i that the α wolf has been assigned to that particular group. This entire process is repeated in each operative period, which occurs every 50 iterations.

Figure 5.6 provides an illustrative example of the order control strategy during an operative period. In this example, group $\mathbf{g}_3^t$ has a value of $\upsilon_3 = 30$, $\mathbf{g}_2^t$ has a value of $\upsilon_2 = 5$, and $\mathbf{g}_1^t$ has a value of $\upsilon_1 = 15$. Using these values, the probabilities for each group are calculated as follows: $p_1 = 15/50$, $p_2 = 5/50$ and $p_3 = 30/50$. Based on the roulette selection method, the order of the groups is determined through the sampling process of the roulette for three different cases. As a result, $\mathbf{g}_3^t$ is more likely to become $\mathbf{g}_1^t$, $\mathbf{g}_1^t$ has a higher chance of becoming $\mathbf{g}_2^t$, and $\mathbf{g}_2^t$ is more likely to transform into $\mathbf{g}_3^t$ since it has the lowest probability.

5.4.6 Diversity Change

After establishing the order of the groups, a diversity-enhancing operation is performed. In this operation, either $\mathbf{g}_2^t$ or $\mathbf{g}_3^t$ is randomly chosen from the last two groups. Subsequently, a random element from the selected group is extracted and transferred to the first group, $\mathbf{g}_1^t$. Figure 5.7 provides an example of this operation. In the given example, after changing the group order through the order control strategy, $\mathbf{g}_3^t$ is randomly selected. From $\mathbf{g}_3^t$, the element $\mathbf{g}_2^t$ is randomly chosen and moved to become part of $\mathbf{g}_1^t$.

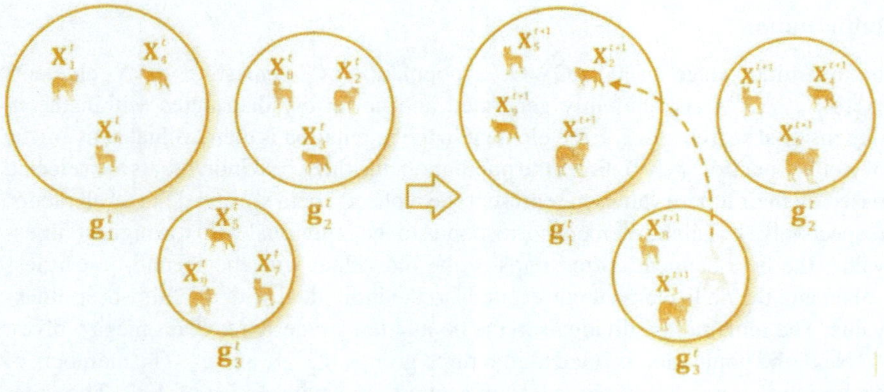

Fig. 5.7 Diversity change mechanism

We have set the minimum number of elements in the last two groups, g_2^t and g_3^t, to be 1. This limit is in place to avoid computational errors in case a group disappears due to the diversity jump strategy. Additionally, it is desirable for elements from g_2^t and g_3^t to gradually merge with g_1^t over the iterations, leading to stronger exploitation. However, it is worth noting that the diversity jump occurs once per operative period. The maximum number of jumps can be calculated as $(t_{\max}/operativeperiod) - 1$. For example, with a maximum of 1000 iterations and an operative period of 50, there would be 19 diversity jumps during the search process. Considering a population size of $N = 50$, the minimum initial size of the groups is 16. Given these conditions, it would be challenging for both g_2^t and g_3^t to contain only 1 element at the end of the iterations since they have an equal probability of being chosen to contribute one of their elements to g_1^t.

5.4.7 Computational Procedure

The computational procedure consists of two primary stages: initialization and the iterative process. During the initialization stage, the population and groups are generated. In the iterative process, the order of the groups is determined using the order control strategy. Based on this order, the groups are processed, starting with the first group and ending with group 3. The parameter a is updated accordingly, based on the order of each group. A detailed explanation of these stages is presented below.

Initialization

In the initial stage of GSAGWO, a population $\mathbf{X}^0$ consisting of N elements $\{\mathbf{x}_1^0, \mathbf{x}_2^0, \ldots, \mathbf{x}_N^0\}$ is randomly generated and uniformly distributed within the n-dimensional search space. Each element in the population is then evaluated using the objective function $f(\mathbf{X}^0)$. From the population, the three best individuals are selected based on their fitness values to represent the alpha $\mathbf{x}_\alpha^0$, beta $\mathbf{x}_\beta^0$, and delta $\mathbf{x}_\delta^0$ elements, respectively. The alpha element corresponds to the individual with the highest fitness value, the beta element corresponds to the individual with the second-best fitness value, and the delta element corresponds to the individual with the third-best fitness value. The remaining individuals in the population are designated as omega wolves.

Next, the population is divided into three groups: $\mathbf{g}_1^0$, $\mathbf{g}_2^0$, and $\mathbf{g}_3^0$. The members of each group are randomly chosen, with each group having a size of $N/3$. The order of these groups is also randomly determined. For each group, the parameter a is initialized using Eq. 5.16. Additionally, the variables $\mathbf{A}_1, \mathbf{A}_2, \mathbf{A}_3, \mathbf{C}_1, \mathbf{C}_2,$ and$\mathbf{C}_3$ are initialized. Finally, the current positions of the elements in each group $\mathbf{g}_i^0$ are updated using Eq. 5.6.

In the iterative process, the order in which the groups are processed is determined based on the order control strategy, taking into account the current iteration t. Initially, when t is less than or equal to 50, the order of the groups is randomly assigned.

However, when t exceeds 50, the order of the groups is determined during the operative phase by considering the information collected in the last 50 iterations. This information is used to assign probabilities to each group based on the number of times the α element has been present in each group. A roulette wheel is then constructed using these probabilities, and the order of the groups is established using the proportional method.

Once the order of the groups has been determined, one of the last two groups is randomly selected. From the selected group, an element is randomly chosen and moved to the first group. This operation helps to increase the diversity of the groups and promotes exploration in the search process.

Subsequently, the GWO operations are performed on each group following the predetermined order. Firstly, the individuals within the current group are evaluated using the objective function $f(\mathbf{X}^t)$. Then, the alpha ($\mathbf{x}_\alpha^t$), beta ($\mathbf{x}_\beta^t$), and delta ($\mathbf{x}_\delta^t$) wolves are updated according to their positions. The parameter a is also updated based on the current group. Additionally, the values of $\mathbf{A}_1, \mathbf{A}_2, \mathbf{A}_3, \mathbf{C}_1, \mathbf{C}_2,$ and $\mathbf{C}_3$ are updated. Finally, the positions of the population members within the current group are updated.

This procedure continues iteratively until the maximum number of iterations t_{max} is reached. The complete GSAGWO process is summarized in Algorithm 5.2.

Algorithm 5.2 The Synchronous-Asynchronous algorithm.

GSAGWO algorithm

Line	Step	Section	Part
1.	Initialize population of N wolves $\mathbf{X}^0 = \{\mathbf{x}_1^0, \mathbf{x}_2^0, \ldots, \mathbf{x}_N^0\}$	I	A
2.	Evaluate population in the objective function $f(\mathbf{X}^0)$		
3.	Set α, β, and δ wolves as:	II	
4.	$\mathbf{x}_\alpha^0 \leftarrow$ best$(f(\mathbf{X}^0))$ $\mathbf{x}_\beta^0 \leftarrow$ second_best$(f(\mathbf{X}^0))$ $\mathbf{x}_\delta^0 \leftarrow$ third_best$(f(\mathbf{X}^0))$		
5.	Create randomly three groups $\mathbf{g}_1^0$, $\mathbf{g}_2^0$, and $\mathbf{g}_3^0$ of $N/3$ elements	III	
6.	Randomly select the order of the groups		
7.	**for** every group	IV	
8.	Initialize a according to the chosen group using **Equation 4.1**		
9.	Set $\mathbf{A}_1, \mathbf{A}_2, \mathbf{A}_3, \mathbf{C}_1, \mathbf{C}_2,$ and $\mathbf{C}_3$		
10.	Update the current position of each element in $\mathbf{g}_i^0$ using **Equation 2.6**		
11.	**end for**		
12.	**while** $(t < t_{max})$		B
13.	**if** $t \leq 50$	V	
14.	Randomly select the order of the groups for evaluation		
15.	**else**		
16.	Use a roulette proportional process to select the group order		
17.	Randomly select one of the two last groups		
18.	Take an element from the selected group and give it to the first group		
19.	**end if**		
20.	**for** every group chosen in the previously established order		
21.	Evaluate each wolf of the chosen group in the objective function $f(\mathbf{g}_i^t)$	VI	
22.	Update $\mathbf{x}_\alpha^t$, $\mathbf{x}_\beta^t$, and $\mathbf{x}_\delta^t$ wolves	VII	
23.	Update a according to the chosen group $\mathbf{g}_i^t$ using **Equation 4.1**	VIII	
24.	Update $\mathbf{A}_1, \mathbf{A}_2, \mathbf{A}_3, \mathbf{C}_1, \mathbf{C}_2,$ and $\mathbf{C}_3$		
25.	Update the current position of each wolf in $\mathbf{g}_i^t$ using **Equation 2.6**	IX	
26.	**end for**		
27.	$t \leftarrow t + 1$		
28.	**end while**		
29.	Best solution $\mathbf{x}_\alpha^{t_{max}}$		

5.4.8 Computational Complexity

In this section, we define the computational complexity of the GSAGWO algorithm. Metaheuristic schemes, being stochastic processes, are complex systems, making it impractical to perform a complexity study from a deterministic perspective. Therefore, we employ the Big-O notation.

To determine the computational complexity of the GSAGWO method, we divide the optimization task into two parts: A and B, as outlined in Algorithm 5.2. Part A (lines 1–11) covers the initialization section, while part B corresponds to the main loop (lines 12–28). Part A consists of four distinct sections: I, II, III, and IV. On the other hand, part B is further divided into sections V, VI, VIII, VIIII, and IX.

When determining the computational complexity, we make certain assumptions: *N* represents the population size, *n* represents the dimension of the optimization problem, and $t_{\max}$ denotes the maximum number of iterations. With these assumptions, we can calculate the computational cost of each section using the following formulas:

Part **A**	Initialization	Total complexity
Section I	Lines 1-2 $O(Nn + N)$	$O(4Nn + 7 + N) + O(N \cdot \log(N))$
Section II	Lines 3-4 $O(N \cdot \log(N))$	
Section III	Lines 5-6 $O(Nn)$	
Section IV	Lines 7-11 $O(Nn + 7)$	
Part **B**	Main loop	
Section V	Lines 13-19 $O(12 \cdot \frac{t_{max}}{50})$	$O\left(t_{max}\left(\frac{12}{50} + N + N \cdot \log(N) + 10 + Nn\right)\right)$
Section VI	Line 21 $O(N \cdot t_{max})$	
Section VII	Line 22 $O(t_{max} \cdot (N \cdot \log(N)))$	
Section VIII	Line 23 $O(10 \cdot t_{max})$	
Section IX	Line 25 $O(Nn \cdot t_{max})$	

Since the initialization process occurs only once, the overall computational cost can be expressed using the following equation:

$$O\left(t_{\max}\left(\frac{12}{50} + N + N \cdot \log(N) + 10 + Nn\right)\right) \tag{5.17}$$

Under these computations, it can be concluded that the GSAGWO has polynomial complexity.

GSAGWO incorporates several desirable properties in its search strategy through the implemented mechanisms. The synchronous mechanism facilitates the exploitation of the search space by sharing information about the best elements among all candidate solutions. On the other hand, the asynchronous scheme promotes stronger exploration and diversity by utilizing partial information. In GSAGWO, we combine both strategies to achieve a better balance between exploration and exploitation.

The division and sequential processing of elements into groups enable the implementation of different search patterns within the same process. In GSAGWO, each group is processed independently, considering only its own elements and employing different values of the parameter a. These varying values allow for experimentation with different exploration–exploitation behaviors. Consequently, the complete search strategy can be divided into diverse search patterns, thereby enhancing its effectiveness.

While GSAGWO divides its search strategy into different search patterns, it also incorporates a global information-sharing aspect. Throughout the optimization process, the search results from each group are collected to determine the global order in which the groups will be processed. In this new order, the first group retains the best element observed thus far, and the remaining groups utilize this information to guide their respective search patterns. Through this mechanism, the effectiveness of each search pattern is significantly enhanced, as they can benefit from the knowledge acquired by the other groups.

The diversity jump process serves as an effective mechanism for artificially increasing diversity within the population. By exploring information from other groups using distinct search patterns, the algorithm can avoid getting trapped in local minima present within the groups.

Despite its advantageous outcomes, the GSAGWO algorithm does have a drawback. This drawback stems from the numerous mechanisms involved in its search strategy, which results in a larger set of operations compared to other metaheuristic schemes that employ a smaller set of operations. However, the optimization results of GSAGWO justify the use of these multiple mechanisms, particularly in complex search scenarios where the algorithm excels.

5.5 Experiments

To assess the effectiveness of GSAGWO, it has been subjected to a comparative analysis with other relevant techniques. The evaluation involved considering recent and enhanced versions of well-known metaheuristic algorithms, namely Synchronous-Asynchronous Particle Swarm Optimization (SAPSO), Ensemble Particle Swarm Optimizer (EPSO), Heterogeneous Comprehensive Learning Particle Swarm Optimization (HCLPSO), L-SHADE, DBL-SHADE, LSHADE-cnEpSin, and LSHADE-SPACMA. Furthermore, the performance of GSAGWO was also compared against the original Grey Wolf Optimizer (GWO) and the Modified Grey Wolf Optimizer (mGWO).

To conduct a comprehensive assessment, all the compared algorithms were rigorously evaluated using the CEC2017 benchmark in its entirety. Additionally, these algorithms were further tested on various engineering optimization problems to gain insights into their performance in practical scenarios. The engineering problems encompassed the design of pressure vessels [68], gear trains [68], tension/compression springs [69], three-bar trusses [70], welded beams [71], and parameter estimation for FM synthesizers [72]. For a detailed description of the engineering problems, please refer to Appendix 1.

To ensure a fair comparison between different metaheuristic algorithms, all experiments were conducted under identical conditions. The metaheuristic algorithms were configured with the same initial parameter values, which are as follows:

- Population Size (N): A population size of 50 individuals was utilized in all experiments.
- Maximum Number of Iterations (t_{max}): The maximum number of iterations was set to 1000 for all algorithms.
- Dimensionality of Test Functions (n): The test functions were implemented in three different dimensions: 30, 50, and 100.
- Total Function Evaluations: The total number of function evaluations was determined based on the dimensionality of the problem. Specifically, 30,000 function evaluations were allowed for 30-dimensional problems, 50,000 for 50-dimensional problems, and 100,000 for 100-dimensional problems.
- Repetition of Experiments: Each experiment was repeated 30 times independently to ensure the consistency and reliability of the results.

By maintaining consistent settings across all experiments, the comparisons between the metaheuristic algorithms were conducted in a controlled and unbiased manner.

The tuning parameters for algorithms SAPSO, EPSO, HCLPSO, L-SHADE, Dbl-SHADE, GWO, and mGWO were determined based on the results provided by the iRice software, which is an automatic parameter configuration tool [73]. iRice is known for efficiently determining the optimal parameter configuration for a specific set of benchmark functions. Hence, we utilized the parameter values recommended by iRice for the aforementioned metaheuristic algorithms.

As for LSHADE-cnEpSin and LSHADE-SPACMA, their performance was evaluated using the CEC2017 benchmark. Consequently, we adopted the parameter values reported by the respective authors, representing the configurations that yielded the best performance for their methods. Detailed information about these parameter settings can be found in Table 5.1.

The performance of each method is evaluated using three different indices: Average Best-so-far (AB), Median Best-so-far (MB), and Standard Deviation (SD) of the best-so-far solutions. These indices are calculated based on the results obtained from 30 independent runs. AB and MB assess the accuracy of the solutions, while SD evaluates the dispersion and robustness of the algorithm. The results for the CEC2017 benchmark functions are presented in the following order: functions in 30 dimensions, 50 dimensions, and 100 dimensions.

5.5.1 Functions from CEC2017 with 30 Dimensions

The experimental results of CEC2017 functions considering 30 dimensions are listed in Table 5.2. Indicators AB, MB, and SD summarize the performance of the results. The best outcomes are highlighted in boldface.

A careful examination of the AB indicator in Table 5.2 reveals that both GSAGWO and DbL-SHADE outperformed the other techniques in a majority of the functions. Specifically, GSAGWO achieved the best AB values in 15 out of 30 functions,

Table 5.1 The parameter settings for the metaheuristic algorithms

Settings configuration	
GSAGWO	a nonlinearly decreased from 3.5 to 1, 2.2 to 0, and 1.1 to 0 for $\mathbf{g}_1$, $\mathbf{g}_2$, and $\mathbf{g}_3$ respectively
[40] SAPSO	$c_1 = 2$, $c_2 = 2$, velocity clamp = 4, minW = 0.35, maxW = 0.95
[24] EPSO	$c_1 = 2$, $c_2 = 2$, minW = 0.25, maxW = 0.95
[25] HCLPSO	$c_1 = 2.5 - 0.5$, $c_2 = 0.5 - 2.5$, $c = 3 - 1.5$, minW = 0.3, maxW = 0.95
[32] L-SHADE	$r^{N^{init}} = 18$, $r^{arc} = 1.5$, $p = 0.11$, $H = 5$
[33] LSHADE-cnEpSin	$\mu F = 0.5$, $\mu CR = 0.5$, $\mu Freq = 0.5$, $ps = 0.5$, $pc = 0.4$, $r^{N^{init}} = 18$, $r^{arc} = 1.4$, $p = 0.11$,$H = 5$
[34] LSHADE-SPACMA	$M_F = 0.5$, $M_{CR} = 0.5$, $M_{FCP} = 0.5$ $r^{arc} = 1.4$, $p = 0.11$,$H = 5$
[35] Dbl-SHADE	$r^{N^{init}} = 18$, $r^{arc} = 1.4$, $p = 0.11$,$H = 5$, $M_F = 0.5$, $M_{CR} = 0.5$
[18] GWO	a linearly decreased from 2.5 to 0
[55] mGWO	a nonlinearly decreased from 2.5 to 0

while DbL-SHADE obtained the best values in 7 out of 30 functions. The remaining functions were dominated by HCLPSO, L-SHADE, and other variants of L-SHADE.

The results presented in Table 5.2 have undergone a statistical analysis to validate their significance. To ensure consistency with other related works, a Wilcoxon test, which is a non-parametric technique, was conducted for this validation. The analysis focused on comparing the performance difference between GSAGWO and the other metaheuristic algorithms. The significance level for the Average Best-so-far (AB) was set at 5% (0.05). The results of the Wilcoxon test, comparing GSAGWO against each of the metaheuristic algorithms, are presented in pairs, such as GSAGWO versus GWO, GSAGWO versus mGWO, GSAGWO versus SAPSO, GSAGWO versus EPSO, GSAGWO versus HCLPSO, GSAGWO versus L-SHADE, GSAGWO vs. LSHADE-cnEpSin, GSAGWO versus LSHADE-SPACMA, and GSAGWO versus DbL-SHADE.

The Wilcoxon test is employed to evaluate the null hypothesis, which states that there is no observable difference between two approaches, against the alternative hypothesis, which suggests a significant difference between the methods being compared. The p-values obtained from the Wilcoxon test indicate whether the null hypothesis can be rejected or accepted. When the p-value is lower than the significance level set for the AB values, the null hypothesis is rejected, indicating a significant difference between the two methods. In such cases, the symbol "+" is used to denote that the null hypothesis is rejected, and our algorithm performs better in terms of AB values compared to its competitor. Conversely, if the null hypothesis is rejected but our algorithm's AB value is worse than that of its competitor, the symbol "−" is used. If the null hypothesis is accepted, indicating no significant difference between the methods, the symbol "=" is used.

Table 5.2 Optimal results achieved by both the metaheuristic algorithms and the discussed technique on the CEC2017 benchmark functions in 30 dimensions

Function		GWO	mGWO	SAPSO	EPSO	HCLPSO	LSHADE	LSHADE cnEpSin	LSHADE SPACMA	DbL SHADE	GSAGWO
$f_1(\mathbf{x})$	AB	2.20E+09	2.37E+09	7.14E+03	3.83E+03	2.97E+03	2.27E+03	1.63E+04	8.68E+03	**2.15E+03**	**2.15E+03**
	MD	1.56E+09	2.07E+09	4.95E+03	2.49E+03	1.92E+03	1.95E+03	**1.59E+03**	2.25E+03	2.10E+03	1.72E+03
	SD	2.02E+09	1.58E+09	7.67E+03	3.88E+03	3.23E+03	1.75E+03	7.51E+04	2.19E+04	3.43E+03	**1.74E+03**
		+	+	+	+	+	+	=	+	=	
$f_2(\mathbf{x})$	AB	4.70E+32	7.22E+31	3.43E+08	2.10E+11	1.13E+14	5.05E+08	3.83E+08	7.14E+08	1.79E+08	**1.21E+08**
	MD	1.02E+27	4.57E+25	6.89E+06	5.89E+08	4.61E+12	**2.52E+06**	1.09E+17	6.74E+22	3.44E+06	8.97E+06
	SD	2.48E+33	3.26E+32	1.18E+09	6.97E+11	2.32E+14	2.10E+09	1.87E+23	1.92E+25	**1.08E+08**	5.02E+08
		+	+	=	+	+	=	+	+	**=**	
$f_3(\mathbf{x})$	AB	5.08E+04	4.69E+04	4.00E+02	1.48E+04	1.56E+04	5.82E+02	2.28E+03	1.86E+04	**3.43E+02**	3.70E+02
	MD	5.10E+04	4.76E+04	3.65E+02	1.33E+04	1.45E+04	3.46E+02	**3.01E+02**	2.58E+03	3.17E+02	7.28E+02
	SD	1.23E+04	1.23E+04	**1.17E+02**	7.01E+03	6.53E+03	7.96E+02	5.43E+03	3.76E+04	3.53E+02	1.86E+02
		+	+	=	+	+	=	+	+	−	
$f_4(\mathbf{x})$	AB	5.83E+02	6.03E+02	4.94E+02	4.84E+02	4.90E+02	4.86E+02	4.99E+02	4.95E+02	4.52E+02	**4.51E+02**
	MD	5.76E+02	5.87E+02	4.90E+02	4.85E+02	4.92E+02	4.86E+02	5.00E+02	4.90E+02	**4.52E+02**	4.65E+02
	SD	4.91E+01	6.10E+01	2.42E+01	2.81E+01	3.04E+01	**1.35E+00**	1.50E+01	1.79E+01	3.32E+00	1.50E+00
		+	+	=	=	=	=	=	=	=	
$f_5(\mathbf{x})$	AB	6.01E+02	6.29E+02	7.70E+02	5.79E+02	5.55E+02	6.07E+02	5.44E+02	5.60E+02	5.64E+02	**5.38E+02**
	MD	5.95E+02	6.06E+02	7.58E+02	5.76E+02	5.55E+02	6.08E+02	5.40E+02	5.61E+02	5.58E+02	**5.22E+02**
	SD	2.41E+01	5.47E+01	5.72E+01	2.38E+01	1.55E+01	1.30E+01	1.49E+01	1.89E+01	2.33E+01	**1.03E+01**
		+	+	+	=	=	+	=	=	=	
$f_6(\mathbf{x})$	AB	6.08E+02	6.09E+02	6.57E+02	6.08E+02	**6.00E+02**	**6.00E+02**	6.05E+02	6.04E+02	6.06E+02	**6.00E+02**

(continued)

Table 5.2 (continued)

Function		GWO	mGWO	SAPSO	EPSO	HCLPSO	LSHADE	LSHADE cnEpSin	LSHADE SPACMA	DbL SHADE	GSAGWO
	MD	6.07E+02	6.09E+02	6.58E+02	6.08E+02	**6.00E+02**	**6.00E+02**	6.04E+02	6.03E+02	6.05E+02	**6.00E+02**
	SD	4.67E+00	1.97E+00	8.97E+00	3.38E+00	5.26E−01	7.20E−03	3.61E+00	3.69E+00	4.35E+00	**1.24E−03**
		=	=	+	=	=	=	=	=	=	
$f_7(\mathbf{x})$	AB	8.89E+02	9.01E+02	1.28E+03	8.51E+02	7.84E+02	8.47E+02	7.80E+02	7.92E+02	7.55E+02	**7.11E+02**
	MD	8.73E+02	8.89E+02	1.26E+03	8.50E+02	7.81E+02	8.47E+02	7.78E+02	7.87E+02	7.36E+02	**7.01E+02**
	SD	4.80E+01	5.29E+01	1.45E+02	3.85E+01	1.59E+01	1.20E+01	1.35E+01	2.47E+01	1.30E+01	**5.71E+00**
		+	+	+	+	+	+	+	+	=	
$f_8(\mathbf{x})$	AB	8.95E+02	9.01E+02	1.00E+03	8.73E+02	8.54E+02	9.08E+02	**8.38E+02**	8.54E+02	8.53E+02	8.53E+02
	MD	8.86E+02	8.95E+02	1.00E+03	8.71E+02	8.51E+02	9.06E+02	**8.36E+02**	8.53E+02	8.42E+02	8.53E+02
	SD	3.91E+01	3.17E+01	3.26E+01	2.45E+01	1.90E+01	**1.20E+01**	1.25E+01	1.85E+01	1.36E+01	1.83E+01
		=	+	+	=	=	+	−	=	−	
$f_9(\mathbf{x})$	AB	2.06E+03	1.74E+03	5.86E+03	1.42E+03	1.01E+03	9.02E+02	1.02E+03	1.14E+03	**9.01E+02**	9.20E+02
	MD	1.73E+03	1.65E+03	5.77E+03	1.30E+03	9.45E+02	**9.01E+02**	1.02E+03	1.12E+03	**9.01E+02**	9.68E+02
	SD	8.60E+02	6.13E+02	1.15E+03	5.13E+02	2.86E+02	**2.56E−01**	4.30E+01	1.06E+02	1.18E+00	1.79E+01
		+	+	+	+	+	−	+	+	−	
$f_{10}(\mathbf{x})$	AB	4.64E+03	5.98E+03	4.85E+03	4.36E+03	4.18E+03	6.14E+03	4.38E+03	4.37E+03	4.41E+03	**3.67E+03**
	MD	4.40E+03	5.20E+03	4.92E+03	4.24E+03	4.37E+03	6.06E+03	4.17E+03	4.29E+03	4.37E+03	**3.79E+03**
	SD	1.27E+03	1.83E+03	5.00E+02	5.30E+02	5.42E+02	**4.14E+02**	7.83E+02	6.47E+02	6.65E+02	5.53E+02
		+	+	+	+	+	+	+	+	+	
$f_{11}(\mathbf{x})$	AB	1.89E+03	1.78E+03	1.25E+03	1.22E+03	1.21E+03	1.17E+03	1.23E+03	1.24E+03	1.29E+03	**1.13E+03**
	MD	1.56E+03	1.47E+03	1.24E+03	1.22E+03	1.20E+03	1.18E+03	1.23E+03	1.24E+03	1.26E+03	**1.11E+03**

(continued)

Table 5.2 (continued)

Function		GWO	mGWO	SAPSO	EPSO	HCLPSO	LSHADE	LSHADE cnEpSin	LSHADE SPACMA	DbL SHADE	GSAGWO
	SD	7.44E+02	7.28E+02	5.52E+01	4.17E+01	4.34E+01	3.33E+01	3.86E+01	4.63E+01	1.24E+02	**2.12E+01**
		+	+	=	=	=	=	=	=	=	
$f_{12}(\mathbf{x})$	AB	7.01E+07	1.05E+08	7.33E+05	2.28E+05	3.90E+05	9.87E+03	9.89E+03	**9.73E+03**	9.98E+03	4.17E+04
	MD	4.82E+07	8.32E+07	5.84E+05	1.57E+05	3.06E+05	8.60E+03	**8.49E+03**	8.57E+03	**8.49E+03**	2.66E+04
	SD	7.24E+07	9.18E+07	5.74E+05	1.82E+05	3.63E+05	4.99E+03	4.99E+03	**4.22E+03**	6.36E+03	4.78E+04
		+	+	+	+	+	−	−	−	−	
$f_{13}(\mathbf{x})$	AB	4.39E+06	6.61E+06	1.58E+04	1.26E+04	1.24E+04	1.59E+03	1.48E+04	1.23E+04	1.46E+03	**1.43E+03**
	MD	1.18E+05	2.51E+05	1.39E+04	8.85E+03	9.19E+03	1.58E+03	1.08E+04	1.08E+04	2.11E+03	**1.50E+03**
	SD	1.38E+07	2.26E+07	1.25E+04	1.06E+04	9.05E+03	7.88E+01	1.28E+04	9.35E+03	3.94E+02	**6.40E+01**
		+	+	+	+	+	=	+	+	=	
$f_{14}(\mathbf{x})$	AB	7.00E+05	2.72E+05	2.06E+04	2.33E+04	2.20E+04	1.77E+03	1.68E+03	2.17E+03	**1.45E+03**	1.48E+03
	MD	2.75E+05	7.71E+04	1.43E+04	2.00E+04	1.45E+04	1.68E+03	1.65E+03	1.92E+03	**1.35E+03**	1.57E+03
	SD	1.01E+06	3.61E+05	1.55E+04	1.54E+04	1.94E+04	2.74E+02	2.15E+02	1.04E+03	**7.56E+00**	8.07E+02
		+	+	+	+	+	=	=	+	−	
$f_{15}(\mathbf{x})$	AB	1.93E+06	4.47E+05	9.43E+03	8.71E+03	4.89E+03	1.60E+03	1.61E+03	1.60E+03	1.61E+03	**1.59E+03**
	MD	8.75E+04	1.06E+05	5.61E+03	5.59E+03	4.11E+03	**1.60E+03**	4.13E+03	3.79E+03	4.71E+03	1.62E+03
	SD	6.80E+06	1.06E+06	8.61E+03	6.73E+03	2.95E+03	3.84E+01	7.33E+03	8.55E+03	6.86E+03	**3.33E+01**
		+	+	+	+	+	=	=	=	=	
$f_{16}(\mathbf{x})$	AB	2.61E+03	2.52E+03	2.96E+03	2.50E+03	2.48E+03	2.54E+03	2.42E+03	2.72E+03	2.70E+03	**2.41E+03**
	MD	2.60E+03	2.47E+03	2.97E+03	2.48E+03	2.43E+03	2.57E+03	**2.38E+03**	2.74E+03	2.67E+03	2.43E+03
	SD	3.42E+02	3.47E+02	2.47E+02	2.51E+02	2.81E+02	**2.13E+02**	2.31E+02	3.06E+02	3.24E+02	2.45E+02

(continued)

Table 5.2 (continued)

Function		GWO	mGWO	SAPSO	EPSO	HCLPSO	LSHADE	LSHADE cnEpSin	LSHADE SPACMA	DbL SHADE	GSAGWO
		=	=	+	=	=	=	=	=	=	
$f_{17}(\mathbf{x})$	AB	2.02E+03	2.05E+03	2.54E+03	2.11E+03	1.98E+03	1.97E+03	1.97E+03	2.22E+03	1.92E+03	**1.91E+03**
	MD	1.99E+03	2.00E+03	2.53E+03	2.10E+03	1.92E+03	1.96E+03	1.96E+03	2.24E+03	**1.90E+03**	2.01E+03
	SD	1.57E+02	1.32E+02	3.32E+02	2.11E+02	1.51E+02	8.37E+01	1.88E+02	1.78E+02	7.56E+01	**2.23E+01**
		+	+	+	+	=	=	=	+	=	
$f_{18}(\mathbf{x})$	AB	1.88E+06	1.54E+06	2.08E+05	2.13E+05	1.58E+05	1.93E+03	1.92E+03	1.99E+03	**1.87E+03**	2.07E+03
	MD	9.90E+05	1.06E+06	1.49E+05	1.32E+05	1.21E+05	1.91E+03	1.91E+03	3.00E+04	**1.87E+03**	1.88E+03
	SD	2.29E+06	1.66E+06	1.56E+05	2.00E+05	1.26E+05	4.50E+01	1.83E+02	1.37E+02	**2.06E+01**	2.17E+02
		+	+	+	+	+	–	–	–	–	
$f_{19}(\mathbf{x})$	AB	2.33E+06	5.63E+06	1.05E+04	9.77E+03	1.01E+04	1.94E+03	1.93E+03	1.98E+03	1.92E+03	**1.91E+03**
	MD	6.32E+05	7.89E+05	7.10E+03	5.45E+03	7.15E+03	1.94E+03	1.92E+03	1.97E+03	**1.92E+03**	1.93E+03
	SD	7.27E+06	2.47E+07	1.07E+04	9.55E+03	7.95E+03	1.02E+01	7.29E+01	8.09E+01	**3.67E+00**	1.10E+01
		+	+	+	+	+	=	=	=	=	
$f_{20}(\mathbf{x})$	AB	2.48E+03	2.42E+03	2.74E+03	2.49E+03	2.37E+03	2.38E+03	2.38E+03	2.48E+03	**2.24E+03**	2.74E+03
	MD	2.43E+03	2.38E+03	2.70E+03	2.49E+03	2.31E+03	2.38E+03	2.31E+03	2.45E+03	**2.25E+03**	2.77E+03
	SD	2.06E+02	1.60E+02	2.17E+02	1.91E+02	1.57E+02	9.28E+01	2.01E+02	2.03E+02	**8.68E+01**	2.98E+02
		–	–	=	–	–	–	–	–	–	
$f_{21}(\mathbf{x})$	AB	2.39E+03	2.41E+03	2.58E+03	2.38E+03	**2.35E+03**	2.40E+03	2.38E+03	**2.35E+03**	**2.35E+03**	2.43E+03
	MD	2.39E+03	2.40E+03	2.57E+03	2.38E+03	2.35E+03	2.40E+03	**2.34E+03**	2.35E+03	2.35E+03	2.41E+03
	SD	2.04E+01	3.92E+01	7.46E+01	2.32E+01	1.65E+01	1.22E+01	1.81E+01	2.26E+01	**9.81E+00**	4.61E+01
		–	–	=	–	–	–	–	–	–	

(continued)

Table 5.2 (continued)

Function		GWO	mGWO	SAPSO	EPSO	HCLPSO	LSHADE	LSHADE cnEpSin	LSHADE SPACMA	DbL SHADE	GSAGWO
$f_{22}(\mathbf{x})$	AB	4.67E+03	5.32E+03	6.46E+03	2.57E+03	2.55E+03	2.30E+03	2.30E+03	2.31E+03	2.30E+03	**2.25E+03**
	MD	4.91E+03	5.20E+03	6.68E+03	2.30E+03	2.30E+03	2.30E+03	2.30E+03	2.30E+03	2.30E+03	**2.25E+03**
	SD	2.12E+03	2.36E+03	8.97E+02	8.30E+02	9.44E+02	4.49E−01	1.93E+00	6.39E+00	**2.60E−04**	3.63E−01
		+	+	+	+	+	=	=	=	=	
$f_{23}(\mathbf{x})$	AB	2.77E+03	2.80E+03	3.43E+03	2.77E+03	2.73E+03	2.76E+03	**2.70E+03**	2.71E+03	**2.70E+03**	2.78E+03
	MD	2.76E+03	2.79E+03	3.37E+03	2.76E+03	2.73E+03	2.76E+03	2.71E+03	**2.70E+03**	**2.70E+03**	2.77E+03
	SD	4.94E+01	5.65E+01	1.80E+02	3.65E+01	2.57E+01	1.37E+01	2.35E+01	2.19E+01	**9.79E+00**	4.22E+01
		−	=	+	−	−	−	−	−	−	
$f_{24}(\mathbf{x})$	AB	2.94E+03	2.96E+03	3.42E+03	2.93E+03	2.90E+03	2.93E+03	2.87E+03	2.89E+03	2.89E+03	**2.63E+03**
	MD	2.92E+03	2.93E+03	3.40E+03	2.92E+03	2.89E+03	2.93E+03	2.87E+03	2.89E+03	2.88E+03	**2.62E+03**
	SD	5.24E+01	7.00E+01	1.49E+02	3.61E+01	3.06E+01	**1.30E+01**	1.55E+01	1.79E+01	2.66E+01	1.58E+01
		=	=	+	=	=	=	=	=	=	
$f_{25}(\mathbf{x})$	AB	2.99E+03	2.98E+03	2.90E+03	2.90E+03	2.89E+03	2.89E+03	2.89E+03	2.89E+03	2.90E+03	**2.66E+03**
	MD	3.00E+03	2.98E+03	2.89E+03	2.89E+03	2.89E+03	2.89E+03	2.89E+03	2.89E+03	2.89E+03	**2.68E+03**
	SD	4.09E+01	2.84E+01	1.68E+01	1.51E+01	1.22E+01	4.56E−01	8.80E+00	9.29E+00	1.30E+01	**3.59E−01**
		=	=	=	=	=	=	=	=	=	
$f_{26}(\mathbf{x})$	AB	4.74E+03	4.81E+03	7.99E+03	4.82E+03	**3.87E+03**	4.56E+03	4.18E+03	4.27E+03	4.32E+03	5.18E+03
	MD	4.72E+03	4.69E+03	8.24E+03	5.03E+03	**4.11E+03**	4.60E+03	**4.11E+03**	4.26E+03	4.26E+03	5.06E+03
	SD	3.37E+02	3.68E+02	2.45E+03	1.16E+03	8.27E+02	**1.50E+02**	2.27E+02	2.27E+02	3.35E+02	7.78E+02
		−	−	+	−	−	−	−	−	−	
$f_{27}(\mathbf{x})$	AB	3.25E+03	3.25E+03	3.75E+03	3.28E+03	3.24E+03	3.30E+03	**3.23E+03**	3.24E+03	3.25E+03	**3.23E+03**

(continued)

Table 5.2 (continued)

Function		GWO	mGWO	SAPSO	EPSO	HCLPSO	LSHADE	LSHADE cnEpSin	LSHADE SPACMA	DbL SHADE	GSAGWO
	MD	3.25E+03	3.25E+03	3.72E+03	3.27E+03	3.25E+03	3.30E+03	3.23E+03	3.24E+03	3.24E+03	**3.22E+03**
	SD	2.60E+01	1.93E+01	2.83E+02	2.75E+01	1.41E+01	**6.40E+00**	1.15E+01	1.58E+01	1.26E+01	2.76E+01
		=	=	+	=	=	=	=	=	=	
$f_{28}(\mathbf{x})$	AB	3.43E+03	3.41E+03	3.22E+03	3.21E+03	3.22E+03	3.22E+03	3.20E+03	3.23E+03	**3.16E+03**	3.20E+03
	MD	3.42E+03	3.39E+03	3.22E+03	3.21E+03	3.22E+03	3.22E+03	3.21E+03	3.23E+03	**3.17E+03**	3.19E+03
	SD	7.58E+01	7.51E+01	2.77E+01	2.72E+01	1.66E+01	2.73E+01	4.92E+01	1.74E+01	4.76E+01	**1.54E+00**
		=	=	=	=	=	=	=	=	−	
$f_{29}(\mathbf{x})$	AB	3.86E+03	3.83E+03	4.32E+03	3.80E+03	3.65E+03	3.61E+03	3.67E+03	3.71E+03	3.46E+03	**3.36E+03**
	MD	3.83E+03	3.83E+03	4.36E+03	3.81E+03	3.63E+03	3.62E+03	3.69E+03	3.68E+03	3.46E+03	**3.44E+03**
	SD	2.07E+02	1.90E+02	2.63E+02	2.06E+02	1.63E+02	5.60E+01	1.84E+02	1.83E+02	**2.57E+01**	2.72E+01
		=	=	+	=	=	=	=	=	=	
$f_{30}(\mathbf{x})$	AB	9.13E+06	7.30E+06	2.36E+04	8.35E+03	1.01E+04	5.75E+03	1.82E+04	1.01E+04	**5.38E+03**	7.38E+04
	MD	7.70E+06	6.70E+06	2.02E+04	7.83E+03	9.82E+03	5.77E+03	1.28E+04	7.70E+03	**5.23E+03**	5.55E+04
	SD	6.01E+06	4.58E+06	1.10E+04	1.70E+03	2.84E+03	3.27E+02	1.86E+04	5.58E+03	**3.13E+02**	7.47E+04
		+	+	−	−	−	−	−	−	−	
Total	+	18	19	21	14	13	5	6	9	1	
	−	4	3	1	5	5	8	8	7	12	
	=	8	8	8	11	12	17	16	14	17	

After applying the Wilcoxon test to the AB data from Table 5.2, the results are reported within the same table, along with a summary at the end. In conclusion, the experimental results in 30 dimensions demonstrate that our method outperforms most of its competitors in the test functions. The results also support the hypothesis that our discussed approach performs better than both the original GWO and its modified version.

5.5.2 *Functions from CEC2017 with 50 Dimensions*

Table 5.3 showcases the experimental results for 50 dimensions. It provides a comparison between GSAGWO and other metaheuristic algorithms. The table includes indicators such as AB, MB, and SD, which reflect the accuracy and dispersion of the results. The best results are indicated in boldface.

Looking at Table 5.3, it can be observed that L-SHADE performs better in 10 out of 30 functions according to the AB index, while GSAGWO achieves superior results in 13 out of 30 functions. Other functions show competitive performance by GSAGWO, L-SHADE, and its variants. This indicates that the discussed method is competitive and demonstrates its superiority.

Furthermore, the results from Table 5.3 confirm that our discussed approach outperforms both GWO and its modified version in all test functions. This validates our hypothesis and establishes the discussed method as an improved version of GWO and its modified counterpart considered in this study.

The Wilcoxon test was also conducted on the results obtained from Table 5.3. The results of the test, presented in the same table, indicate that the null hypothesis has been accepted for DbL-SHADE in 17 functions and for the other L-SHADE variants in 18 functions. This suggests that there is no significant difference between our discussed approach and these methods. However, the null hypothesis has been rejected for most of the functions when comparing with the remaining algorithms.

From the rejected null hypotheses, it can be observed that DbL-SHADE achieves the best performance in 12 functions, while the other L-SHADE variants exhibit superior results in 6 functions. This analysis confirms that DbL-SHADE is the most competitive algorithm in terms of 50 dimensions. However, the discussed method surpasses DbL-SHADE in 13 functions, indicating its effectiveness. The overall performance of the other algorithms can be summarized in the table, where the symbol "=" denotes the number of functions where the null hypothesis was accepted, and the symbol "+" represents the number of functions where our algorithm outperformed the compared algorithm, while the symbol "−" indicates the number of functions where the compared algorithm achieved better results than our method.

Based on the results of the Wilcoxon test, the experimental outcomes in 50 dimensions demonstrate that our method generally outperforms most of the compared algorithms in the majority of test functions. It is worth noting that DbL-SHADE emerges as the most competitive algorithm. However, it is important to emphasize that our method was specifically designed to surpass GWO and its variants.

Table 5.3 Optimal results achieved by both the metaheuristic algorithms and the discussed technique on the CEC2017 benchmark functions in 50 dimensions

Function		GWO	mGWO	SAPSO	EPSO	HCLPSO	LSHADE	LSHADE cnEpSin	LSHADE SPACMA	DbL SHADE	GSAGWO
$f_1(\mathbf{x})$	AB	6.77E+09	8.89E+09	2.69E+03	3.36E+03	1.43E+02	1.04E+02	1.38E+02	1.48E+02	1.03E+02	**1.02E+02**
	MD	6.67E+09	7.52E+09	1.08E+03	2.27E+03	1.33E+02	1.04E+02	1.38E+02	1.78E+03	1.03E+02	**1.02E+02**
	SD	3.51E+09	4.30E+09	3.43E+03	3.66E+03	1.04E+01	**5.99E−02**	4.77E−01	4.31E+00	7.55E−01	2.14E−01
		+	+	+	+	=	=	=	+	=	
$f_2(\mathbf{x})$	AB	1.65E+53	5.53E+53	3.09E+06	7.82E+10	9.63E+07	7.83E+02	4.83E+02	4.72E+02	3.88E+02	**3.75E+02**
	MD	2.78E+48	2.97E+49	1.37E+04	2.12E+06	6.16E+07	3.20E+02	3.17E+02	4.65E+02	**3.15E+02**	3.18E+02
	SD	8.87E+53	1.89E+54	9.10E+06	3.22E+11	2.06E+06	7.21E+02	2.21E+03	6.21E+02	3.58E+02	**3.57E+02**
		+	+	+	+	+	+	+	+	=	
$f_3(\mathbf{x})$	AB	1.10E+05	1.03E+05	4.51E+02	3.78E+03	4.85E+03	3.51E+02	3.50E+02	3.51E+02	3.49E+02	**3.41E+02**
	MD	1.09E+05	1.01E+05	3.74E+02	3.38E+03	5.37E+03	3.51E+02	3.52E+02	3.80E+02	3.48E+02	**3.42E+02**
	SD	1.94E+04	2.07E+04	5.83E+02	1.75E+03	4.08E+03	**2.33E+00**	4.93E+01	3.29E+02	2.66E+01	1.59E+01
		+	+	+	+	+	=	=	=	=	
$f_4(\mathbf{x})$	AB	1.14E+03	1.18E+03	4.81E+02	4.70E+02	4.96E+02	4.70E+02	4.61E+02	4.84E+02	**4.45E+02**	**4.45E+02**
	MD	1.05E+03	1.12E+03	4.88E+02	4.71E+02	5.09E+02	4.67E+02	4.65E+02	5.00E+02	**4.44E+02**	4.45E+02
	SD	3.51E+02	3.36E+02	3.07E+01	3.47E+01	2.45E+01	1.17E+01	1.14E+01	4.24E+01	4.67E+01	**8.11E+00**
		+	+	+	=	=	=	=	=	−	
$f_5(\mathbf{x})$	AB	7.28E+02	7.34E+02	7.51E+02	5.75E+02	5.50E+02	5.56E+02	5.45E+02	5.41E+02	5.32E+02	**5.29E+02**
	MD	7.28E+02	7.37E+02	7.37E+02	5.71E+02	5.50E+02	5.57E+02	5.45E+02	5.40E+02	5.32E+02	**5.25E+02**
	SD	3.30E+01	5.37E+01	5.94E+01	2.02E+01	1.49E+01	9.56E+00	4.37E+01	2.60E+01	3.07E+01	**3.03E+00**
		+	+	+	+	=	=	=	=	=	
$f_6(\mathbf{x})$	AB	6.18E+02	6.18E+02	6.61E+02	6.05E+02	6.01E+02	6.02E+02	6.01E+02	6.01E+02	**6.00E+02**	**6.00E+02**

(continued)

Table 5.3 (continued)

Function		GWO	mGWO	SAPSO	EPSO	HCLPSO	LSHADE	LSHADE cnEpSin	LSHADE SPACMA	DbL SHADE	GSAGWO
	MD	6.17E+02	6.18E+02	6.61E+02	6.04E+02	6.01E+02	6.02E+02	6.02E+02	6.02E+02	**6.00E+02**	**6.00E+02**
	SD	5.03E+00	4.23E+00	7.91E+00	3.38E+00	4.44E−02	5.60E−02	5.95E+00	3.00E+00	4.45E−02	**1.28E−02**
		=	=	=	=	=	=	=	=	=	
$f_7(\mathbf{x})$	AB	1.09E+03	1.08E+03	1.29E+03	8.29E+02	7.86E+02	7.92E+02	7.88E+02	7.49E+02	7.27E+02	**7.23E+02**
	MD	1.07E+03	1.06E+03	1.31E+03	8.28E+02	7.82E+02	7.90E+02	7.88E+02	7.45E+02	7.27E+02	**7.25E+02**
	SD	7.80E+01	7.91E+01	1.51E+02	2.87E+01	1.81E+01	8.45E+00	3.11E+01	2.09E+01	2.88E+01	**3.38E+00**
		+	+	+	+	+	+	+	=	=	
$f_8(\mathbf{x})$	AB	1.02E+03	1.06E+03	1.00E+03	8.70E+02	8.54E+02	9.02E+02	8.58E+02	8.73E+02	8.56E+02	**8.53E+02**
	MD	1.03E+03	1.03E+03	1.00E+03	8.67E+02	**8.51E+02**	8.93E+02	8.93E+02	8.73E+02	8.54E+02	9.09E+02
	SD	3.85E+01	7.50E+01	4.34E+01	2.07E+01	1.74E+01	3.67E+01	3.67E+01	1.94E+01	**7.13E+00**	2.63E+01
		+	+	+	=	=	+	=	=	=	
$f_9(\mathbf{x})$	AB	8.07E+03	8.62E+03	6.32E+03	1.34E+03	9.68E+02	9.40E+02	**9.00E+02**	9.38E+02	**9.00E+02**	9.53E+02
	MD	7.99E+03	8.30E+03	6.16E+03	1.23E+03	9.17E+02	9.30E+02	**9.00E+02**	9.37E+02	**9.00E+02**	9.63E+02
	SD	3.24E+03	4.23E+03	1.78E+03	3.34E+02	1.50E+02	**2.27E−02**	4.47E−02	1.64E+01	3.79E−02	1.15E+00
		+	+	+	+	=	−	−	−	−	
$f_{10}(\mathbf{x})$	AB	7.22E+03	7.88E+03	5.01E+03	4.15E+03	4.27E+03	4.97E+03	4.39E+03	4.40E+03	**3.77E+03**	3.88E+03
	MD	7.07E+03	7.30E+03	5.08E+03	4.19E+03	4.36E+03	4.97E+03	3.90E+03	4.39E+03	3.75E+03	**3.63E+03**
	SD	1.30E+03	2.27E+03	6.71E+02	5.16E+02	5.57E+02	4.53E+02	4.02E+02	8.11E+02	**1.35E+02**	3.74E+02
		+	+	+	=	+	+	+	+	−	
$f_{11}(\mathbf{x})$	AB	4.94E+03	4.71E+03	1.24E+03	1.22E+03	1.19E+03	1.14E+03	1.57E+03	1.28E+03	1.14E+03	**1.13E+03**
	MD	4.64E+03	4.65E+03	1.25E+03	1.22E+03	1.18E+03	**1.13E+03**	1.46E+03	1.27E+03	1.14E+03	**1.13E+03**

(continued)

Table 5.3 (continued)

Function		GWO	mGWO	SAPSO	EPSO	HCLPSO	LSHADE	LSHADE cnEpSin	LSHADE SPACMA	DbL SHADE	GSAGWO
	SD	2.15E+03	1.79E+03	4.90E+01	3.94E+01	3.33E+01	3.19E+01	3.12E+02	4.45E+01	3.21E+01	**2.28E+01**
		+	+	=	=	=	=	+	=	=	
$f_{12}(\mathbf{x})$	AB	1.61E+09	9.23E+08	4.59E+05	7.36E+04	2.66E+05	3.78E+04	2.18E+04	2.64E+04	**1.93E+04**	3.65E+04
	MD	8.74E+08	7.00E+08	3.49E+05	5.49E+04	1.94E+05	3.30E+04	3.93E+04	2.26E+04	**1.12E+04**	2.09E+04
	SD	1.75E+09	7.13E+08	3.48E+05	5.50E+04	2.09E+05	1.76E+04	3.75E+03	3.62E+03	**1.89E+03**	4.30E+04
		+	+	+	+	+	=	−	−	−	
$f_{13}(\mathbf{x})$	AB	1.57E+08	1.61E+08	1.58E+04	1.49E+04	1.20E+04	1.48E+03	1.47E+03	1.48E+04	**1.43E+03**	1.44E+03
	MD	1.24E+08	1.26E+08	9.10E+03	1.17E+04	9.94E+03	1.48E+03	1.47E+03	1.48E+03	**1.46E+03**	1.47E+03
	SD	1.54E+08	1.66E+08	1.65E+04	1.22E+04	8.40E+03	6.08E+01	5.74E+01	6.27E+01	**2.20E+01**	4.19E+01
		+	+	+	+	+	=	=	=	−	
$f_{14}(\mathbf{x})$	AB	7.51E+05	1.12E+06	2.32E+04	1.37E+04	2.09E+04	1.46E+03	1.45E+03	1.47E+03	**1.44E+03**	**1.44E+03**
	MD	4.87E+05	8.36E+05	2.03E+04	1.12E+04	1.23E+04	1.45E+03	1.45E+03	1.46E+03	1.44E+03	**1.42E+03**
	SD	7.14E+05	9.19E+05	1.77E+04	1.07E+04	2.02E+04	4.84E+00	4.47E+00	5.39E+00	**4.28E+00**	5.62E+00
		+	+	+	+	+	=	=	=	=	
$f_{15}(\mathbf{x})$	AB	1.31E+07	1.28E+07	1.37E+04	5.06E+03	6.27E+03	1.60E+03	1.60E+03	1.62E+03	**1.58E+03**	1.59E+03
	MD	8.96E+04	1.88E+05	1.00E+04	3.41E+03	4.69E+03	1.60E+03	1.60E+03	1.61E+03	**1.58E+03**	1.59E+03
	SD	2.36E+07	2.49E+07	1.15E+04	5.12E+03	5.24E+03	1.87E+01	1.56E+02	3.17E+01	**2.52E+00**	5.05E+00
		+	+	+	+	+	=	=	=	−	
$f_{16}(\mathbf{x})$	AB	3.20E+03	3.27E+03	2.84E+03	2.46E+03	2.33E+03	2.20E+03	2.18E+03	2.19E+03	1.85E+03	**1.79E+03**
	MD	3.18E+03	3.11E+03	2.91E+03	2.44E+03	2.29E+03	2.18E+03	2.16E+03	2.19E+03	1.82E+03	**1.80E+03**
	SD	3.49E+02	5.54E+02	3.77E+02	2.55E+02	2.70E+02	1.88E+02	1.81E+02	1.79E+02	2.27E+02	**1.72E+02**

(continued)

Table 5.3 (continued)

Function		GWO	mGWO	SAPSO	EPSO	HCLPSO	LSHADE	LSHADE cnEpSin	LSHADE SPACMA	DbL SHADE	GSAGWO
		+	+	+	+	+	+	=	=	=	
$f_{17}(\mathbf{x})$	AB	2.84E+03	2.93E+03	2.46E+03	2.07E+03	1.96E+03	1.89E+03	1.84E+03	1.87E+03	1.83E+03	**1.81E+03**
	MD	2.86E+03	2.97E+03	2.48E+03	2.08E+03	1.96E+03	1.86E+03	1.99E+03	1.83E+03	1.83E+03	**1.82E+03**
	SD	3.03E+02	2.85E+02	3.29E+02	1.64E+02	1.52E+02	6.71E+01	5.24E+01	3.21E+01	**3.27E+00**	2.18E+01
		+	+	+	+	+	=	=	=	=	
$f_{18}(\mathbf{x})$	AB	5.54E+06	8.87E+06	2.17E+05	1.55E+05	1.57E+05	7.18E+03	2.98E+03	6.63E+04	**1.87E+03**	1.97E+03
	MD	4.06E+06	4.00E+06	1.65E+05	1.01E+05	1.21E+05	5.76E+03	2.71E+03	3.72E+04	**1.86E+03**	1.96E+03
	SD	5.19E+06	1.43E+07	1.79E+05	1.66E+05	1.06E+05	5.95E+03	3.77E+02	7.39E+04	**4.17E+01**	1.44E+02
		+	+	+	+	+	+	=	+	–	
$f_{19}(\mathbf{x})$	AB	3.77E+06	6.11E+06	9.09E+03	1.02E+04	7.11E+03	1.93E+03	**1.92E+03**	1.93E+03	**1.92E+03**	**1.92E+03**
	MD	1.34E+06	2.18E+06	6.84E+03	4.96E+03	3.27E+03	1.93E+03	**1.92E+03**	1.57E+04	1.93E+03	**1.92E+03**
	SD	7.74E+06	1.14E+07	7.19E+03	9.66E+03	7.47E+03	9.06E+00	**2.94E+00**	6.42E+03	4.06E+00	7.81E+00
		+	+	+	+	+	=	=	=	=	
$f_{20}(\mathbf{x})$	AB	3.07E+03	3.02E+03	2.82E+03	2.40E+03	2.30E+03	**2.24E+03**	3.11E+03	2.90E+03	3.08E+03	2.26E+03
	MD	2.92E+03	2.93E+03	2.91E+03	2.43E+03	2.29E+03	**2.21E+03**	3.15E+03	2.91E+03	2.99E+03	2.23E+03
	SD	3.89E+02	4.03E+02	2.56E+02	1.49E+02	1.16E+02	8.10E+01	3.06E+02	3.09E+02	3.17E+02	**1.83E+01**
		+	+	+	=	=	–	+	+	+	
$f_{21}(\mathbf{x})$	AB	2.50E+03	2.54E+03	2.57E+03	2.37E+03	**2.35E+03**	2.36E+03	2.42E+03	2.37E+03	2.38E+03	2.41E+03
	MD	2.50E+03	2.53E+03	2.57E+03	2.37E+03	**2.35E+03**	2.36E+03	2.42E+03	2.37E+03	2.38E+03	2.41E+03
	SD	3.49E+01	6.64E+01	6.12E+01	1.85E+01	2.10E+01	8.69E+00	5.44E+01	1.73E+01	2.42E+01	2.28E+01
		=	=	=	–	–	–	=	–	–	

(continued)

Table 5.3 (continued)

Function		GWO	mGWO	SAPSO	EPSO	HCLPSO	LSHADE	LSHADE cnEpSin	LSHADE SPACMA	DbL SHADE	GSAGWO
$f_{22}(\mathbf{x})$	AB	9.43E+03	1.02E+04	6.41E+03	**2.30E+03**	2.39E+03	**2.30E+03**	**2.30E+03**	**2.30E+03**	**2.30E+03**	2.31E+03
	MD	8.87E+03	9.78E+03	6.63E+03	**2.30E+03**	2.31E+03	**2.30E+03**	**2.30E+03**	**2.30E+03**	**2.30E+03**	2.31E+03
	SD	2.18E+03	2.45E+03	1.57E+03	1.35E+01	4.84E+02	6.49E−08	3.48E−08	7.43E−09	**1.24E−09**	3.79E−03
		+	+	+	−	=	−	−	−	−	
$f_{23}(\mathbf{x})$	AB	2.98E+03	3.01E+03	3.48E+03	2.75E+03	2.72E+03	2.71E+03	2.68E+03	2.70E+03	2.61E+03	**2.60E+03**
	MD	2.98E+03	2.99E+03	3.44E+03	2.75E+03	2.72E+03	2.70E+03	2.67E+03	2.68E+03	**2.61E+03**	**2.61E+03**
	SD	5.90E+01	9.37E+01	2.47E+02	3.37E+01	**2.41E+01**	5.08E+01	3.78E+01	2.68E+01	3.58E+01	4.24E+01
		=	+	+	=	=	=	=	=	=	
$f_{24}(\mathbf{x})$	AB	3.15E+03	3.19E+03	3.47E+03	2.92E+03	2.88E+03	2.88E+03	2.87E+03	2.86E+03	2.81E+03	**2.80E+03**
	MD	3.13E+03	3.15E+03	3.47E+03	2.91E+03	2.88E+03	2.88E+03	2.86E+03	2.86E+03	2.81E+03	**2.80E+03**
	SD	9.46E+01	1.22E+02	2.25E+02	3.10E+01	1.87E+01	1.15E+01	4.07E+01	2.28E+01	3.07E+01	**3.45E+00**
		+	+	+	=	=	=	=	=	=	
$f_{25}(\mathbf{x})$	AB	3.54E+03	3.64E+03	2.90E+03	2.90E+03	2.89E+03	2.89E+03	2.78E+03	2.76E+03	**2.63E+03**	**2.63E+03**
	MD	3.45E+03	3.59E+03	2.89E+03	2.89E+03	2.89E+03	2.89E+03	2.76E+03	2.74E+03	2.63E+03	**2.62E+03**
	SD	3.63E+02	2.14E+02	1.91E+01	1.47E+01	1.04E+01	7.05E−02	3.95E−02	2.15E−02	**2.29E−02**	3.33E−01
		+	+	+	+	+	+	=	=	=	
$f_{26}(\mathbf{x})$	AB	6.21E+03	6.52E+03	6.92E+03	3.97E+03	**3.78E+03**	4.00E+03	4.07E+03	4.71E+03	3.90E+03	4.82E+03
	MD	6.14E+03	6.38E+03	7.81E+03	4.37E+03	4.11E+03	3.99E+03	4.05E+03	4.67E+03	**3.91E+03**	4.87E+03
	SD	4.54E+02	6.11E+02	3.00E+03	9.57E+02	7.97E+02	1.12E+02	1.81E+02	2.94E+01	**2.42E+01**	7.69E+02
		+	+	+	−	−	−	−	−	−	
$f_{27}(\mathbf{x})$	AB	3.60E+03	3.60E+03	3.80E+03	3.26E+03	3.24E+03	**3.21E+03**	**3.21E+03**	3.26E+03	**3.21E+03**	3.27E+03

(continued)

Table 5.3 (continued)

Function		GWO	mGWO	SAPSO	EPSO	HCLPSO	LSHADE	LSHADE cnEpSin	LSHADE SPACMA	DbL SHADE	GSAGWO
	MD	3.60E+03	3.60E+03	3.71E+03	3.25E+03	3.24E+03	**3.21E+03**	**3.21E+03**	3.26E+03	**3.21E+03**	3.26E+03
	SD	9.90E+01	8.18E+01	3.09E+02	2.56E+01	1.46E+01	5.54E+00	4.78E−01	6.53E+01	**1.48E−01**	2.93E+01
		+	+	+	−	−	−	−	−	−	
$f_{28}(\mathbf{x})$	AB	4.39E+03	4.17E+03	3.22E+03	3.19E+03	3.22E+03	3.16E+03	3.42E+03	3.04E+03	3.03E+03	**3.00E+03**
	MD	4.38E+03	4.06E+03	3.21E+03	3.20E+03	3.21E+03	3.20E+03	3.40E+03	3.04E+03	**3.02E+03**	3.14E+03
	SD	3.64E+02	4.31E+02	2.13E+01	4.00E+01	**2.11E+01**	4.82E+01	8.61E+01	2.73E+01	3.34E+01	5.13E+01
		+	+	+	+	+	+	+	=	=	
$f_{29}(\mathbf{x})$	AB	4.54E+03	4.69E+03	4.37E+03	3.72E+03	3.57E+03	3.49E+03	3.36E+03	3.19E+03	3.18E+03	**3.11E+03**
	MD	4.50E+03	4.64E+03	4.38E+03	3.67E+03	3.57E+03	3.48E+03	3.35E+03	3.19E+03	3.16E+03	**3.09E+03**
	SD	4.07E+02	3.08E+02	2.86E+02	1.96E+02	1.59E+02	3.30E+01	3.70E+02	2.81E+02	3.80E+02	**2.93E+01**
		+	+	+	+	+	+	=	=	=	
$f_{30}(\mathbf{x})$	AB	1.11E+08	1.17E+08	1.73E+04	8.49E+03	8.29E+03	5.11E+03	5.11E+03	6.06E+03	**5.07E+03**	5.29E+03
	MD	9.84E+07	1.11E+08	1.29E+04	7.77E+03	7.36E+03	5.10E+03	5.09E+03	6.00E+03	**5.07E+03**	5.98E+03
	SD	3.67E+07	4.32E+07	9.34E+03	2.29E+03	2.84E+03	8.43E+01	8.76E+01	2.65E+02	**4.97E+01**	5.85E+03
		+	+	+	+	+	−	−	+	−	
Total	+	27	28	27	18	16	9	6	6	1	
	−	0	0	0	4	3	7	6	6	12	
	=	3	2	3	8	11	14	18	18	17	

5.5.3 Functions from CEC2017 with 100 Dimensions

Table 5.4 displays the experimental results of CEC2017 test functions in 100 dimensions. It showcases the outcomes of the GSAGWO algorithm compared to other metaheuristic algorithms. The performance of the results is evaluated using indicators AB, MB, and SD. The best results are emphasized in boldface to highlight their superiority.

The analysis of Table 5.4 reveals that the enhanced versions of L-SHADE and GSAGWO demonstrate the best performance. Among the L-SHADE variants, DbL-SHADE stands out as the most competitive, achieving the best results in 7 functions. On the other hand, GSAGWO outperforms the other algorithms by attaining the best AB values in 8 functions. Consequently, our discussed approach exhibits superior performance compared to its competitors in the majority of the benchmark functions.

The results from Table 5.4 were also subjected to the Wilcoxon test, and the corresponding outcomes are presented in the same table. Notably, the null hypothesis was accepted for DbL-SHADE in 17 functions and rejected in 13 functions. Among the rejected cases, DbL-SHADE achieved the best performance in 10 functions, while our discussed method outperformed it in 3 functions. This analysis confirms that DbL-SHADE emerges as the most competitive algorithm for the 100-dimensional case. However, our discussed method surpasses DbL-SHADE in a greater number of functions. The performance of the remaining algorithms is summarized at the end of Table 5.4, indicating the number of functions for which the null hypothesis was rejected (=), the number of functions where our algorithm outperformed the comparison algorithm (+), and the number of functions where the comparison algorithm obtained better results than our method (−).

Based on Table 5.4, the null hypothesis was found to be accepted for GWO in function f_4, indicating no significant difference between this algorithm and our discussed method. Conversely, the null hypothesis was rejected for the remaining functions. The symbol '+' associated with these functions signifies that our algorithm exhibits better performance than GWO. Summarizing the results obtained from the Wilcoxon test, we can conclude that among the enhanced versions of classical metaheuristics, DbL-SHADE and our discussed approach emerge as the top-performing algorithms in experiments conducted with 30, 50, and 100 dimensions. However, our method consistently demonstrates superiority when compared to all the other algorithms. Furthermore, when comparing with GWO and mGWO in particular, our method consistently outperforms them across all dimensions.

We performed the Kruskal–Wallis test to analyze the differences among multiple algorithms and conduct a robust statistical analysis. The Kruskal–Wallis test is a non-parametric statistical test that allows us to determine the differences among three or more independent and non-normally distributed groups [75]. It is an extension of the Wilcoxon test and tests the null hypothesis that the groups are subsets of the same population, while the alternative hypothesis considers that there is a significant difference among the groups. The p-values obtained from the Kruskal–Wallis test indicate whether the null hypothesis is accepted or rejected. If the p-value is below

Table 5.4 Optimal results achieved by both the metaheuristic algorithms and the discussed technique on the CEC2017 benchmark functions in 100 dimensions

Function		GWO	mGWO	SAPSO	EPSO	HCLPSO	LSHADE	LSHADE cnEpSin	LSHADE SPACMA	DbL SHADE	GSAGWO
$f_1(\mathbf{x})$	AB	4.26E+10	3.88E+10	5.01E+03	3.08E+03	3.72E+03	3.25E+02	1.90E+02	2.15E+02	**1.45E+02**	1.89E+02
	MD	4.16E+10	3.79E+10	3.67E+03	1.67E+03	2.72E+03	2.89E+02	1.79E+03	2.01E+02	1.45E+02	**1.24E+02**
	SD	8.37E+09	7.29E+09	4.66E+03	3.54E+03	3.44E+03	2.22E+02	2.02E+01	1.62E+01	**4.18E−01**	6.72E+01
		+	+	+	+	+	+	=	+	−	
$f_2(\mathbf{x})$	AB	2.6E+128	2.00E+134	2.80E+09	1.59E+08	3.58E+11	3.54E+03	2.88E+02	2.91E+02	2.25E+02	**2.15E+02**
	MD	2.7E+118	9.10E+120	1.43E+05	1.68E+04	9.47E+07	3.49E+02	2.75E+02	2.80E+02	2.18E+02	**2.12E+02**
	SD	1.0E+129	1.10E+135	9.17E+09	8.53E+08	1.83E+12	7.31E+00	1.50E+01	1.40E+01	5.35E+00	**1.22E+00**
		+	+	+	+	+	+	+	+	=	
$f_3(\mathbf{x})$	AB	2.70E+05	2.56E+05	**3.00E+02**	3.05E+02	6.22E+02	**3.00E+02**	**3.00E+02**	**3.00E+02**	**3.00E+02**	**3.00E+02**
	MD	2.70E+05	2.54E+05	**3.00E+02**	3.01E+02	5.12E+02	**3.00E+02**	**3.00E+02**	**3.00E+02**	**3.00E+02**	**3.00E+02**
	SD	2.90E+04	1.99E+04	5.98E−02	6.89E+00	3.27E+02	5.34E−10	1.49E−11	3.22E−10	**2.32E−12**	1.06E−08
		+	+	+	=	+	=	=	=	=	
$f_4(\mathbf{x})$	AB	3.80E+03	3.69E+03	4.81E+02	4.50E+02	4.89E+02	4.59E+02	4.76E+02	4.84E+02	4.49E+02	**4.42E+02**
	MD	3.75E+03	3.66E+03	4.85E+02	4.68E+02	4.89E+02	4.59E+02	4.69E+02	4.60E+02	4.46E+02	**4.41E+02**
	SD	9.23E+02	5.83E+02	3.12E+01	3.37E+01	2.34E+01	1.41E+00	2.29E+02	4.24E+01	4.74E+01	**3.56E−01**
		=	=	=	=	+	=	=	+	=	
$f_5(\mathbf{x})$	AB	1.11E+03	1.15E+03	7.64E+02	5.59E+02	5.53E+02	5.21E+02	5.19E+02	5.55E+02	5.32E+02	**5.05E+02**
	MD	1.12E+03	1.14E+03	7.52E+02	5.55E+02	5.48E+02	5.21E+02	5.10E+02	5.48E+02	5.13E+02	**5.04E+02**
	SD	6.43E+01	9.99E+01	6.54E+01	1.90E+01	1.77E+01	2.71E+00	7.13E+01	4.07E+01	4.46E+01	**2.88E−01**
		+	+	+	+	+	=	=	+	=	
$f_6(\mathbf{x})$	AB	6.36E+02	6.37E+02	6.58E+02	6.01E+02	**6.00E+02**	**6.00E+02**	**6.00E+02**	**6.00E+02**	**6.00E+02**	**6.00E+02**

(continued)

Table 5.4 (continued)

Function		GWO	mGWO	SAPSO	EPSO	HCLPSO	LSHADE	LSHADE cnEpSin	LSHADE SPACMA	DbL SHADE	GSAGWO
	MD	6.36E+02	6.36E+02	6.60E+02	6.01E+02	**6.00E+02**	**6.00E+02**	**6.00E+02**	**6.00E+02**	**6.00E+02**	6.01E+02
	SD	4.50E+00	3.85E+00	1.06E+01	8.60E−01	7.53E−04	1.13E−05	7.11E−01	3.96E−01	**4.88E−06**	1.02E−02
		+	+	+	=	=	=	=	=	=	
$f_7(\mathbf{x})$	AB	1.92E+03	1.98E+03	1.24E+03	8.12E+02	7.89E+02	7.54E+02	7.90E+02	7.46E+02	7.23E+02	**7.13E+02**
	MD	1.91E+03	1.98E+03	1.22E−03	8.12E−02	7.82E−02	7.53E−02	7.65E−02	7.36E−02	7.23E−02	**7.14E−02**
	SD	1.07E+02	1.33E−02	1.27E−02	2.38E−01	2.02E−01	4.21E−00	3.28E−01	1.03E−01	6.23E−00	**2.87E−00**
		+	−	−	−	−	−	−	=	=	
$f_8(\mathbf{x})$	AB	1.43E+03	1.48E−03	1.02E−03	8.61E−02	8.52E−02	8.44E−02	8.19E−02	8.20E−02	**8.18E−02**	8.58E−02
	MD	1.42E+03	1.46E−03	1.01E−03	8.57E−02	8.47E−02	8.41E−02	**8.15E−02**	8.17E−02	8.16E−02	8.40E−02
	SD	1.11E+02	9.28E−01	6.11E−01	1.69E−01	1.80E−01	3.21E−00	4.97E−00	3.48E−00	**3.21E−00**	2.09E−01
		+	−	−	=	−	−	−	−	−	
$f_9(\mathbf{x})$	AB	3.50E+04	3.93E−04	6.57E−03	1.03E−03	9.17E−02	**9.00E−02**	**9.00E−02**	**9.00E−02**	**9.00E−02**	9.15E−02
	MD	3.71E+04	4.04E−04	6.22E−03	1.01E−03	9.08E−02	**9.00E−02**	**9.00E−02**	**9.00E−02**	**9.00E−02**	9.17E−02
	SD	1.22E+04	7.85E−03	1.76E−03	9.41E−01	1.71E−01	7.00E−14	1.79E−16	5.19E−16	**7.00E−19**	2.50E−00
		+	−	−	−	=	−	−	−	−	
$f_{10}(\mathbf{x})$	AB	1.60E+04	1.83E−04	4.97E−03	3.72E−03	3.78E−03	3.65E−03	3.51E−03	3.57E−03	**3.35E−03**	3.85E−03
	MD	1.55E+04	1.68E−04	5.01E−03	3.82E−03	3.73E−03	3.65E−03	3.47E−03	3.49E−03	**3.27E−03**	3.76E−03
	SD	3.06E+03	4.98E−03	5.08E−02	4.60E−02	4.84E−02	2.98E−02	2.07E−02	1.27E−02	**9.80E−01**	1.13E−02
		+	−	−	−	−	−	−	−	−	
$f_{11}(\mathbf{x})$	AB	5.72E+04	5.31E−04	1.24E−03	1.21E−03	1.18E−03	1.13E−03	1.12E−03	**1.11E−03**	1.13E−03	1.12E−03
	MD	5.42E+04	5.33E−04	1.23E−03	1.22E−03	1.18E−03	1.12E−03	1.11E−03	**1.10E−03**	**1.10E−03**	1.13E−03

(continued)

Table 5.4 (continued)

Function		GWO	mGWO	SAPSO	EPSO	HCLPSO	LSHADE	LSHADE cnEpSin	LSHADE SPACMA	DbL SHADE	GSAGWO
	SD	1.35E+04	1.27E+04	3.81E−01	4.40E−01	3.53E−01	2.45E−01	2.85E−01	**2.50E−00**	1.78E−02	1.58E−01
		+	+	=	=	=	=	−	−	=	
$f_{12}(\mathbf{x})$	AB	9.33E+09	6.44E−09	2.60E−05	3.12E−04	5.58E−04	**2.26E+03**	2.81E+03	2.51E+03	2.42E+03	3.88E+04
	MD	8.29E+09	5.64E+09	1.90E+05	3.08E+04	4.15E+04	**2.29E+03**	2.52E+03	2.48E+03	2.40E+03	2.47E+04
	SD	5.59E+09	2.92E+09	3.34E+05	1.41E+04	3.93E+04	**3.62E+02**	4.82E+02	4.39E+02	2.22E+03	3.10E+04
		+	+	+	−	+	−	−	−	−	
$f_{13}(\mathbf{x})$	AB	9.05E+08	6.97E+08	1.74E+04	1.29E+04	1.44E+04	1.35E+03	1.34E+03	1.34E+03	**1.32E+03**	1.39E+03
	MD	6.51E+08	6.50E+08	8.27E+03	6.54E+03	1.19E+04	1.36E+03	1.34E+03	1.36E+03	**1.33E+03**	1.38E+03
	SD	8.27E+08	5.16E+08	1.89E+04	1.27E+04	1.15E+04	3.71E+01	2.81E+01	5.92E+01	**2.24E+01**	4.63E+01
		+	+	+	+	+	=	=	=	−	
$f_{14}(\mathbf{x})$	AB	7.37E+06	6.53E+06	9.63E+03	8.04E+03	1.07E+04	**1.42E+03**	**1.42E+03**	**1.42E+03**	**1.42E+03**	**1.42E+03**
	MD	7.04E+06	5.62E+06	7.78E+03	7.96E+03	6.86E+03	**1.42E+03**	**1.42E+03**	**1.42E+03**	**1.42E+03**	1.43E+03
	SD	4.70E+06	2.90E+06	7.52E+03	4.51E+03	1.18E+04	5.78E+00	3.47E+00	1.12E+00	**5.78E−01**	3.96E+01
		+	+	+	+	+	=	=	=	=	
$f_{15}(\mathbf{x})$	AB	1.49E+08	1.35E+08	1.25E+04	8.03E+03	3.61E+03	**1.51E+03**	**1.51E+03**	**1.51E+03**	**1.51E+03**	**1.51E+03**
	MD	3.72E+07	7.77E+07	7.85E+03	5.34E+03	3.31E+03	**1.51E+03**	**1.51E+03**	**1.51E+03**	**1.51E+03**	**1.51E+03**
	SD	2.81E+08	2.05E+08	1.15E+04	6.28E+03	1.83E+03	5.12E+02	5.51E+01	4.15E+02	**3.71E+00**	4.48E+00
		+	+	+	+	+	=	=	=	=	
$f_{16}(\mathbf{x})$	AB	5.74E+03	5.79E+03	2.86E+03	2.35E+03	2.31E+03	2.13E+03	1.98E+03	2.00E+03	**1.94E+03**	2.66E+03
	MD	5.77E+03	5.81E+03	2.83E+03	2.30E+03	2.32E+03	2.18E+03	1.97E+03	1.98E+03	**1.95E+03**	2.66E+03
	SD	7.09E+02	5.11E+02	3.31E+02	2.34E+02	1.96E+02	5.78E+02	**1.18E+02**	1.35E+02	1.39E+02	3.13E+02

(continued)

Table 5.4 (continued)

Function		GWO	mGWO	SAPSO	EPSO	HCLPSO	LSHADE	LSHADE cnEpSin	LSHADE SPACMA	DbL SHADE	GSAGWO
		+	+	+	−	−	−	−	−	−	
$f_{17}(\mathbf{x})$	AB	4.63E+03	5.17E+03	2.47E+03	1.95E+03	1.89E+03	1.77E+03	1.74E+03	1.76E+03	1.71E+03	**1.70E+03**
	MD	4.68E+03	5.12E+03	2.46E+03	1.91E+03	1.87E+03	1.78E+03	1.73E+03	1.75E+03	1.72E+03	**1.71E+03**
	SD	5.48E+02	7.63E+02	2.47E+02	1.43E+02	1.16E+02	9.80E+00	7.40E+00	1.13E+01	8.60E+00	**4.89E+00**
		+	+	+	+	+	=	=	=	=	
$f_{18}(\mathbf{x})$	AB	5.45E+06	6.25E+06	1.36E+05	1.28E+05	1.27E+05	1.86E+03	1.85E+03	1.87E+03	**1.84E+03**	1.85E+03
	MD	5.18E+06	4.84E+06	9.73E+04	1.11E+05	9.78E+04	1.87E+03	1.86E+03	1.89E+03	**1.83E+03**	1.86E+03
	SD	3.36E+06	5.07E+06	9.81E+04	8.15E+04	9.58E+04	8.99E+00	1.24E+01	9.90E+00	**8.17E+00**	4.11E+01
		+	+	+	+	+	=	=	=	−	
$f_{19}(\mathbf{x})$	AB	1.61E+08	8.83E+07	1.08E+04	7.58E+03	5.24E+03	1.95E+03	1.92E+03	1.94E+03	**1.91E+03**	**1.91E+03**
	MD	5.69E+07	7.16E+07	5.18E+03	6.63E+03	4.00E+03	1.94E+03	1.94E+03	1.95E+03	**1.91E+03**	**1.91E+03**
	SD	2.58E+08	8.13E+07	1.31E+04	6.07E+03	4.86E+03	3.45E+00	7.78E+00	8.45E+00	2.30E+00	**1.58E+00**
		+	+	+	+	+	=	=	=	=	
$+f_{20}(\mathbf{x})$	AB	4.82E+03	5.47E+03	2.72E+03	2.28E+03	2.23E+03	2.17E+03	2.14E+03	2.16E+03	2.09E+03	**2.08E+03**
	MD	4.67E+03	5.08E+03	2.75E+03	2.28E+03	2.21E+03	2.19E+03	2.12E+03	2.17E+03	2.09E+03	**2.08E+03**
	SD	7.18E+02	1.19E+03	1.74E+02	1.24E+02	7.96E+01	4.20E+01	**3.60E+01**	7.80E+01	3.70E+01	2.13E+02
		+	+	+	+	+	+	+	+	=	
$f_{21}(\mathbf{x})$	AB	2.96E+03	2.96E+03	2.58E+03	2.36E+03	2.35E+03	2.32E+03	2.57E+03	2.48E+03	2.52E+03	**2.31E+03**
	MD	2.95E+03	2.95E+03	2.57E+03	2.36E+03	2.35E+03	2.32E+03	2.57E+03	2.47E+03	2.52E+03	**2.31E+03**
	SD	6.17E+01	9.30E+01	6.66E+01	1.73E+01	1.52E+01	3.08E+00	8.82E+01	3.54E+01	4.02E+01	**2.51E−01**
		+	+	+	=	=	=	+	+	+	

(continued)

Table 5.4 (continued)

Function		GWO	mGWO	SAPSO	EPSO	HCLPSO	LSHADE	LSHADE cnEpSin	LSHADE SPACMA	DbL SHADE	GSAGWO
$f_{22}(\mathbf{x})$	AB	1.85E+04	2.18E+04	6.19E+03	2.40E+03	**2.30E+03**	**2.30E+03**	**2.30E+03**	**2.30E+03**	**2.30E+03**	**2.30E+03**
	MD	1.86E+04	2.02E+04	6.55E+03	2.30E+03	**2.30E+03**	**2.30E+03**	**2.30E+03**	**2.30E+03**	**2.30E+03**	**2.30E+03**
	SD	1.44E+03	5.43E+03	1.70E+03	5.22E+02	1.14E+00	4.22E−13	**3.42E−15**	1.55E−11	2.22E−07	2.85E−08
		+	+	+	+	=	=	=	=	=	
$f_{23}(\mathbf{x})$	AB	3.56E+03	3.55E+03	3.44E+03	2.73E+03	2.70E+03	2.67E+03	2.66E+03	2.67E+03	2.66E+03	**2.65E+03**
	MD	3.55E+03	3.52E+03	3.44E+03	2.73E+03	2.70E+03	2.67E+03	**2.66E+03**	**2.66E+03**	**2.66E+03**	2.67E+03
	SD	7.35E+01	1.40E+02	2.09E+02	2.51E+01	1.39E+01	**5.60E+00**	4.29E+01	4.51E+01	4.01E+01	2.85E+01
		+	+	+	+	=	=	=	=	=	
$f_{24}(\mathbf{x})$	AB	4.14E+03	4.12E+03	3.40E+03	2.90E+03	2.87E+03	2.84E+03	2.83E+03	2.83E+03	2.81E+03	**2.80E+03**
	MD	4.11E+03	4.10E+03	3.39E+03	2.90E+03	2.87E+03	2.84E+03	2.83E+03	2.84E+03	2.82E+03	**2.81E+03**
	SD	1.11E+02	1.15E+02	1.25E+02	3.67E+01	2.42E+01	**3.49E+00**	3.53E+01	4.97E+01	5.99E+01	6.43E+00
		+	+	+	+	=	=	=	=	=	
$f_{25}(\mathbf{x})$	AB	6.06E+03	6.04E+03	2.90E+03	2.90E+03	**2.89E+03**	**2.89E+03**	**2.89E+03**	2.91E+03	**2.89E+03**	2.97E+03
	MD	6.17E+03	6.01E+03	2.89E+03	2.89E+03	**2.89E+03**	**2.89E+03**	**2.89E+03**	2.93E+03	**2.89E+03**	2.96E+03
	SD	6.01E+02	6.59E+02	1.63E+01	1.99E+01	1.03E+01	4.20E−02	2.69E+01	5.34E+01	**5.93E−03**	3.94E+01
		+	+	−	−	−	−	−	−	−	
$f_{26}(\mathbf{x})$	AB	1.41E+04	1.44E+04	7.86E+03	3.84E+03	3.70E+03	3.67E+03	2.93E+03	2.98E+03	2.72E+03	**2.68E+03**
	MD	1.40E+04	1.44E+04	8.21E+03	4.32E+03	3.93E+03	3.67E+03	2.79E+03	2.83E+03	2.71E+03	**2.71E+03**
	SD	1.23E+03	1.30E+03	2.12E+03	9.10E+02	5.91E+02	**5.65E+01**	2.62E+02	3.22E+02	8.47E+02	7.22E+02
		+	+	+	+	+	+	+	+	=	
$f_{27}(\mathbf{x})$	AB	4.05E+03	4.00E+03	3.74E+03	3.23E+03	3.23E+03	3.22E+03	3.90E+03	3.67E+03	**3.20E+03**	3.26E+03

(continued)

Table 5.4 (continued)

Function		GWO	mGWO	SAPSO	EPSO	HCLPSO	LSHADE	LSHADE cnEpSin	LSHADE SPACMA	DbL SHADE	GSAGWO
	MD	4.04E+03	3.99E+03	3.77E+03	3.23E+03	3.22E+03	3.23E+03	3.88E+03	3.68E+03	**3.20E+03**	3.25E+03
	SD	1.33E+02	1.35E+02	2.77E+02	1.21E+01	1.03E+01	7.22E+00	1.45E+02	7.44E+01	**5.25E+00**	1.97E+01
		+	+	+	–	–	–	+	+	–	
$f_{28}(\mathbf{x})$	AB	7.69E+03	7.98E+03	3.21E+03	3.16E+03	3.20E+03	3.13E+03	3.32E+03	3.47E+03	3.04E+03	**3.01E+03**
	MD	7.62E+03	7.64E+03	3.21E+03	3.11E+03	3.20E+03	3.10E+03	3.11E+03	3.47E+03	3.03E+03	**3.00E+03**
	SD	1.22E+03	1.19E+03	3.34E+01	6.09E+01	2.70E+01	4.86E+01	5.52E+02	4.72E+01	6.38E+01	**9.46E+00**
		+	+	+	+	+	=	+	+	=	
$f_{29}(\mathbf{x})$	AB	8.30E+03	8.21E+03	4.26E+03	3.59E+03	3.48E+03	3.39E+03	3.28E+03	3.38E+03	3.35E+03	**3.16E+03**
	MD	8.31E+03	8.13E+03	4.32E+03	3.58E+03	3.45E+03	3.39E+03	3.37E+03	3.48E+03	3.37E+03	**3.15E+03**
	SD	7.81E+02	5.91E+02	3.56E+02	1.10E+02	1.05E+02	**1.42E+01**	5.94E+02	6.27E+02	5.53E+01	2.32E+01
		+	+	+	+	+	+	+	+	+	
$f_{30}(\mathbf{x})$	AB	7.76E+08	7.55E+08	9.75E+03	8.44E+03	8.27E+03	**5.05E+03**	8.38E+03	6.05E+03	6.83E+03	5.86E+03
	MD	5.54E+08	5.51E+08	9.42E+03	7.69E+03	7.76E+03	5.05E+03	9.98E+03	6.27E+03	**4.69E+03**	5.02E+03
	SD	7.10E+08	6.83E+08	3.19E+03	2.28E+03	2.00E+03	**7.92E+01**	1.31E+02	7.43E+02	1.15E+03	5.20E+02
		+	+	+	+	+	–	+	+	+	
Total	+	29	29	27	19	18	6	9	11	3	
	–	0	0	1	5	5	8	7	7	10	
	=	1	1	2	6	7	16	14	12	17	

Table 5.5 The p-values resulting from the Kruskal–Wallis test, which compares GSAGWO with various metaheuristic algorithms for the CEC2017 benchmark functions in 100 dimensions

Functions	GSAGWO and metaheuristic algorithms	Functions	GSAGWO and metaheuristic algorithms
$f_1(\mathbf{x})$	4.96E−67	$f_{16}(\mathbf{x})$	1.59E−55
$f_2(\mathbf{x})$	4.64E−67	$f_{17}(\mathbf{x})$	6.68E−60
$f_3(\mathbf{x})$	8.39E−67	$f_{18}(\mathbf{x})$	9.10E−58
$f_4(\mathbf{x})$	1.59E−57	$f_{19}(\mathbf{x})$	3.20E−61
$f_5(\mathbf{x})$	1.97E−65	$f_{20}(\mathbf{x})$	1.69E−52
$f_6(\mathbf{x})$	1.07E−67	$f_{21}(\mathbf{x})$	9.63E−66
$f_7(\mathbf{x})$	6.79E−66	$f_{22}(\mathbf{x})$	3.35E−61
$f_8(\mathbf{x})$	2.40E−64	$f_{23}(\mathbf{x})$	1.58E−66
$f_9(\mathbf{x})$	2.71E−66	$f_{24}(\mathbf{x})$	1.45E−65
$f_{10}(\mathbf{x})$	1.05E−62	$f_{25}(\mathbf{x})$	2.81E−57
$f_{11}(\mathbf{x})$	3.29E−64	$f_{26}(\mathbf{x})$	4.78E−56
$f_{12}(\mathbf{x})$	2.58E−65	$f_{27}(\mathbf{x})$	5.03E−63
$f_{13}(\mathbf{x})$	5.26E−61	$f_{28}(\mathbf{x})$	5.71E−65
$f_{14}(\mathbf{x})$	2.19E−57	$f_{29}(\mathbf{x})$	2.15E−63
$f_{15}(\mathbf{x})$	1.75E−61	$f_{30}(\mathbf{x})$	5.05E−65

the confidence level of 5% (0.05) for the Average Best-so-far (AB) values of each metaheuristic, it means that at least one method significantly differs in its results. The p-values resulting from the test are presented in Table 5.5.

According to the p-values reported in Table 5.5, the null hypothesis has been rejected for all functions. This indicates that there is a significant difference among the algorithms being compared.

For the Kruskal–Wallis statistical analysis, we chose two specific functions, namely $f_7(\mathbf{x})$, and $f_{28}(\mathbf{x})$, to represent a multimodal and composite function in 100 dimensions. Figure 5.8 displays the boxplots generated from the Kruskal–Wallis test, which consider these selected functions.

Upon examining Fig. 5.8, it can be deduced that the null hypothesis is not upheld for functions f_7 and f_{28}. The Kruskal–Wallis test provides substantial evidence to refute the null hypothesis, as corroborated by the p-values presented in Table 5.5 for these specific functions.

5.5.4 Analysis of Convergence

In this section, the evaluation of optimization algorithms goes beyond comparing the accuracy of the final fitness value. Instead, a convergence test is carried out

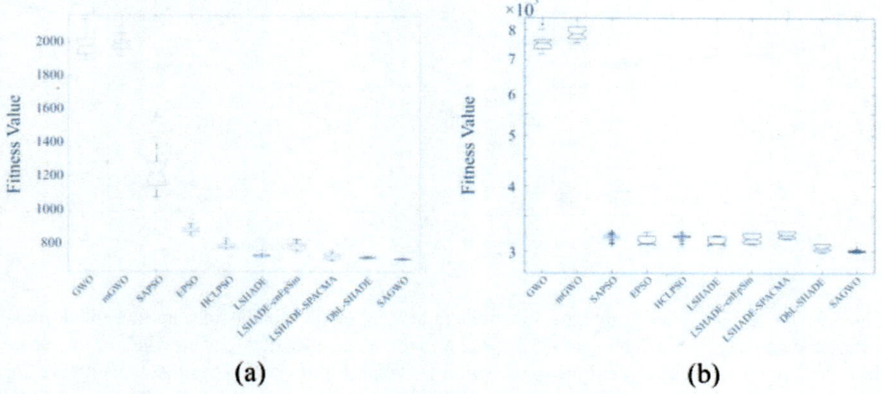

Fig. 5.8 The Kruskal–Wallis test was performed on functions f_7 and f_{28} in 100 dimensions. The results are illustrated through boxplots, showing the comparison between GSAGWO and other metaheuristic algorithms for each function. Specifically, **a** displays the boxplot for function f_7, while **b** presents the boxplot for function f_{28}

to assess the speed at which each algorithm approaches the optimal solution [76–80]. It is recognized that solely examining the final fitness value accuracy may not provide a comprehensive assessment of an optimization algorithm's performance. Hence, the focus of this study is to analyze the convergence characteristics of the compared algorithms. The performance evaluation of each algorithm is conducted based on a single execution in the experiments. The obtained results are presented in Fig. 5.9. The analysis focuses on two benchmark functions, namely f_7 and f_{28}, which have been chosen to represent a multimodal and a composite function, respectively, based on the classification provided in CEC2017. These evaluations are performed considering 100 dimensions. Upon inspecting Fig. 5.9, it becomes evident that our discussed method exhibits a quicker and more accurate convergence compared to the other algorithms.

Based on the observations from Fig. 5.9, it is evident that our algorithm outperforms the other methods. While it may not have achieved the best results in all functions, it has demonstrated its competitiveness with state-of-the-art metaheuristic algorithms.

5.5.5 *Analysis of Diversity*

This section focuses on examining the transition from exploration to exploitation in order to analyze the diversity of GSAGWO. To understand how the exploration and exploitation change over time, the average distance between all solutions is calculated for each iteration [81–84]. The analysis specifically considers two benchmark functions: f_6 and f_{26}, which represent a multimodal and composite function,

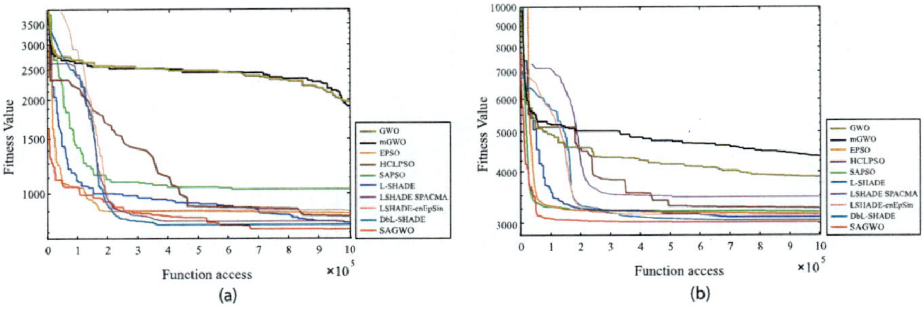

Fig. 5.9 The convergence graphs for functions f_7 and f_{28} in 100 dimensions are presented, illustrating the convergence behavior of GSAGWO and the metaheuristic algorithms. Figure **a** shows the convergence of these algorithms for function f_7, while Figure **b** displays the convergence for function f_{28}

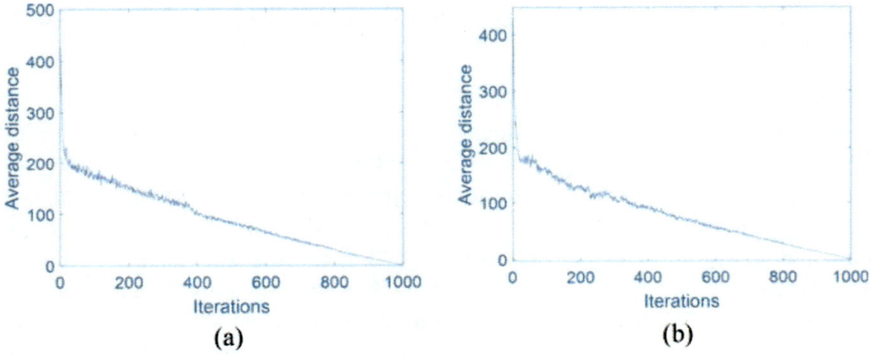

Fig. 5.10 Diversity plots for 30 dimensions: **a** f_6, and **b** f_{26}

respectively. The test is conducted with 30 dimensions. The diversity of GSAGWO, considering these functions, is illustrated in Fig. 5.10.

Figure 5.10 clearly depicts the behavior of the transition from exploration to exploitation. Initially, during the early stages of the search process, the average distance between particles is noticeably higher, indicating a more exploratory behavior. However, as the number of iterations increases, the average distance nonlinearly decreases, indicating a shift towards more exploitative behavior.

5.5.6 Computational Cost

As an additional analysis, the speed performance of the algorithms is examined. The experiments involve running the algorithms independently in 30 dimensions for each benchmark function. The average elapsed real-time (in seconds) for unimodal, multimodal, and composite test functions is recorded and presented in Table 5.6. The best results are highlighted in bold. The experiments were conducted on a PC with 8 GB of memory and an Intel® Core i7 Processor (3.4 GHz).

Based on the information in Table 5.6, it is clear that DbL-SHADE demonstrates the fastest execution time, whereas L-SHADE exhibits the slowest performance. The remaining algorithms display comparable speed results.

5.5.7 Application in Engineering Problems

In order to expand our analysis, we have chosen several engineering optimization problems to assess the performance of the compared optimization methods. These problems include the pressure vessel design problem f_{A1} [68], the gear train problem f_{A2} [68], the tension/compression spring design problem f_{A3} [69], the three-bar truss design problem f_{A4} [70], the welded beam design problem f_{A5} [71], and the parameter estimation for the FM synthesizer problem f_{A6} [72]. All of these problems are formulated as minimization problems. For a more detailed description of these engineering optimization problems, please refer to Appendix 1.

The experimental settings, including the configuration parameters, are provided in Table 5.1. The population size was set to 50 individuals, and the maximum number of iterations was set to 1000. The results of the experiments, based on 30 independent runs, are presented in Table 5.7. The best values for the AB, MB, and SD metrics are highlighted in bold for each engineering problem.

Based on the results presented in Table 5.7, LSHADE-cnEpSin and DbL-SHADE performed the best for problem f_{A1}. GSAGWO showed the best performance for problem f_{A2}. Multiple algorithms, including GSAGWO, L-SHADE, HCLPSO, and EPSO, achieved the best results for problem f_{A3}. Most algorithms demonstrated similar performance for problem f_{A4}. GSAGWO, L-SHADE, and various L-SHADE variants yielded the best results for problem f_{A5}. L-SHADE emerged as the top-performing algorithm for problem f_{A6}. In summary, our discussed algorithm demonstrated competitive performance by outperforming other methods in four out of five engineering problems.

Table 5.8 presents the best solution **x** obtained by the GSAGWO algorithm from 30 independent runs for each engineering optimization problem. Here, **x** represents the set of design variables $\{x_1, \ldots, x_n\}$.

5.6 Conclusions

This research introduces a modified version of the Grey Wolf Optimizer (GWO) called Group-based Synchronous-Asynchronous Grey Wolf Optimizer (GSAGWO). The discussed method divides the population into three groups, each containing an equal number of candidate solutions. These groups are processed individually in

Table 5.6 Elapsed real-time in seconds for the metaheuristic algorithms and the discussed technique when considering the CEC2017 test functions in 30 dimensions

	GWO	mGWO	SAPSO	EPSO	HCLPSO	LSHADE	LSHADE cnEpSin	LSHADE SPACMA	DbL SHADE	GSAGWO
$f_1(\mathbf{x})$	1.167	1.191	1.976	1.509	1.239	7.350	1.93	1.89	**1.04**	1.529
$f_2(\mathbf{x})$	1.193	1.215	2.049	1.602	1.288	7.482	1.40	1.24	**1.07**	1.569
$f_3(\mathbf{x})$	1.183	1.197	2.196	1.733	1.296	7.278	1.32	1.38	**0.89**	1.545
$f_4(\mathbf{x})$	1.192	1.207	2.034	1.626	1.268	7.446	1.20	1.46	**0.92**	1.562
$f_5(\mathbf{x})$	1.233	1.265	2.043	1.592	1.322	7.367	1.14	1.46	**0.92**	1.598
$f_6(\mathbf{x})$	1.374	1.431	2.158	1.706	1.467	7.626	1.32	1.44	**0.94**	1.730
$f_7(\mathbf{x})$	1.263	1.280	2.116	1.575	1.345	7.511	1.34	1.55	**0.89**	1.611
$f_8(\mathbf{x})$	1.272	1.287	2.098	1.617	1.363	7.458	1.26	1.64	**0.93**	1.634
$f_9(\mathbf{x})$	1.282	1.295	2.150	1.574	1.371	7.399	1.39	1.49	**0.91**	1.609
$f_{10}(\mathbf{x})$	1.324	1.337	2.149	1.709	1.500	7.383	1.37	1.40	**0.94**	1.697
$f_{11}(\mathbf{x})$	1.288	1.310	2.123	1.667	1.377	7.551	1.38	1.45	**0.90**	1.676
$f_{12}(\mathbf{x})$	1.320	1.333	2.170	1.695	1.419	7.520	1.17	1.50	**0.96**	1.678
$f_{13}(\mathbf{x})$	1.304	1.322	2.163	1.718	1.386	7.486	1.25	1.42	**0.95**	1.663
$f_{14}(\mathbf{x})$	1.368	1.387	2.211	2.047	1.465	7.492	1.41	1.51	**0.93**	1.761
$f_{15}(\mathbf{x})$	1.317	1.330	2.163	1.738	1.421	7.519	1.20	1.46	**0.92**	1.667
$f_{16}(\mathbf{x})$	1.346	1.372	2.185	1.695	1.445	7.706	1.27	1.47	**0.98**	1.672
$f_{17}(\mathbf{x})$	1.455	1.472	2.322	1.769	1.550	7.663	1.35	1.42	**1.00**	1.867
$f_{18}(\mathbf{x})$	1.361	1.379	2.184	1.860	1.489	7.529	1.19	1.52	**0.93**	1.703
$f_{19}(\mathbf{x})$	1.977	2.044	2.831	2.281	2.084	8.108	1.44	1.65	**1.14**	2.318
$f_{20}(\mathbf{x})$	1.512	1.558	2.352	2.148	1.620	7.686	1.44	1.61	**1.01**	1.853

(continued)

Table 5.6 (continued)

	GWO	mGWO	SAPSO	EPSO	HCLPSO	LSHADE	LSHADE cnEpSin	LSHADE SPACMA	DbL SHADE	GSAGWO
$f_{21}(\mathbf{x})$	1.575	1.604	2.408	1.901	1.679	7.805	1.37	1.45	**1.06**	1.914
$f_{22}(\mathbf{x})$	1.631	1.661	2.462	1.965	1.758	7.888	1.43	1.62	**0.99**	2.018
$f_{23}(\mathbf{x})$	1.696	1.725	2.511	2.039	1.812	7.941	1.55	1.54	**1.07**	2.031
$f_{24}(\mathbf{x})$	1.746	1.768	2.622	2.112	1.845	7.898	1.38	1.63	**1.08**	2.086
$f_{25}(\mathbf{x})$	1.713	1.718	2.563	2.129	1.799	7.906	1.33	1.75	**1.05**	2.061
$f_{26}(\mathbf{x})$	1.847	1.838	2.658	2.138	1.915	8.130	1.54	1.66	**1.11**	2.167
$f_{27}(\mathbf{x})$	1.926	1.925	2.730	2.355	1.995	8.162	1.50	1.65	**1.10**	2.243
$f_{28}(\mathbf{x})$	1.831	1.858	2.709	2.163	1.885	8.204	1.39	1.77	**1.08**	2.174
$f_{29}(\mathbf{x})$	1.780	1.756	2.569	2.164	1.834	7.973	1.44	2.02	**1.07**	2.076
$f_{30}(\mathbf{x})$	2.273	2.267	3.105	2.916	2.384	8.517	1.59	1.91	**1.18**	2.632

Table 5.7 The optimal results of both the metaheuristic algorithms and the discussed technique were evaluated for the engineering optimization problems

Function		GWO	mGWO	SAPSO	EPSO	HCLPSO	LSHADE	LSHADEcnEpSin	LSHADESPACMA	DbLSHADE	GSAGWO
$f_{A1}(\mathbf{x})$	AB	1.03E+04	9.79E+03	1.39E+04	1.19E+04	1.29E+04	1.29E+04	**8.81E+03**	8.98E+03	**8.81E+03**	1.09E+04
	MD	8.82E+03	8.82E+03	1.45E+04	1.37E+04	1.37E+04	8.82E+03	**8.81E+03**	**8.81E+03**	**8.81E+03**	**8.81E+03**
	SD	2.28E+03	1.99E+03	3.06E+03	2.40E+03	3.86E+03	2.69E+03	**3.70E−12**	8.92E+02	**3.70E−12**	2.46E+03
$f_{A2}(\mathbf{x})$	AB	1.27E−10	1.38E−10	3.64E−09	2.14E−10	4.36E−10	4.32E−10	5.49E−10	2.70E−10	2.23E−10	**2.03E−10**
	MD	2.32E−11	2.33E−11	1.86E−09	2.32E−11	6.98E−11	2.70E−11	1.17E−10	**2.31E−11**	**2.31E−11**	**2.31E−11**
	SD	3.07E−10	3.13E−10	5.92E−09	2.57E−10	6.72E−10	**1.76E−10**	6.47E−10	3.57E−10	2.73E−10	5.04E−10
$f_{A3}(\mathbf{x})$	AB	6.67E+298	1.00E+299	7.28E−03	**7.27E−03**	**7.27E−03**	**7.27E−03**	5.00E+29	5.33E+29	3.67E+29	**7.27E−03**
	MD	7.28E−03	7.29E−03	8.28E−03	7.29E−03	**7.27E−03**	**7.27E−03**	5.00E+29	1.00E+30	8.96E−03	**7.27E−03**
	SD	6.55E+04	6.55E+04	6.55E+04	6.55E+04	6.55E+04	3.05E+29	5.09E+29	5.07E+29	4.90E+29	**6.02E−07**
$f_{A4}(\mathbf{x})$	AB	2.65E+02	2.66E+02	**2.64E+02**	**2.64E+02**	**2.64E+02**	**2.64E+02**	**2.64E+02**	**2.64E+02**	**2.64E+02**	**2.64E+02**
	MD	2.65E+02	2.66E+02	2.66E+02	**2.64E+02**	**2.64E+02**	**2.64E+02**	**2.64E+02**	**2.64E+02**	**2.64E+02**	**2.64E+02**
	SD	1.83E−03	3.58E−03	1.42E−01	3.23E−04	1.76E−03	6.25E−04	1.49E−14	1.06E−14	**0.00E+00**	6.23E−04
$f_{A5}(\mathbf{x})$	AB	2.33E+299	1.33E+299	2.01E+01	1.10E+01	1.53E+01	**1.73E+00**	**1.73E+00**	**1.73E+00**	**1.73E+00**	**1.73E+00**
	MD	1.74E+00	1.74E+00	2.00E+01	1.00E+01	1.21E+01	**1.73E+00**	**1.73E+00**	**1.73E+00**	**1.73E+00**	**1.73E+00**
	SD	6.55E+04	6.55E+04	6.55E+04	6.55E+04	6.55E+04	2.50E−04	2.26E−03	**7.52E−09**	2.16E−06	3.72E−04
$f_{A6}(\mathbf{x})$	AB	3.09E+01	3.08E+01	3.10E+01	3.09E+01	3.10E+01	**3.07E+01**	3.09E+01	3.10E+01	3.09E+01	3.10E+01
	MD	3.11E+01	3.12E+01	**3.10E+01**	**3.10E+01**	**3.10E+01**	**3.10E+01**	**3.10E+01**	**3.10E+01**	**3.10E+01**	**3.10E+01**
	SD	2.56E−01	3.67E−01	**2.17E−14**	3.02E−01	2.00E−01	4.23E−01	2.56E−01	2.15E−01	3.18E−01	2.02E−01

Table 5.8 The GSAGWO algorithm has achieved the best solutions for each engineering optimization problem described in Appendix 1

Problem	x_1	x_2	x_3	x_4	x_5	x_6	$f_{Ai}(\mathbf{x})$ for $i = 1, \ldots, 6$
$f_{A1}(\mathbf{x})$	0.8213	1.3710	51.0045	91.4286			1.08E+04
$f_{A2}(\mathbf{x})$	19.2021	15.5090	42.8588	48.5356			2.02E−10
$f_{A3}(\mathbf{x})$	0.0500	0.4800	4.0565				7.25E−03
$f_{A4}(\mathbf{x})$	0.7886	0.4085					2.60E+02
$f_{A5}(\mathbf{x})$	0.2057	3.4724	9.0361	0.2058			1.71E+00
$f_{A6}(\mathbf{x})$	1.0340	5.0113	−1.0967	5.3284	1.8701	4.6197	3.08E+01

a specific order, following a control strategy that adjusts the exploration–exploitation balance during optimization. The algorithm also incorporates a mechanism to enhance population diversity by exchanging elements between the groups. Unlike other synchronous-asynchronous approaches, our scheme employs a simple mechanism to determine the number of elements and the processing order (synchronous or asynchronous), without relying on complex methods to select agents and strategies based on the population's current state.

GSAGWO incorporates several properties in its search strategy. By combining synchronous and asynchronous strategies, we aim to achieve a better balance between exploration and exploitation. The division and sequential processing of elements into groups allow for different search patterns to be implemented within the same process, enhancing the overall search strategy's effectiveness. Despite the division, results are shared globally. Information from each group's search results is used to determine the global processing order of the groups, significantly enhancing the potential of each search pattern. Lastly, the diversity jump process serves as an effective mechanism to artificially increase population diversity during the optimization procedure.

The GSAGWO algorithm has undergone testing on both CEC2017 benchmark functions and engineering problems. In addition, its performance has been compared to that of GWO, various modified versions of GWO, and other similar metaheuristic algorithms. The results affirm the effectiveness and accuracy of the discussed technique.

Appendix 1

This appendix provides a description of the engineering optimization problems that were implemented in the experiments.

Pressure Vessel Design Problem

The primary goal of the pressure vessel problem is to minimize the cost by considering specific variables: $\mathbf{x} = [x_1, x_2, x_3, x_4]$. Here, x_1 represents the shell's thickness T_s, x_2 corresponds to the head's thickness T_h, x_3 represents the internal radius of the vessel R, and x_4 represents the length of the vessel L. Further details about the problem can be found in Problem 5.1. The mathematical model for this problem is described below.

Problem 5.1 Pressure vessel design problem

Problem: Pressure Vessel Design

Minimize:

$$f_{B1}(\mathbf{x}) = 0.6224x_1x_2x_4 + 1.7781x_2x_3^2 + 3.1661x_1^2x_4 + 19.84x_1^2x_3$$

Subject to:

$$g_1(\mathbf{x}) = -x_1 + 0.0193x_3 \le 0$$
$$g_2(\mathbf{x}) = -x_2 + 0.00954x_3 \le 0$$
$$g_3(\mathbf{x}) = -\pi x_3^2x_4 - (4/3)\pi x_3^3 + 1{,}296{,}000 \le 0$$
$$g_4(\mathbf{x}) = x_4 - 240 \le 0$$

$$0 \le x_i \le 100, \quad i = 1,2$$
$$0 \le x_i \le 200, \quad i = 3,4$$

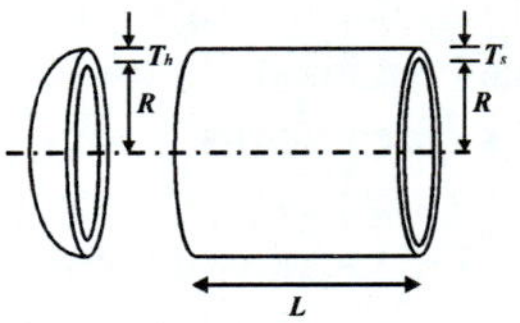

Gear Train Design Problem

In this case, the objective is to minimize the squared difference between the gear's teeth ratio and a specified scalar value, which is set as $1/6.931$. The variables considered in this problem are the number of teeth for each gear: A, B, D, and F. These variables are represented as $x_1 = A, x_2 = B, x_3 = D$ and $x_4 = F$ in the mathematical model. For further details, refer to Problem 5.2.

Problem 5.2 Pressure vessel design problem

Problem: Gear Train Design

Minimize:

$$f_{B2}(\mathbf{x}) = \left(\left(\frac{1}{6.931}\right) - \left(\frac{x_3x_2}{x_1x_4}\right)\right)^2$$

Subject to:

$$0 \le x_i \le 600, \quad i = 1,2,3,4$$

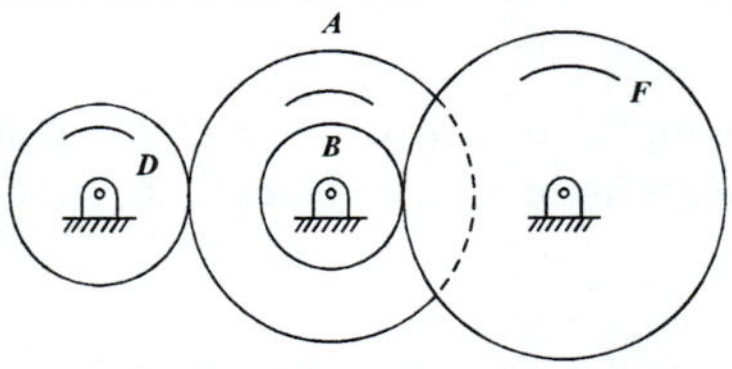

Tension/Compression Spring Design Problem

In this optimization problem, the objective is to minimize the tension or compression on a load represented by ***P*** applied to a spring. The variables in this problem are defined as follows: the wire diameter as ***d*** (x_1), the coil diameter as ***D*** (x_2), and the number of active coils as n (x_3). The mathematical model for this problem can be found in Problem 5.3.

Problem 5.3 Tension/compression spring design problem

Problem: Tension/Compression Spring Design

Minimize:

$$f_{B3}(\mathbf{x}) = (x_3 + 2)x_2x_1^2$$

Subject to:

$$g_1(\mathbf{x}) = 1 - \frac{x_2^3 x_3}{71{,}785x_1^4} \le 0$$

$$g_2(\mathbf{x}) = \frac{4x_2^2 - x_1x_2}{12{,}566(x_2x_1^3 - x_1^4)} + \frac{1}{5{,}108x_1^2} - 1 \le 0$$

$$g_3(\mathbf{x}) = 1 - \frac{140.45x_1}{x_2^2 + x_3} \le 0$$

$$g_4(\mathbf{x}) = \frac{x_1 + x_2}{1.5} \le 0$$

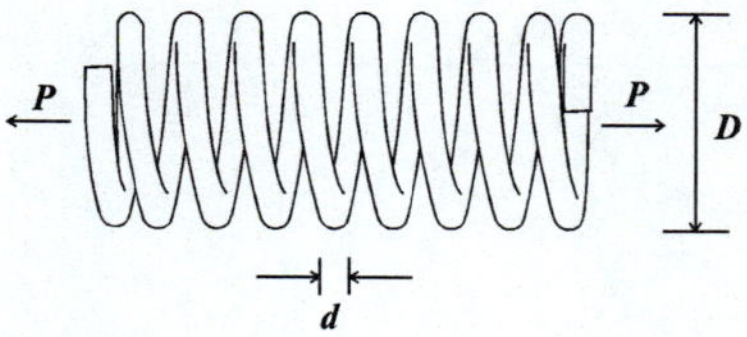

$$0.5 \le x_1 \le 2$$
$$0.25 \le x_2 \le 1.3$$
$$2 \le x_3 \le 15$$

The Three-Bar Truss Design Problem

The objective of this problem is to minimize the manufacturing cost of a three-bar system while taking into account specific load and stress constraints imposed on the system. The variables in this problem, denoted as x_1 and x_2, are described in Problem 5.4. The static parameters include the distance between the axes (l), the static load (P), and the external force (stress) (σ). The mathematical model for this problem is outlined as follows.

Problem 5.4 Tension/compression spring design problem

Problem: Three-Bar Truss Design

Minimize:

$$f_{B4}(\mathbf{x}) = l(2\sqrt{2x_1} + x_2)$$

Subject to:

$$g_1(\mathbf{x}) = \frac{\sqrt{2x_1} + x_2}{\sqrt{2x_1^2} + 2x_1x_2}P - \sigma \leq 0$$

$$g_2(\mathbf{x}) = \frac{x_2}{\sqrt{2x_1^2} + 2x_1x_2}P - \sigma \leq 0$$

$$g_3(\mathbf{x}) = \frac{1}{\sqrt{2x_2^2} + x_1}P - \sigma \leq 0$$

$$0 \leq x_i \leq 1, \quad i = 1,2$$

$$l = 100 \text{ cm}, \quad P = 2 \text{ kN/cm}^2, \quad \sigma = 2 \text{ kN/cm}^2$$

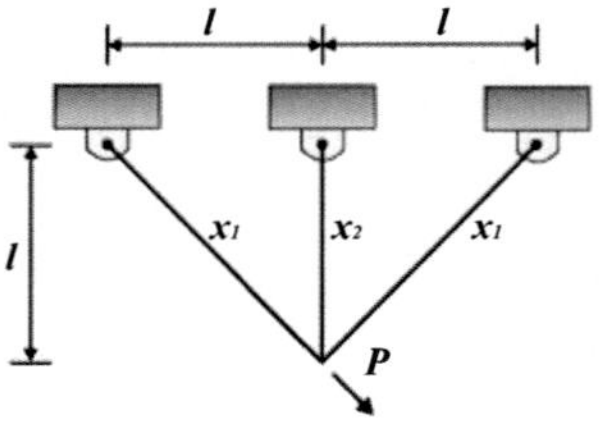

The Welded Beam Design Problem

The goal of this problem is to minimize the cost associated with welding applied to a beam placed on another surface, while ensuring that the welding meets specific stress criteria such as shear stress (τ), buckling load (P_c), bending stress (σ), and deflection (δ). The amount of welding is limited by the thickness (h) and length (l) assigned to the variables. Additionally, the design of the beam involves other important variables, such as width (t) and thickness (b). In the mathematical model, the variables are represented as $x_1 = h$, $x_2 = l$, $x_3 = t$, and $x_4 = b$. The model is formulated as follows.

Problem 5.5 Welded beam design problem

Problem: Welded Beam Design

Minimize:

$f_{B5}(\mathbf{x}) = 1.10471x_1^2x_2 + 0.04811x_3x_4(14 + x_2)$

Subject to:

$g_1(\mathbf{x}) = \tau(\mathbf{x}) - \tau_{\max} \geq 0$

$g_2(\mathbf{x}) = \sigma(\mathbf{x}) - \sigma_{\max} \leq 0$

$g_3(\mathbf{x}) = x_1 - x_4 \leq 0$

$g_4(\mathbf{x}) = 0.1047x_1^2 + 0.04811x_3x_4(14 + x_2) - 5 \leq 0$

$g_5(\mathbf{x}) = 0.125 - x_1 \leq 0$

$g_6(\mathbf{x}) = \delta(\mathbf{x}) - \delta_{\max} \leq 0$

$g_7(\mathbf{x}) = P - P_c(\mathbf{x})$

$0.1 \leq x_i \leq 2, \quad i = 1,4$

$0.1 \leq x_i \leq 10, \quad i = 2,3$

$\sigma(\mathbf{x}) = \frac{6PL}{x_4x_3}, \quad P_c(\mathbf{x})$

$= \frac{4.013E\sqrt{\frac{x_3^2x_4^6}{36}}}{L^2}\left(1 - \frac{x_3}{2L}\sqrt{\frac{E}{4G}}\right)$

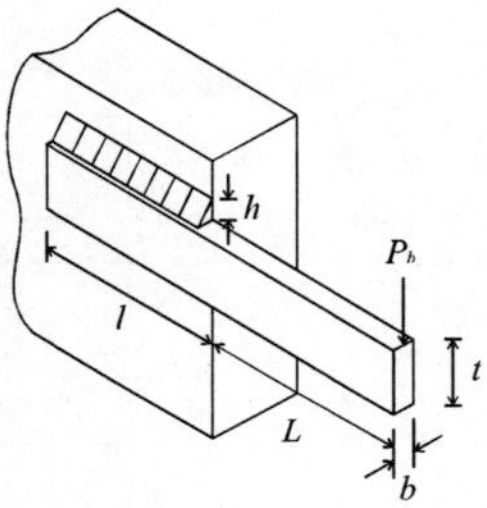

$P = 6000$ lb, $L = 14$ in, $E = 30 \times 10^6$ psi, $G = 12 \times 10^6$ psi

$\tau_{\max} = 13{,}600$ psi, $\sigma_{\max} = 30{,}000$ psi, $\delta_{\max} = 0.25$ in

Parameter Estimation for FM Synthesizers

The primary goal of this problem is to generate a sound signal, denoted as $y(\mathbf{x}, t)$, that matches another given sound signal, $y_0(t)$. The variables involved in the problem are represented as $\mathbf{x} = [x_1 = a_1, x_2 = \omega_1, x_3 = a_2, x_4 = \omega_2, x_5 = a_3, x_6 = \omega_3]$, where a_i and ω_i correspond to the finite wave amplitude and frequency, respectively. The objective function aims to minimize the error between the difference of the two signals, $y(\mathbf{x}, t)$ and $y_0(t)$. The mathematical description of the problem is as follows.

Problem 5.6 Parameter estimation for FM synthesizer

Problem: Parameter Estimation for FM Synthesizers

Minimize:

$$f_{B6}(\mathbf{x}) = \sum_{t=0}^{100} \left(y(\mathbf{x}, t) - y_0(t)\right)^2$$

$$y(x, t) = x_1 \sin(x_2 \theta t) + x_2 \sin(x_4 \theta t) + x_5 \sin(x_6 \theta t)$$

Subject to:

$$-6.4 \leq x_i \leq 6.35, \quad i = 1,2,3,4,5,6$$

$$\theta = \frac{2\pi}{100}$$

References

1. Pardalos, P.M., Romeijn, H.E., Tuy, H.: Recent developments and trends in global optimization. J. Comput. Appl. Math. **124**, 209–228 (2000)
2. Yang, X.-S.: Engineering Optimization: An Introduction with Metaheuristic Application. Wiley, USA (2010)
3. Arpan Kumar Kar: Bio inspired computing—a review of algorithms and scope of applications. Expert Syst. Appl. **59**, 20–32 (2016)
4. Karaboga, D., Ozturk, C.: A novel clustering approach: artificial bee colony (ABC) algorithm. Appl. Soft Comput. **11**(1), 652–657 (2011)
5. Bhandari, A.K., Singh, V.K., Kumar, A., Singh, G.K.: Cuckoo search algorithm and wind driven optimization based study of satellite image segmentation for multilevel thresholding using Kapur's entropy. Expert Syst. Appl. **41**(7), 3538–3560 (2014)
6. Yang, X.S., Hosseini, S.S.S., Gandomi, A.H.: Firefly algorithm for solving non-convex economic dispatch problems with valve loading effect. Appl. Soft Comput. **12**(3), 1180–1186 (2012)
7. Kavousi-Fard, A., Samet, H., Marzbani, F.: A new hybrid modified firefly algorithm and support vector regression model for accurate short term load forecasting. Expert Syst. Appl. **41**(13), 6047–6056 (2014)
8. Hong, W.C.: Chaotic particle swarm optimization algorithm in a support vector regression electric load forecasting model. Energy Convers. Manage. **50**(1), 105–117 (2009)
9. Kumar, M., Rawat, T.K.: Optimal design of FIR fractional order differentiator using cuckoo search algorithm. Expert Syst. Appl. **42**(7), 3433–3449 (2015)
10. Long, N.C., Meesad, P., Unger, H.: A highly accurate firefly based algorithm for heart disease prediction. Expert Syst. Appl. **42**(21), 8221–8231 (2015)
11. Gogna, A., Tayal, A.: Metaheuristics: review and application. J. Exp. Theor. Artif. Intell. **25**(4), 503–526 (2013)
12. Grefenstette, J.J.: Genetic algorithms and their applications. In: Proceedings of the Second International Conference on Genetic Algorithms. Psychology Press, Hillsdale, New Jersey (2013)
13. Karaboga, D., Gorkemli, B., Ozturk, C., Karaboga, N.: A comprehensive survey: Artificial bee colony (ABC) algorithm and applications. Artif. Intell. Rev. **42**(1), 21–57 (2014)

14. Yang, X.-S.: Nature-Inspired Optimization Algorithms. Elsevier (2014)
15. Leboucher, C., et al.: An enhanced particle swarm optimization method integrated with evolutionary game theory. IEEE Trans. Games **10**(2), 221–230 (2018)
16. Fausto, F., Cuevas, E., Valdivia, A., González, A.: A global optimization algorithm inspired in the behavior of selfish herds. Biosystems **160**, 39–55 (2017)
17. Gao, H., Shi, Y., Pun, C.M., Kwong, S.: An improved artificial bee colony algorithm with its application. IEEE Trans. Ind. Informatics **15**(4), 1853–1865 (2019)
18. Mirjalili, S., Mirjalili, S.M., Lewis, A.: Grey wolf optimizer. Adv. Eng. Softw. **69**, 46–61 (2014)
19. Shi, Z., Li, Q., Zhang, S., Huang, X.: Improved crow search algorithm with inertia weight factor and roulette wheel selection scheme. In: Proceedings of 2017 10th International Symposium on Computational Intelligence and Design (ISC), vol. 1, 2018-January, pp. 205–209 (2017)
20. Mirjalili, S., Lewis, A.: The whale optimization algorithm. Adv. Eng. Softw. **95**, 51–67 (2016)
21. Cai, Z., Niu, J., Yang, X.: A multi measure improved firefly algorithm. In: Proceedings of 2018 2nd IEEE Advanced Information Management Communicates Electronic and Automation Control Conference IMCEC 2018, no. Imcec, pp. 20–26 (2018)
22. Xu, L., et al.: Enhanced moth-flame optimization based on cultural learning and gaussian mutation. J. Bionic. Eng. **15**(4), 751–763
23. Zaldívar, D., Morales, B., Rodríguez, A., Valdivia, A., Cuevas, G.E., Pérez-Cisneros, M.: A novel bio-inspired optimization model based on Yellow Saddle Goatfish behavior. BioSystems **174**(March), 1–21 (2018)
24. Lynn, N., Suganthan, P.N.: Ensemble particle swarm optimizer. Appl. Soft Comput. J. **55**, 533–548 (2017)
25. Lynn, N., Suganthan, P.N.: Heterogeneous comprehensive learning particle swarm optimization with enhanced exploration and exploitation. Swarm Evol. Comput. **24**, 11–24 (2015)
26. Meng, X.B., Gao, X.Z., Lu, L., Liu, Y., Zhang, H.: A new bio-inspired optimisation algorithm: Bird Swarm Algorithm. J. Exp. Theor. Artif. Intell. **28**(4), 673–687 (2016)
27. Dhiman, G., Kumar, V.: Spotted hyena optimizer: a novel bio-inspired based metaheuristic technique for engineering applications. Adv. Eng. Softw. **114**, 48–70 (2017)
28. Wang, G.-G., Deb, S., Coelho, L.S.D.S.: Earthworm optimization algorithm: a bioinspired metaheuristic algorithm for global optimization problems. Int. J. BioInspired Comput. (2015)
29. Wang, G.G.: Moth search algorithm: a bio-inspired metaheuristic algorithm for global optimization problems. Memet. Comput. **10**(2), 151–164 (2018)
30. Dhiman, G., Kumar, V.: Emperor penguin optimizer: a bio-inspired algorithm for engineering problems. Knowl.-Based Syst. **159**(May), 20–50 (2018)
31. Kaveh, A., Kooshkebaghi, M.: Artificial coronary circulation system: a new bio-inspired metaheuristic algorithm. Sci. Iran. **26**(5A), 2731–2747 (2019)
32. Tanabe, R., Fukunaga, A.S.: Improving the search performance of SHADE using linear population size reduction. In: Proceedings on 2014 IEEE Congress Evolutionary Computation (CEC), pp. 1658–1665 (2014)
33. Awad, N.H., Ali, M.Z., Suganthan, P.N.: Ensemble sinusoidal differential covariance matrix adaptation with Euclidean neighborhood for solving CEC2017 benchmark problems. In: 2017 IEEE Congress Evolutionary Computation (CEC), pp. 372–379 (2017)
34. Mohamed, A.W., Hadi, A.A., Fattouh, A.M., Jambi, K.M.: LSHADE with semi-parameter adaptation hybrid with CMA-ES for solving CEC 2017 benchmark problems. In: 2017 IEEE Congress on Evolutionary Computation (CEC), pp. 145–152 (2017)
35. Viktorin, A., Senkerik, R., Pluhacek, M., Kadavy, T., Zamuda, A.: Distance based parameter adaptation for Success-History based differential evolution. Swarm Evol. Comput. **50** (2019)
36. Chen, P., Li, Q., Zhang, C., Cui, J., Zhou, H.: Hybrid chaos-based particle swarm optimization-ant colony optimization algorithm with asynchronous pheromone updating strategy for path planning of landfill inspection robots. Int. J. Adv. Rob. Syst. (2019). https://doi.org/10.1177/1729881419859083
37. Aziz, NAA, Ibrahim, Z., Aziz, N.H.A., Rahman, T.A.: Asynchronous simulated kalman filter optimization algorithm. Int. J. Eng. Technol. 7.4.27:44–49. Web. 12 Mar. 2020 (2018)

38. Aziz, N.A.A., Ibrahim, Z.: Asynchronous particle swarm optimization for swarm robotics. Procedia Eng. **41**, 951–957 (2012)
39. Aziz, N.A.A., et al.: Synchronous vs asynchronous gravitational search algorithm. In: 2013 1st International Conference on Artificial Intelligence, Modelling and Simulation, Kota Kinabalu, pp. 37–42 (2013). https://doi.org/10.1109/AIMS.2013.14
40. Ab Aziz, N.A., Mubin, M., Mohamad, M.S., Ab Aziz, K.: A synchronous-asynchronous particle swarm optimisation algorithm. Sci. World J. 17 (2014)
41. Aziz, N.A.A.A., Ibrahim, Z., Aziz, N.H.A., Mubin, M., Mokhtar, N., Shapiai, M.I.: A fitness-based adaptive synchronous-asynchronous switching in simulated kalman filter optimizer. In: 2019 International Conference on Computer and Information Sciences (ICCIS), Sakaka, Saudi Arabia, pp. 1–5 (2019)
42. Ab Aziz, N.A., Ibrahim, Z., Mubin, M., et al.: Adaptive switching gravitational search algorithm: an attempt to improve diversity of gravitational search algorithm through its iteration strategy. Sādhanā **42,** 1103–1121 (2017). https://doi.org/10.1007/s12046-017-0674-0
43. Aziz, N.A.A., Ibrahim, Z., Mubin, M., Nawawi, S.W., Mohamad, M.S.: Improving particle swarm optimization via adaptive switching asynchronous—synchronous update. Appl. Soft Comput. **72**, 298–311 (2018)
44. Lu, C., Xiao, S., Li, X., Gao, L.: An effective multi-objective discrete grey wolf optimizer for a real-world scheduling problem in welding production. Adv. Eng. Softw. **99**, 161–176 (2016)
45. Li, S.X., Wang, J.S.: Dynamic modeling of steam condenser and design of pi controller based on grey wolf optimizer. Math. Probl. Eng. **2015**, 9 (2015)
46. Wong, L.I., Sulaiman, M.H., Mohamed, M.R., Hong, M.S.: Grey wolf optimizer for solving economic dispatch problems. In: 2014 IEEE International Conference on Power and Energy (PECon), pp. 150–154 . IEEE (2014)
47. Devi, E.M., Suganthe, R.C.: Feature selection in intrusion detection grey wolf optimizer. Asian J. Res. Soc. Sci. Hum. **7**(3), 671–682 (2017)
48. Al-Aboody, N.A., Al-Raweshidy, H.S.: Grey wolf optimization-based energy-efficient routing protocol for heterogeneous wireless sensor networks. In: 2016 4th International Symposium on Computational and Business Intelligence (ISCBI), pp. 101–107. IEEE (2016)
49. Faris, H., Aljarah, I., Al-Betar, M.A., Mirjalili, S.: Grey wolf optimizer: a review of recent variants and applications. Neural Comput. Appl. **30**(2), 413–435 (2018)
50. Mirjalili, S., Saremi, S., Mirjalili, S.M., Coelho, L.S.: Multi-objective grey wolf optimizer: a novel algorithm for multi-criterion optimization. Expert Syst. Appl. **47**, 106–119 (2016)
51. Pan, T.S., Dao, T.K., Chu, S.C., et al.: A communication strategy for paralleling grey wolf optimizer. In: International Conference on Genetic and Evolutionary Computing. Springer, pp. 253–262 (2015)
52. Kamboj, V.K.: A novel hybrid pso-gwo approach for unit commitment problem. Neural Comput. Appl. **27**(6), 1643–1655 (2016)
53. Jitkongchuen, D.: A hybrid differential evolution with grey wolf optimizer for continuous global optimization. In: 2015 7th International Conference on Information Technology and Electrical Engineering (ICITEE). IEEE, pp. 51–54 (2015)
54. Zhu, A., Xu, C., Li, Z., Wu, J., Liu, Z.: Hybridizing grey wolf optimization with differential evolution for global optimization and test scheduling for 3d stacked soc. J. Syst. Eng. Electron. **26**(2), 317–328 (2015)
55. Mittal, N., Singh, U., Sohi, B.S.: Modified grey wolf optimizer for global engineering optimization. Appl. Comput. Intell. Soft Comput. 8 (2016)
56. Al-Betar, M.A., Awadallah, M.A., Faris, H., Aljarah, I., Hammouri, A.I.: Natural selection methods for Grey Wolf Optimizer. Expert Syst. Appl. **113**, 481–498 (2018)
57. Malik, M.R.S., Mohideen, E.R., Ali, L.: Weighted distance grey wolf optimizer for global optimization problems. In: 2015 IEEE International Conference on Computational Intelligence and Computing Research (ICCIC). IEEE, pp. 1–6 (2015)
58. Luo, Q., Zhang, S., Li, Z., Zhou, Y.: A novel complex-valued encoding grey wolf optimization algorithm. Algorithms **9**(1), 4 (2015)

59. Saremi, S., Mirjalili, S.Z., Mirjalili, S.M.: Evolutionary population dynamics and grey wolf optimizer. Neural Comput. Appl. **26**(5), 1257–1263 (2015)
60. Mech, L.D.: Alpha status, dominance, and division of labor in wolf packs. Can. J. Zool. **77**, 1196–1203 (1999)
61. Muro, C., Escobedo, R., Spector, L., Coppinger, R.: Wolf-pack (Canis lupus) hunting strategies emerge from simple rules in computational simulations. Behav. Process. **88**, 192–197 (2011)
62. Back, T.: Selective pressure in evolutionary algorithms: a characterization of selection mechanisms. In: IEEE World Congress on Computational Intelligence, Proceedings of the First IEEE Conference on Evolutionary Computation, pp. 57–62. IEEE (1994)
63. Miller, B.L., Goldberg, D.E.: Genetic algorithms, selection schemes, and the varying effects of noise. Evol. Comput. **4**(2), 113–131 (1996)
64. Holland, J.H.: Adaptation in natural and artificial systems. The University of Michigan Press, Ann Arbor (1975)
65. Goldberg, D., Deb, K., Korb, B.: Messy genetic algorithms: motivation, analysis, and first results. Complex Syst. **3**, 493–530 (1989)
66. Long, W., Liang, X., Cai, S., Jiao, J., Zhang, W.: A modified augmented Lagrangian with improved grey wolf optimization to constrained optimization problems. Neural Comput. Appl. **28**(Suppl), S421–S438 (2016)
67. Qin, A.K., Suganthan, P.N.: Self-adaptive differential evolution algorithm for numerical optimization. In: 2005 IEEE Congress on Evolutionary Computer, vol. 2, pp. 1785–1791
68. Sandgren, E.: Nonlinear integer and discrete programming in mechanical design optimization J. Mech. Des. **112**(2), 223 (1990)
69. Arora, J.S.: Introduction to Optimum Design. New York (1989)
70. Rao, S.S.: Engineering Optimization: Theory and Practice. John Wiley & Sons (2009)
71. Koski, J.: Defectiveness of weighting method in multicriterion optimization of structures. Commun. Appl. Numer. Methods **1**(6), 333–337 (1985)
72. Das, S., Suganthan, P.N.: Problem Definitions and Evaluation Criteria for CEC 2011 Competition on Testing Evolutionary Algorithms on Real World Optimization Problems (2010)
73. López-Ibáñez, M., Dubois-Lacoste, J., Pérez Cáceres, L., Birattari, M., Stützle, T.: The irace package: Iterated racing for automatic algorithm configuration. Oper. Res. Perspect. **3**, 43–58 (2016)
74. Wilcoxon, F.: Individual comparisons by ranking methods. Biom. Bull.
75. Van Hecke, T.: Power study of Anova versus Kruskal-Wallis test. J. Stat. Manag. Syst. **15**(2–3), 241–247 (2012)
76. Cuevas, E., Gálvez, J., Hinojosa, S., Avalos, O., Zaldívar, D., Pérez-Cisneros, M.: A comparison of evolutionary computation techniques for IIR model identification. J. Appl. Math. (2014)
77. Zaldivar, D., Morales, B., Rodríguez, A., Valdivia-G, A., Cuevas, E., Pérez-Cisneros, M.: A novel bio-inspired optimization model based on Yellow Saddle Goatfish behavior. Biosystems **174**, 1–21 (2018)
78. Cuevas, E., Zaldivar, D., Pérez-Cisneros, M.: Seeking multi-thresholds for image segmentation with learning automata. Mach. Vis. Appl. **22**, 805–818 (2011)
79. Cuevas, E., Wario, F., Osuna-Enciso, V., Zaldivar, D., Pérez-Cisneros, M.: Fast algorithm for multiple-circle detection on images using learning automata. IET Image Proc. **6**(8), 1124–1135 (2012)
80. Hinojosa, S., Dhal, K.G., Abd Elaziz, M., Oliva, D., Cuevas, E.: Entropy-based imagery segmentation for breast histology using the stochastic fractal search. Neurocomputing **321**, 201–215 (2018)
81. Bandyopadhyay, R., Basu, A., Cuevas, E., Sarkar, R.: Harris Hawks optimisation with Simulated Annealing as a deep feature selection method for screening of COVID-19 CT-scans. Appl. Soft Comput. **111**, 107698 (2021)
82. Cuevas, E., González, A., Fausto, F., Zaldívar, D., Pérez-Cisneros, M.: Multithreshold segmentation by using an algorithm based on the behavior of locust swarms. Math. Probl. Eng. (2015)

83. Ibrahim, R.A., Elaziz, M.A., Oliva, D., Cuevas, E., Lu, S.: An opposition-based social spider optimization for feature selection. Soft. Comput. **23**, 13547–13567 (2019)
84. Ramírez-Ortegón, M.A., Tapia, E., Ramírez-Ramírez, L.L., Rojas, R., Cuevas, E.: Transition pixel: a concept for binarization based on edge detection and gray-intensity histograms. Pattern Recogn. **43**(4), 1233–1243 (2010)

Chapter 6
Efficient Image Contrast Enhancement by Using the Moth Swarm Algorithm

Image Contrast Enhancement (ICE) plays a vital role in various applications related to image processing and computer vision. Its primary aim is to enhance the quality of visual information in processed images. However, the presence of noise and small pixel sets in images not only hampers their visualization but also hinders the effectiveness of ICE algorithms. The inclusion of irrelevant information disrupts the appropriate distribution of significant pixel intensities, leading to issues such as undesirable artifacts, noise amplification, over saturation, and compromised human visual perception in most existing ICE methods. This chapter introduces a novel ICE method for grayscale and color images that addresses these problems. The presented approach effectively eliminates noisy and irrelevant information to improve the distribution of significant pixel intensities in the enhanced image. It achieves this by removing multiple groups of pixels that do not represent any objects or important image details. This elimination process utilizes the Mean-shift algorithm, which replaces these irrelevant pixel sets in the original histogram with local maxima representing significant pixel densities. Subsequently, the Moth Swarm Algorithm (MSA) is employed to redistribute the pixel intensities of the reduced histogram, maximizing the value derived from Kullback–Leibler entropy (KL-entropy). To evaluate the effectiveness of the presented approach, we conducted experiments using various publicly available datasets commonly used in the literature. We also compared our results with those obtained from other well-known ICE techniques. The experimental results demonstrate that the presented approach successfully highlights important image details while improving the overall visual appearance for human observers.

E. Cuevas et al., *New Metaheuristic Schemes: Mechanisms and Applications*, Intelligent Systems Reference Library 246, https://doi.org/10.1007/978-3-031-45561-2_6

6.1 Introduction

Knowledge-Based Systems (KBS) in the field of artificial intelligence involve the extraction and representation of engineering knowledge. In computer vision, the core of KBS lies in the extraction of information through optimization principles. Many image analysis problems within KBS can be simplified as optimization processes. In this approach, the image's information is assessed based on a knowledge base comprising essential quality characteristics. Through a search strategy, the optimal image is identified as the solution to the KBS problem.

Image enhancement (IE) is a computer vision task that can be viewed as a Knowledge-Based Systems (KBS) problem. Its broad range of applications in fields like medicine, security, and transportation has garnered significant interest from the computer vision community [1, 2, 21, 31, 35]. IE aims to enhance the visual information within an image, emphasizing differences among object features. The primary objective is to improve human interpretability or enable further processing within automated computer vision systems [33, 35].

Commonly used IE methods include histogram equalization (HE), quadratic transformation, and fuzzy logic operations [1, 2, 21]. Among these techniques, HE is the most widely employed due to its simplicity and effectiveness in image enhancement. HE considers the statistical properties of pixels and redistributes them across the entire intensity scale, regardless of the number of pixels corresponding to a particular intensity value [24, 30, 31, 33, 34]. However, these methods face challenges when confronted with noise or irrelevant small pixel sets, resulting in inadequate redistribution of pixel data. Consequently, these approaches often produce enhanced images with undesired artifacts and amplified noise [9, 11, 25].

In recent years, the problem of Image Contrast Enhancement (ICE) has been addressed using metaheuristic techniques as an alternative approach to designing ICE schemes. Unlike histogram equalization (HE) techniques, metaheuristic methods tackle contrast enhancement as an optimization problem. These methods consider certain aspects of the enhancement process as an objective function, with the optimal solution representing a high-quality enhanced image. Generally, contrast enhancement algorithms based on metaheuristic schemes have shown better results in terms of pixel intensity redistribution compared to HE [2].

Several examples of these approaches include the utilization of metaheuristic algorithms such as Artificial Bee Colony (ABC) [17], Genetic Algorithms (GA) [11], Particle Swarm Optimization (PSO) [18], Social Spider Optimization (SSO) [9, 25], Cuckoo Search Algorithm (CS) [4], Firefly Algorithm (FFA) [37], Gravitational Search Algorithm (GSA) [29], Differential Evolution (DE) [8], among others. While these methods have demonstrated interesting results in image enhancement, they suffer from several limitations, such as premature convergence to local optima and a lack of population diversity. In these methods, the redistribution of pixel data is evaluated using objective functions [11] that primarily consider heterogeneity. The higher the variation in pixel intensities in the enhanced image, the better the quality of distribution. However, these approaches have limited effectiveness in improving contrast enhancement in the presence of noise and irrelevant small sets of pixels. To address this issue, it is crucial to remove irrelevant visual information prior to redistribution. Recently, hybrid approaches that combine metaheuristic techniques with other computational schemes have been introduced in the literature for image contrast enhancement. These methods aim to enhance the capabilities of the metaheuristic approach by incorporating specific computational elements that provide additional information on the contrast enhancement process. For instance, the Gamma-PSO algorithm proposed by Mahmood et al. [21] determines the optimal values of a Gamma correction using PSO to enhance the contrast of a previously transformed image with wavelet transform. Another interesting approach is the Fuzzy-IPSO method described by Subrahmanyeswara-Rao [32], which combines fuzzy logic and PSO to enhance image contrast. This method utilizes a fuzzy system to obtain histogram information, and the histogram is then adjusted using weights computed through an improved version of the PSO algorithm.

The Moth Swarm Algorithm (MSA) [26] is a metaheuristic approach inspired by the orientation behavior of moths towards moonlight. Unlike other metaheuristic schemes, MSA incorporates three distinct sub-populations. Each individual in the sub-population is subjected to a different evolutionary operation, reflecting specific behaviors observed in moth insects. By integrating these operators into the search strategy, MSA addresses critical issues commonly found in other metaheuristic algorithms, such as imbalanced exploration–exploitation trade-off and premature convergence. The unique characteristics of MSA have led to its widespread application in various complex engineering problems, including image processing [5], power systems [26], energy conversion [10], and more.

The presence of noise and small sets of pixel intensities in images often indicates the existence of insignificant or irrelevant details that obscure important image information. These elements refer to sets of pixels that are too small to represent objects or visible details in the image. Such information not only hampers human visualization but also negatively impacts the image enhancement process. Eliminating this irrelevant information becomes essential to improve the redistribution of pixel data in the enhanced image. By removing these unnecessary pixel intensities, the space previously occupied by them can be effectively utilized to highlight relevant pixel intensities during redistribution. The Mean-shift algorithm [3] is a non-parametric technique widely employed to replace sparsely existing features $\mathbf{x}$ with a densely contained element within their neighborhood (local maximum) $\mathbf{x}^*$. This algorithm starts from a point $\mathbf{x}$ and computes a new location in the direction of the greatest increase in the feature space until reaching the local maximum $\mathbf{x}^*$. Mean-shift's simplicity and favorable properties have led to its application in various domains, including regression [3] and clustering [36].

On the contrary, the Kullback–Leibler entropy (KL-entropy) [6, 19], originally introduced by Kullback and Leibler, provides a means to objectively measure the divergence between two probability distributions. Its primary purpose is to quantify the difference between a given probability distribution and a reference distribution. The KL-entropy value is minimal when the two distributions are completely dissimilar, while it reaches its maximum when they closely resemble each other. The

KL-entropy finds applications in various domains, such as assessing randomness in continuous time-series, characterizing relative (Shannon) entropy in information systems, and measuring the information content when comparing different statistical inference models.

Unlike other entropy measures, the KL-entropy not only helps us identify the optimal data distribution in terms of information content but also determines the distribution that best resembles the original image data. Consequently, utilizing the KL-entropy enables us to enhance image contrast while maintaining a close resemblance to the original information, thereby improving the visual appearance of the images without introducing distortions.

This chapter introduces a novel algorithm for Image Contrast Enhancement (ICE) of images. Distinguished from existing approaches, the presented method incorporates an initial step of eliminating irrelevant sets of pixel intensities. This elimination process aims to reduce the number of histogram features, enabling the utilization of the full histogram range during the enhancement process. The Mean-shift method is employed to identify and remove insignificant small sets of pixels by replacing sporadic intensities with more abundant features. Once the reduced histogram is obtained, the contrast improvement process is performed.

In this process, the Moth Swarm Algorithm (MSA) is adopted to redistribute the pixel intensities within the complete range of the reduced histogram. The objective is to maximize the Kullback–Leibler Entropy (KL-entropy) between a candidate distribution and the original image information. This chapter presents three key contributions:

(I) The introduction of an elimination process to reduce the number of histogram features, enhancing the redistribution of pixel intensities. This process is crucial for achieving significant contrast enhancement. To the best of our knowledge, no comprehensive study has previously addressed the incorporation of an elimination process within an ICE scheme to improve the redistribution process.
(II) The integration of KL-entropy for the redistribution process, leading to enhanced images with improved human visual appearance.
(III) The utilization of the Moth Swarm Algorithm (MSA) to find the optimal redistribution that best represents the enhanced image. The search characteristics of MSA facilitate effective exploration of the solution space without encountering the associated issues of other metaheuristic algorithms.

The presented scheme's performance is evaluated on representative datasets commonly used in the literature. Comparative analysis is conducted against other well-known techniques in the field. Experimental results demonstrate that the presented method outperforms existing schemes across multiple performance indexes.

The remaining sections of this chapter are structured as follows: Sect. 6.2 provides a concise overview of the Mean-shift algorithm; Sect. 6.3 discusses the key aspects of the ICE problem; Sect. 6.4 highlights the crucial elements of the MSA; Sect. 6.5

presents the discussed approach; Sect. 6.6 presents the experimental results; and finally, Sect. 6.7 concludes the chapter by summarizing the findings and implications.

6.2 Mean-Shift Scheme

The Mean-shift algorithm [3] is a non-parametric technique commonly employed to eliminate irrelevant information present in a dataset. This method replaces a sparsely represented feature $\mathbf{x}$ in the feature space with a densely located element $\mathbf{x}^*$ that corresponds to a local maximum. The Mean-shift algorithm initiates at a random point $\mathbf{x}$ within the density feature map and computes a new position in the direction of the steepest increase of the density feature space until reaching the local maximum density $\mathbf{x}^*$. The Mean-shift algorithm consists of two main processes. Firstly, a Kernel function is applied to estimate the probability density function (PDF) of the data. Then, an optimization step based on gradient ascent is performed to locate the local maxima (density attractors) within the PDF [13].

6.2.1 Estimation of the Probability Density Function

The mean-shift algorithm estimates the probability density function by defining a symmetric kernel function K, which is non-negative and satisfies the following conditions:

$$K(\mathbf{x}) \geq 0; \quad K(-\mathbf{x}) = K(\mathbf{x}); \quad \int K(\mathbf{x})dx = 1 \tag{6.1}$$

Given a set of n features, denoted as $\{\mathbf{x}_1, \mathbf{x}_2, \ldots, \mathbf{x}_n\}$, the density function $\hat{f}(\mathbf{x})$ can be estimated using the kernel function K as follows:

$$\hat{f}(\mathbf{x}) = \frac{1}{nh}\sum_{i=1}^{n} K\left(\frac{\mathbf{x} - \mathbf{x}_i}{h}\right), \tag{6.2}$$

where h represents the influence area of K.

6.2.2 Attractors

The primary concept behind estimating a probability density function (PDF) for a given dataset is to identify characteristic points $\mathbf{x}^*$ that accurately represent the distribution of the data points being modeled. These points correspond to the local maxima found in the estimated PDF $\hat{f}(\mathbf{x})$. To locate these points, a gradient-based optimization method is employed. The procedure starts by initializing a random point $\mathbf{x}$. Then, a new position is computed by moving in the direction of the density gradient $\nabla\hat{f}(\mathbf{x})$ until the local maximum $\mathbf{x}^*$ is reached. The update rule can be expressed as follows:

$$\mathbf{x}^{t+1} = \mathbf{x}^t + \delta \cdot \nabla\hat{f}(\mathbf{x}^t), \tag{6.3}$$

In the given expression, t represents the current iteration number, and δ represents the step size used in the algorithm. In this formulation, the density gradient at a specific point $\mathbf{x}$ is determined by evaluating the derivative of the probability density function (PDF) using the following equation:

$$\nabla\hat{f}(x) = \frac{\partial}{\partial x}\hat{f}(x) = \frac{1}{nh}\sum_{i=1}^{n}\frac{\partial}{\partial x}K\left(\frac{x - x_i}{h}\right) \tag{6.4}$$

Figure 6.1 provides a visual depiction of the mean-shift process. In this illustration, Fig. 6.1a represents the original dataset, while Fig. 6.1b displays the estimated probability density function (PDF) $\hat{f}(\mathbf{x})$ obtained using Eq. (6.2). The mean-shift procedure involves starting at a specific point $\mathbf{x}$ and iteratively computing a new location that moves in the direction of the steepest increase in the feature space. This process continues until a local maximum $\mathbf{x}^*$ is reached. Consequently, the feature $\mathbf{x}$ with a low density is replaced by $\mathbf{x}^*$, which serves as a representative feature from its neighborhood.

6.3 Image Contrast Enhancement Considering as an Optimization Problem

An optimization-based approach is employed for contrast enhancement, where the contrast enhancement process is linked to an objective function. This objective function is used to evaluate and compare different histogram distributions as potential solutions to the contrast enhancement problem. Each histogram is represented by a 256-dimensional vector, and the objective function measures the quality of the enhancement achieved by each histogram. To find the optimal histogram, a metaheuristic approach is applied, iteratively improving the solutions (histograms) in each generation of the optimization process based on the values of the objective function. Typically, the goal is to maximize the objective function, meaning that lower values indicate better-improved results in terms of contrast enhancement.

Figure 6.2 depicts the enhancement process within an optimization framework. In the figure, a solution labeled as A undergoes modifications using the operators of the optimization approach. This modification leads to the creation of a new solution

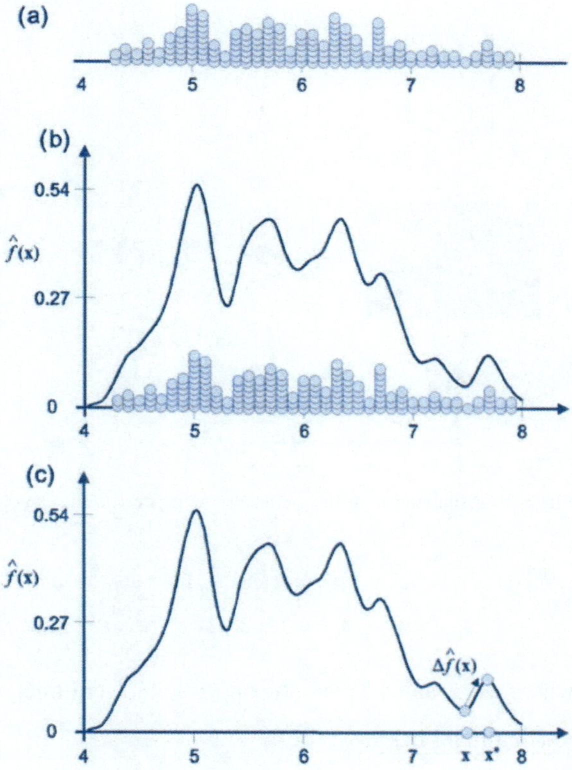

Fig. 6.1 Explanation of the mean-shift process: **a** An example of a dataset, **b** The estimated probability density function $\hat{f}(\mathbf{x})$ obtained using Eq. (6.2), and **c** The replacement of $\mathbf{x}$ by $\mathbf{x}^*$ as a result of the density gradient operation

labeled as B. As a result of these modifications, solution B exhibits an improved enhancement index when evaluated by the objective function.

6.4 Moth Swarm Algorithm (MSA)

The MSA [26] is a modern swarm algorithm specifically developed for addressing global optimization problems. It introduces a division of the swarm population into three distinct groups: Pathfinders, Prospectors, and Onlookers. The subsequent subsections provide a concise overview of the complete MSA algorithm. For a more comprehensive understanding, it is recommended to refer to the original work by [26].

Initial Phase

In the MSA methodology, the population consists of a group of p individual moths represented as $\mathbf{M} = \{\mathbf{m}_1, \mathbf{m}_2, \ldots \mathbf{m}_p\}$, where each $\mathbf{m}_i = [m_{i,1}, m_{i,2}, \ldots m_{i,n}]$ denotes a potential solution. In the initialization phase, the search agents randomly assume positions based on the following model:

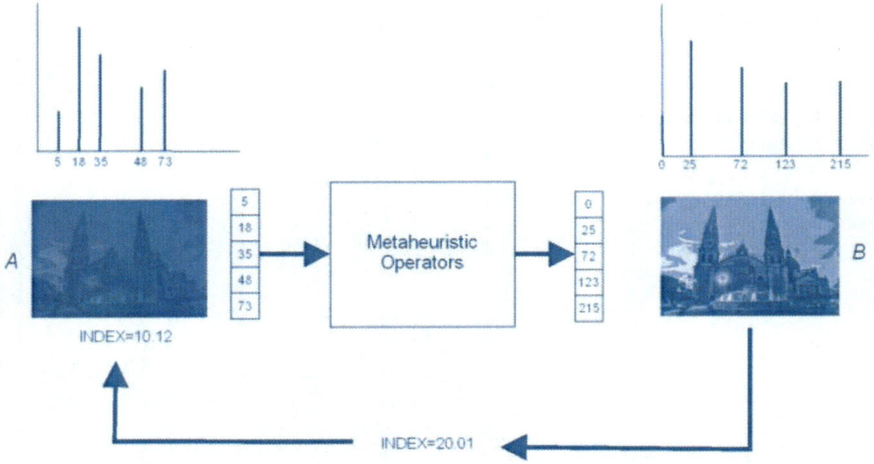

Fig. 6.2 Visualization of the enhancement process within an optimization framework

$$m_{ij} = \text{rand}(0, 1) \cdot (m_j^{\max} - m_j^{\min}) + m_j^{\min}$$
$$\forall\, i \in \{1, 2, \ldots, p\}, \quad j \in \{1, 2, \ldots, n\} \tag{6.5}$$

where, $m_j^{\max}$ and $m_j^{\min}$ are the upper and lower limits of the search space.

<u>Recognition Phase</u>

During this phase, every candidate solution undergoes two operations: crossover and Lévy perturbation [14]. The crossover point is determined using the following expression:

$$\mu_j^i = \frac{\sqrt{\frac{1}{p_n}\sum_{a=1}^{p_n}\left(m_{aj}^i - \overline{m}_j^i\right)^2}}{\overline{m}_j^i} \qquad \overline{m}_j^i = \frac{1}{p_n}\sum_{a=1}^{p_n} m_{aj}^i \tag{6.6}$$

With p_n representing the number of pathfinder moths, the new solution is calculated using the following equation:

$$\vartheta^i = \frac{1}{n}\sum_{j=1}^{n}\mu_j^i \tag{6.7}$$

<u>Lévy Movements</u>

The MSA utilizes Lévy movements [14] to produce random perturbations. The process of generating a Lévy L_i random sample involves employing Mantegna's algorithm [23], which can be explained as follows:

$$L_i \sim scale \oplus Lévy(\alpha) \sim 0.01 \frac{v}{|z|^{\frac{1}{\alpha}}}, \tag{6.8}$$

where $\alpha \in [0, 2]$, *scale* is the dispersion size, $\oplus$ is the entrywise multiplication, $v = N(0, \mu_v)$ and $z = N(0, \mu_z)$ are two normal distributions.

Lévy Mutation

For *nc* operations ($nc \in cp$), the algorithm creates a sub-trial vector $\overrightarrow{x_p} = [x_{p1}, x_{p2}, \ldots, x_{pn_c}]$ as follows:

$$\begin{aligned} \overrightarrow{x_p^i} &= L_1^i * \left(\overrightarrow{m_{r^2}^i}, -, \overrightarrow{m_{r^3}^i}\right) + L_2^i * \left(\overrightarrow{m_{r^4}^i}, -, \overrightarrow{m_{r^5}^i}\right) \\ \forall r^1 &\neq r^2 \neq r^3 \neq r^4 \neq r^5 \neq p \in \{1,\ 2,\ \ldots,\ np\} m_{r^1}^i, \end{aligned} \tag{6.9}$$

where L_1 and L_2 are two independent Lévy random samples [23].

Adaptive Crossover

Each pathfinder moth updates its position by combining crossover operations and mutated variables. Consequently, the trail or mixed solution Tms_{pj}^i for the element i can be described as follows:

$$Tms_{pj}^i = \begin{cases} \overrightarrow{x_{pj}^i} & if\ j \in cp \\ \overrightarrow{m_{pj}^i} & if\ j \notin cp \end{cases} \tag{6.10}$$

Selection Strategy

After the adaptive crossover operator is finished, the fitness value of the complete trail solution is computed and compared to its corresponding host solution. Subsequently, a set of *np* solutions is selected using the roulette scheme, as described in the work by Mohamed et al. [26].

Transversal Flight

The subset of elements exhibiting the highest luminescence intensities is designated as prospectors for the subsequent iteration. Throughout the optimization process, the number of prospectors, denoted as pf, gradually decreases according to the following pattern:

$$pf = round\left((p - p_n) \times \left(1 - \frac{i}{i_c}\right)\right) \tag{6.11}$$

where i_c represents the current iteration number.

Celestial Navigation

Throughout the optimization process, as the number of prospectors decreases, the number of onlookers, denoted as o_n, increases ($o_n = p - p_n - p_f$). This shift in the population composition can result in a rapid acceleration of the convergence rate. Onlookers are moths within the swarm that possess lower luminescent sources. During this phase, onlookers perform an exhaustive search in close proximity to prospector elements. The onlookers are divided into two equal groups to carry out their exploration tasks.

Gaussian Walks

During this phase, the aim is to explore promising regions in the search space. To achieve this, a stochastic Gaussian distribution is utilized to introduce perturbations to the original positions. As a result, the new positions of the onlookers belonging to the first subset are calculated using the following formulation:

$$m_o^{i+1} = m_o^i + \varepsilon_1 + \left[\varepsilon_2 \times best - \varepsilon_3 \times m_o^i\right]; \forall o \in [1, 2, \ldots, on_G] \tag{6.12}$$

In this equation, ε_1 represents a random sample obtained from a Gaussian distribution, while ε_2 and ε_3 are random numbers uniformly distributed between 0 and 1.

Associative Learning Mechanism

The second subset of onlookers, with a size of $on_A = o_n - on_G$, adjusts its positions towards the global best solution based on immediate memory. The memory is initialized using a continuous uniform Gaussian distribution within the range from $m_o^i - m_o^{\min} to m_o^{\max} - m_o^i$. The calculation for the new positions is as follows:

$$m_o^{i+1} = m_o^i + 0.001 \cdot G\left[m_o^i - m_o^{\min}, m_o^{\max} - m_o^i\right] + \left(1 - \frac{g}{G}\right) \cdot r_1 \cdot \left(pbest^i - m_o^i\right)$$
$$+ \frac{2g}{G} \cdot r_2 \cdot \left(gbest^i - m_o^i\right), \tag{6.13}$$

The equation above is used to compute the new positions for onlookers $o\epsilon$, where o belongs to the range $[1, 2, \ldots, on_A]$. The social factor is represented by $2g/G$, while the cognitive factor is represented by $1 - (\frac{g}{G})$. r_1 and r_2 are random numbers between 0 and 1, and $pbest$ is a light source selected randomly from the new pathfinders group based on the probability value pv_p. The MSA algorithm's flowchart is depicted in Fig. 6.3.

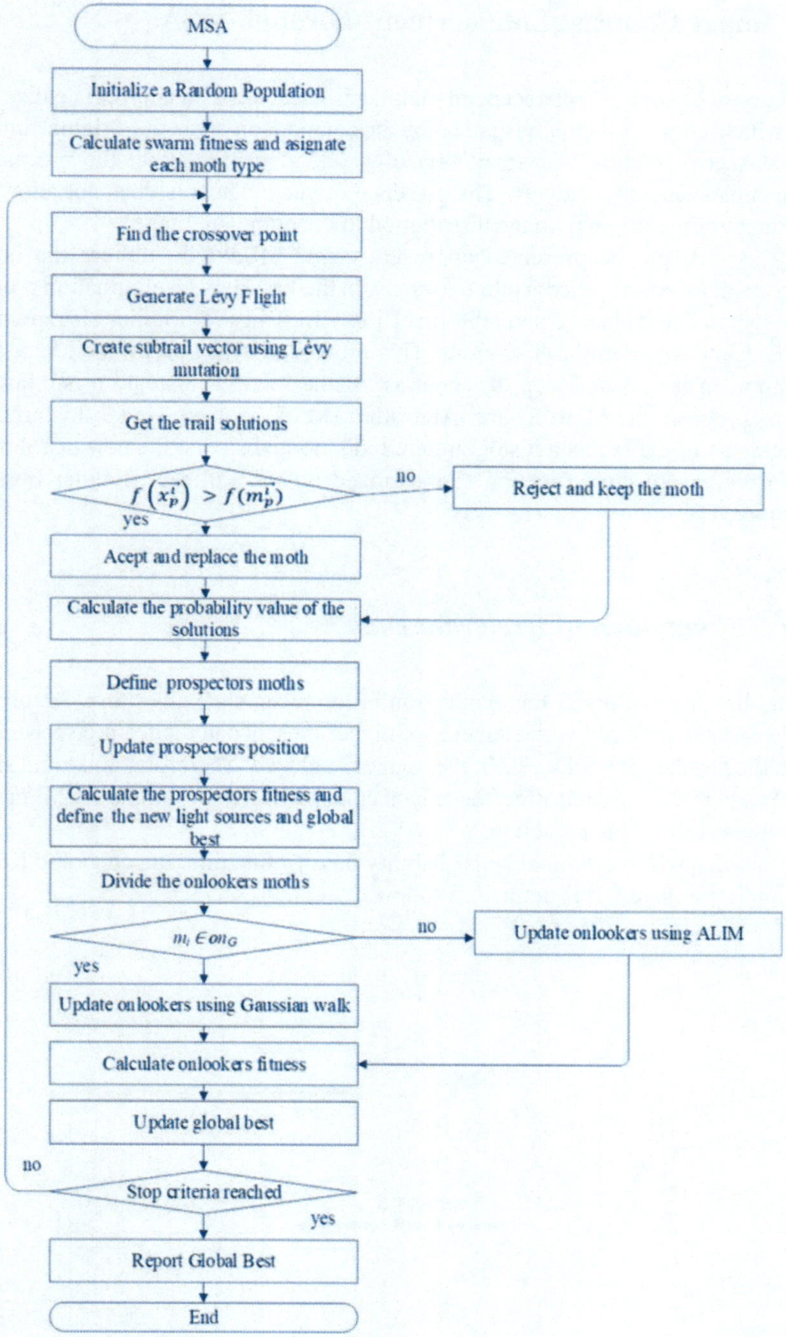

Fig. 6.3 The processing step for the MSA algorithm

6.5 Image Contrast Enhancement Through MSA

The purpose of contrast enhancement on images is to define an appropriate distribution, which allows a better visualization in comparison with the original image. The existence of noise and small sets of pixel intensities affect the process of image enhancement negatively. The presence of such elements does not allow the adjustment of the original image distribution in the complete range.

In this chapter, the presented approach, called MSA-ICE, divides the image contrast enhancement process into two steps. In the first step, an elimination process is conducted for replacing sporadic pixel intensities in the original histogram by abundant intensities in the histogram. This process is achieved by the Mean-shift algorithm. In the second step, the contrast of the reduced histogram is adjusted. For this propose, the Moth Swarm Algorithm (MSA) is used. Under this process, the elements of the reduced histogram are redistributed so that the new distribution maximizes an objective function characterized by the Kullback–Leibler Entropy (KL-entropy).

6.5.1 *Elimination of Irrelevant Data*

During this process, under the application of the mean-shift algorithm, features of low density are replaced with elements of higher presence in the feature space. As a result, the number of gray levels in the image is reduced. Therefore, noise and small sets of pixel intensities that affect the enhancement process are eliminated. Figure 6.4 shows the result of this reduction.

In this chapter, to estimate the probability density function, the quadratic Kernel function is employed. It is defined as follows:

$$K(x) = \left(\frac{15}{16}\right) \cdot \left(1 - x^2\right)^2 \tag{6.14}$$

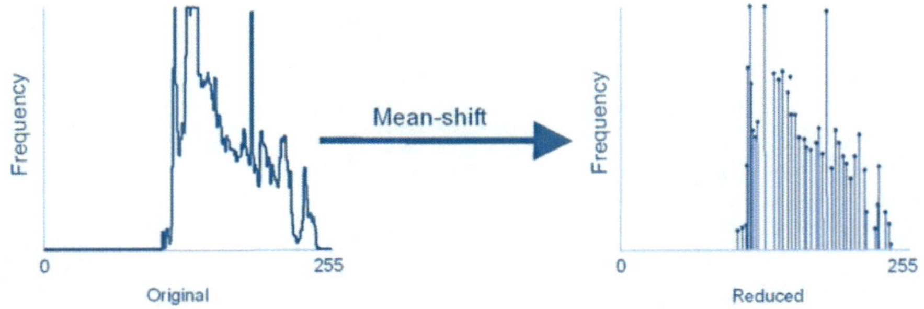

Fig. 6.4 Reduction pixel intensities obtained by the mean-shift algorithm

During this operation, we consider an original histogram H_O, which is defined as follows:

$$H_O = \{p_0, \ p_1, \ \ldots, \ p_{L-1}\}, \ \ p_r = \frac{n_r}{m \cdot n}, \ \ r \in (0, 1, \ldots, L-1) \tag{6.15}$$

The original histogram H_O, with L representing the number of grayscale intensities and n_r referring to the number of pixels with intensity r, is used in the operation. The mean-shift method is applied to H_O to produce the reduced histogram H_R, denoted as $\{q_0, q_1, \ldots, q_{L-1}\}$. In this scheme, for each pixel intensity i, the mean-shift algorithm identifies the pixel j that represents its local maximum. Consequently, the components i and j of the reduced histogram H_R are computed as $q_j = p_j + p_i$, while q_i is set to zero. This operation ensures that the pixels with intensity i in the image are considered as pixels with intensity j. The result of this operation can be seen in Fig. 6.4.

6.5.2 Objective Function

To enhance the contrast, the Moth Swarm Algorithm (MSA) is utilized to adjust the pixel intensities in the reduced histogram H_R. The quality of the contrast enhancement is evaluated using the following objective function:

$$J_P(H_R) = \log(\log \mathrm{e_i(I)}) \cdot \frac{\mathrm{e_p(I)}}{\mathrm{hp} * \mathrm{vp}} \cdot \mathrm{KL_e(I)} \tag{6.16}$$

J_P combines various image characteristics that are deemed significant in the context of image enhancement. It incorporates: (1) the count of edge pixels $\mathrm{e_p(I)}$ identified through the application of a Sobel filter on the image, (2) the total sum of intensities of the edge pixels $\mathrm{e_i(I)}$, (3) the resolution of the image $hp * vp$, and (4) the symmetric Kullback–Leibler entropy $\mathrm{KL_e(I)}$ between the original and reduced images, which is defined as follows:

$$\mathrm{KL_e(I)} = \sum h_1 \cdot \log \frac{h_1}{h_2} + \sum h_2 \cdot \log \frac{h_2}{h_1} \tag{6.17}$$

where h_1 represents the histogram of the original image, and h_2 represents the histogram of the reduced image.

The objective function J_P integrates the number of edge elements $\mathrm{e_p(I)}$, the intensity of edge pixels $\mathrm{e_i(I)}$, and the Kullback–Leibler entropy $\mathrm{KL_e(I)}$ of the image. The quantities $\mathrm{e_p(I)}$ and $\mathrm{e_i(I)}$ reflect the presence and strength of important features in the enhanced image. They indicate the level of fine details captured in the enhanced image. On the other hand, the $\mathrm{KL_e(I)}$ value measures the similarity between the original histogram h_1) and the histogram of the enhanced image (h_2). In this way, J_P

values are used to identify the enhanced image that maximizes the preservation of fine details while maintaining the structural characteristics of the original image.

6.5.3 Penalty Function

Under a contrast enhancement process, the pixel intensities of the reduced histogram H_R are rearranged to improve the visualization of the image. The new positions are assigned in a sequential manner, ensuring coverage of the entire grayscale range. Specifically, the pixel intensity i is always positioned before intensity $i+1$. However, during the optimization process, some candidate solutions may exhibit inconsistencies. These inconsistencies arise when a pixel intensity $i+1$ is relocated to a position preceding pixel intensity i. To address this issue, a penalty function is introduced to penalize solutions that display such inconsistencies. Consequently, the problem of enhancing contrast can be framed as a constrained optimization task, which can be expressed as follows:

$$\begin{aligned} &\textbf{max } \mathrm{Fit}(H_R) \\ &\text{subject to : } q_i < q_{i+1}, \ \forall i \in (0, 1, \ldots, L-1) \end{aligned} \tag{6.18}$$

In this formulation, the function $\mathrm{Fit}(H_R)$ is defined as follows:

$$\mathrm{Fit}(H_R) = \begin{cases} \frac{\mathrm{fit}(H_R)}{P_e(H_R)}, & \text{in case of inconsistency} \\ \mathrm{fit}(H_R), & \text{otherwise} \end{cases} \tag{6.19}$$

where $P_e(H_R)$ is the penalty function that directly influences the fitness function as defined in Eq. (6.25). The penalty function value, $P_e(H_R)$, increases as the number of inconsistencies in the produced solutions increases. $P_e(H_R)$ can be expressed using the following formulation:

$$P_e(H_R) = (\mathrm{NVC} + \mathrm{SVC}) * \mathrm{PC} \tag{6.20}$$

NVC represents the count of produced inconsistencies, while PC is a predefined constant set to a high value to ensure effective penalization. SVC refers to the sum of all pixel elements n_d that are associated with the inconsistency d.

$$\mathrm{SVC} = \sum_{d=1}^{\mathrm{NVC}} \max\{0, n_d\} \tag{6.21}$$

6.6 Experiments

To assess the effectiveness of the MSA-ICE approach, a series of experiments were conducted using representative image datasets, namely CSIQ [20] and TID2013 [27]. The CSIQ dataset consists of 116 images generated by modifying the contrast levels of 30 reference images. On the other hand, the TID2013 dataset contains 250 images grouped into 10 categories, each based on 25 reference images, with each category representing a different type of contrast.

For evaluating the performance of the MSA-ICE approach, six performance indexes were utilized: Structural similarity index measurement (SSIM), Mean squared error (MSE), Edge Preserve Index (EPI), Entropy (E), Relative Enhancement Contrast (REC), Range Redistribution (RR), and Computational Complexity (CC). These indexes were chosen to ensure compatibility with other related works. The first six indexes assess the quality of the enhanced images, while the last index measures the computational effort required by the approach.

To calculate the various indexes, an original image with unsatisfactory contrast, denoted as **R**, is taken as input for an ICE method. This ICE method generates an enhanced image, denoted as **I**. The Structural similarity index measurement (SSIM) [16] is then used to assess the similarity between the enhanced image **I** and the original image **R**. The SSIM is computed based on the pixel data of **I** denoted as $\{p_1, \ldots, p_{M\times N}\}$, and the original data denoted as $\{r_1, \ldots, r_{M\times N}\}$. The calculation of SSIM is as follows:

$$SSIM = \frac{(2\mu_{\mathbf{I}}\mu_{\mathbf{R}} + Q_1)(2\sigma_{\mathbf{IR}} + Q_2)}{(\mu_{\mathbf{I}}^2 + \mu_{\mathbf{R}}^2 + Q_1)(\sigma_{\mathbf{I}}^2 + \sigma_{\mathbf{R}}^2 + Q_2)} \tag{6.22}$$

The equation incorporates two small positive constants denoted as Q_1 and Q_2, typically set to 0.01. The mean values of the segmented data, denoted as $\mu_{\mathbf{I}}$, and the reference data, denoted as $\mu_{\mathbf{R}}$, are calculated. Similarly, the variances of the segmented data, $\sigma_{\mathbf{I}}$, and the reference data, $\sigma_{\mathbf{R}}$, are computed. The covariance between the segmented data **I** and the reference data **R**, denoted as $\sigma_{\mathbf{IR}}$, is also taken into account. The $SSIM$ produces values ranging from 0 to 1.

The MSE measures the resemblance between a segmented image **I** and its reference **R** by calculating the average of the squared differences between the pixel values of the segmented image and its corresponding reference pixels. The MSE is computed using the following formula:

$$MSE = \frac{1}{MN}\sum_{i=1}^{M}\sum_{j=1}^{N}(\mathbf{I}(i,j) - \mathbf{R}(i,j))^2 \tag{6.23}$$

The Edge Preserve Index (EPI) assesses the preservation of edges in the enhanced image **I** compared to the original image **R**. A higher EPI value indicates better performance of the algorithm. The EPI is calculated using the following formula:

$$EPI = \frac{\sum_{i=1}^{M}\sum_{j=1}^{N}|\mathbf{I}(i,j) - \mathbf{I}(i,j+1)| + |\mathbf{I}(i,j) - \mathbf{I}(i+1,j)|}{\sum_{i=1}^{M}\sum_{j=1}^{N}|\mathbf{R}(i,j) - \mathbf{R}(i,j+1)| + |\mathbf{R}(i,j) - \mathbf{R}(i+1,j)|} \tag{6.24}$$

The Entropy (E) assesses the information content of the enhanced image **I**. E is a quality measurement that indirectly evaluates the number of details existent in **I**. A larger value of E corresponds to more information content in **I**. The index E is calculated with the following model:

$$E = -\sum_{i=0}^{L-1} P(q) \cdot \log(P(q)) \tag{6.25}$$

In the equation, $P(q)$ represents the probability density function at intensity level q (where q ranges from 0 to $L-1$) of the image **I**. Here, L denotes the total number of grayscale levels present in the image **I**.

The Relative Enhancement Contrast (REC) index, as proposed by Zhao et al. [38], provides a measure of the contrast disparity between the enhanced image **I** and the original image **R**. REC quantifies the variations in intensity among different objects and structures present in both images. A higher REC value indicates a more improved image in terms of contrast. The REC index is calculated using the following formula:

$$REC = 20 \cdot \log\left[\frac{1}{M \times N}\sum_{i=1}^{M}\sum_{j=1}^{N}(\mathbf{I}(i,j))^2 - \left(\frac{1}{M \times N}\sum_{i=1}^{M}\sum_{j=1}^{N}\mathbf{R}(i,j)\right)^2\right] \tag{6.26}$$

The Range Redistribution (RR) index, as introduced by Chen et al. [7], evaluates the distribution of pixel intensities in the enhanced image **I**. RR provides insights into how the intensities are spread across the image. The RR index is calculated using the following formula:

$$RR = \frac{1}{M \times N \cdot (M \times N - 1)}\sum_{q=1}^{L-1}\sum_{r=q}^{L-1} P(q)P(r)(r-q) \tag{6.27}$$

The Range Redistribution (RR) index, involving the probability density functions $P(q)$ and $P(r)$, assesses the distribution of pixel intensities in the enhanced image **I**. The RR index quantifies how well the histogram of **I** is spread out, without any significant concentration of intensities. The probabilities are computed at the intensity levels q and r, both ranging from 0 to $L-1$, where L represents the total number of grayscale levels in **I**. A higher RR value indicates a more evenly distributed histogram without prominent intensity clusters.

The ICE methods used in the experiments are inherently complex, involving various random operations and stochastic sub-routines. Due to the unpredictable nature of these processes, it is not feasible to analyze their complexity deterministically. Instead, the computational complexity (CC) is assessed by measuring the amount of computational effort in seconds required for each algorithm.

The experimental results are presented in four subsections to evaluate different aspects of the MSA-ICE approach. In Sect. 6.6.1, the performance of MSA-ICE is assessed using metrics such as SSIM, MSE, EPI, E, REC, RR, and CC on grayscale images. Section 6.6.2 compares the contrast enhancement results achieved by the proposed objective function J_P (Eq. 6.16), which is based on KL-entropy, with other objective functions commonly used in contrast enhancement literature. Section 6.6.3 conducts a comparative study among different ICE metaheuristic methods. Lastly, in Sect. 6.6.4, the results of MSA-ICE are extended to color images.

6.6.1 Comparison Over Grayscale Images

In this subsection, we evaluate the performance of the MSA-ICE scheme using grayscale images. The experiments involve applying MSA-ICE to the images in the CSIQ and TID2013 datasets. The results obtained from MSA-ICE are compared with similar schemes based on metaheuristic methods such as the Artificial Bee Colony (ABC) algorithm, Differential Evolution (DE), and Firefly Algorithm (FFA). For all the comparisons, the population size is set to $N = 30$, and the maximum number of iterations is set to $k_{\max} = 100$. This choice of parameters is made to ensure compatibility with other related works reported in the literature.

The specific parameter setup for each metaheuristic scheme is described as follows:

- ABC: The algorithm is implemented by setting the parameter $limit =$ num of Food Sources $*$ dims, where num Of $FoodSources = N$ (population size) and dims $= n$ (dimensionality of the solution space) [17].
- DE: The crossover rate is set to $CR = 0.5$, while the differential weight is given as $F = 0.2$ [28].
- FFA: The randomness factor and light absorption coefficient are set to $\alpha = 0.2$ and $\gamma = 1.0$, respectively [37].
- MSA: The number of sub-trial vectors is set to nc $= 8$.

In addition to the metaheuristic schemes mentioned earlier, two recent hybrid approaches, namely the Gamma-PSO algorithm [22] and the Fuzzy-IPSO method [32], were also considered in the comparisons. Furthermore, two classical algorithms based on histogram equalization, namely Local and Adaptive Histogram Equalization (CLAHE) [30] and Weighted Thresholded Histogram Equalization (WTHE) [34], were included in the analysis. All algorithms were configured using the parameter values specified in their respective references, and these parameter configurations remained unchanged throughout the experimental work. The selected parameter values were reported as the best performing configurations in their respective references. All experiments were conducted using MatLab® R2016a on a computer equipped with an Intel® Core™ i7-3.40 GHz processor and Windows 8 (64-bit, 8 GB of memory) as the operating system.

In cases where the pixel intensities of an image are concentrated in a narrow range, the fine details of the image may be lost, resulting in low contrast. To evaluate the ability of the compared image contrast enhancement techniques to extract information from such images, average results in terms of various indexes including $SSIM$, MSE, EPI, E, REC, and RR are recorded in Table 6.1 for all 116 images from the CSIQ dataset. Higher values indicate better performance for all metrics except MSE, where lower values are desired. The $SSIM$ and EPI indexes indirectly assess the visual perception of an image by evaluating the amount of edge information present. A higher $SSIM$ value indicates better preservation of important details. According to Table 6.1, the MSA-ICE method achieves the highest values for SSIM and EPI, indicating its superior performance in preserving relevant edges and significant features, making it suitable for feature-related processing tasks. The presented method also achieves the lowest MSE values, indicating minimal distortion or error in the processed image. The entropy (E) is an index that quantitatively evaluates the information content of an image and reflects the ability of a method to extract fine details. Table 6.1 shows that the presented scheme achieves comparatively higher entropy values, indicating higher information content in the enhanced image. With a high REC value, the presented algorithm demonstrates its ability to significantly enhance the contrast of images compared to its competitors. The Range Redistribution (RR) index is also important for evaluating an image contrast enhancement scheme's ability to highlight hidden details. A high RR value suggests that more features can be detected and displayed in the image. According to the RR index, the MSA-ICE method achieves the highest value, indicating superior pixel intensity redistribution without concentration. The exceptional performance of the MSA-ICE method can be attributed to its ability to replace sporadic characteristics in the original image with meaningful pixel intensities.

Upon analyzing Table 6.1, it is evident that the MSA-ICE scheme outperforms the other considered algorithms in terms of average values. The results indicate that the MSA-ICE achieves the highest scores across the $SSIM$, MSE, EPI, E, REC, and RR indexes. The Fuzzy-IPSO, Gamma-PSO, and ABC approaches rank second, performing relatively well in most performance indexes. The CLAHE and WTHE techniques fall into the third category. On the other hand, the DE and FFA methods yield the lowest results among all the compared algorithms.

Table 6.2 presents the averaged results of the eight algorithms, including MSA-ICE, for all 250 images in the TID2013 dataset. Once again, the MSA-ICE method achieved the highest performance indexes. Similar to the results obtained from the CSIQ dataset, Fuzzy-IPSO, Gamma-PSO, and ABC algorithms performed well and secured the second position in most of the performance indexes. The CLAHE and WTHE techniques showed slightly lower performance in the third category. On the other hand, the DE and FFA methods yielded the poorest results.

The outstanding performance of the MSA-ICE method can be attributed to two key factors. Firstly, the inclusion of an elimination process enables the removal of irrelevant information during the enhancement process, allowing for the utilization

Table 6.1 The performance of ABC, DE, FA, Gamma-PSO, Fuzzy-IPSO, CLAHE, WTHE, and MSA-ICE algorithms was evaluated using the CSIQ dataset. The numerical values represent the average results in terms of the $SSIM$, MSE, EPI, E, REC, and RR indexes

Index	ABC	DE	FFA	Gamma-PSO	Fuzzy-IPSO	CLAHE	WTHE	MSA-ICE
$SSIM$	0.7412	0.5874	0.6847	0.7502	0.7598	0.7021	0.7157	0.8620
MSE	5.88E+03	6.51E+03	6.31E+03	5.80E+03	5.77E+03	6.22E+03	6.15E+03	5.22E+03
EPI	1.1132	0.9721	1.1031	1.0021	1.0254	1.0014	1.0997	1.2491
E	6.0141	4.8741	5.8741	6.0982	6.0098	5.5721	5.7864	6.3871
REC	1.0964	0.9847	1.0101	1.0782	1.0320	1.0004	1.0089	1.1272
RR	39.4714	27.214	36.4124	39.0088	38.9801	30.0087	30.8714	43.0121

Table 6.2 Average performance values of ABC, DE, FA, Gamma-PSO, Fuzzy-IPSO, CLAHE, WTHE, and MSA-ICE algorithms in terms of the indexes $SSIM, MSE, EPI, E, REC, and RR$

Index	ABC	DE	FFA	Gamma-PSO	Fuzzy-IPSO	CLAHE	WTHE	MSA-ICE
$SSIM$	0.6784	0.4914	0.6210	0.6708	0.6921	0.6107	0.6521	0.7421
MSE	6.21E+03	7.18E+03	6.91E+03	6.20E+03	6.19E+03	7.07E+03	6.73E+03	6.08E+03
EPI	1.0974	0.9974	1.0087	1.0021	1.0198	1.0004	1.0241	1.1987
E	5.8041	5.4170	5.6127	5.8421	5.9001	5.5496	5.7410	6.1047
REC	1.0021	0.9784	0.9974	0.9997	1.0087	0.9870	1.0007	1.0975
RR	37.4120	21.0745	25.7412	37.2104	38.0078	31.7847	27.8741	41.2101

of the entire histogram range. This is evident from the higher RR value, indicating better redistribution of pixel intensities. Secondly, the incorporation of the MSA algorithm significantly enhances the exploration–exploitation capabilities, enabling effective avoidance of local minima. Consequently, the presented method achieves the best possible enhanced image in terms of KL-entropy.

To conduct a detailed analysis of the quantitative and qualitative results, a set of six representative images $I_1 - I_6$ were selected from both the CSIQ and TID2013 datasets. These images, which exhibit complexity, are displayed in Figs. 6.5, 6.6, 6.7, 6.8, 6.9 and 6.10.

Table 6.3 provides the quantitative results of the algorithms for the selected images $I_1 - I_6$. The table presents the performance indexes, including SSIM, MSE, EPI, E, REC, and RR. The results clearly indicate that the MSA-ICE method outperforms its competitors in terms of performance indexes. In general, the DE and GSA methods demonstrate poor performance, suggesting the need for multiple executions to obtain consistent results. When evaluated only once, these methods often yield suboptimal solutions and produce distorted images. On the other hand, the ABC,

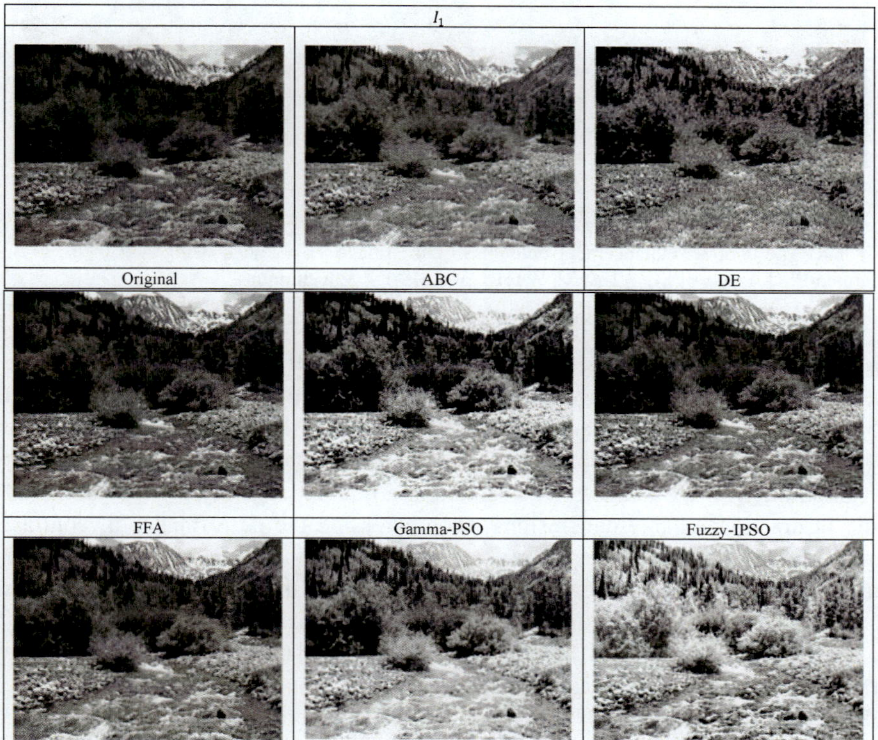

Fig. 6.5 Contrast enhancement outcomes of the image I_1 achieved by ABC, DE, FFA, Gamma-PSO, Fuzzy-IPSO, CLAHE, WTHE, and MSA-ICE techniques

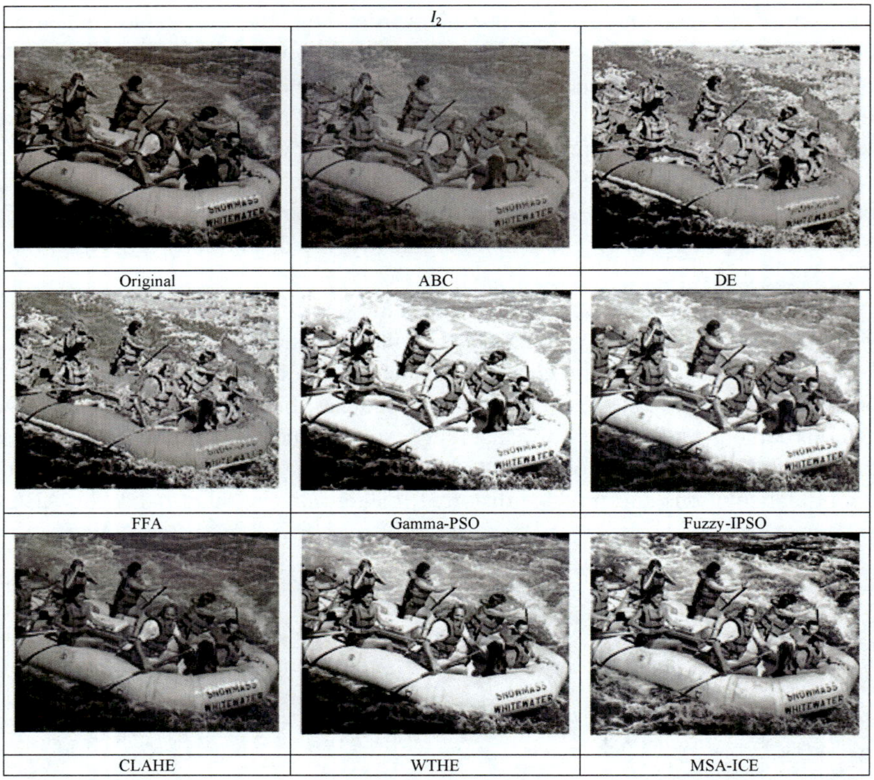

Fig. 6.6 The contrast enhancement outcomes of the image I_2 obtained using ABC, DE, FFA, Gamma-PSO, Fuzzy-IPSO, CLAHE, WTHE, and MSA-ICE techniques

WTHE, CLAHE, and FFA algorithms exhibit good performance across most of the performance indexes. However, their inability to correctly redistribute intensities in the enhanced results is evident from the RR index and the visual perception of the produced images when compared to the MSA-ICE method.

In addition to the quantitative evaluation, a qualitative analysis plays a crucial role in drawing reliable conclusions when assessing the performance of contrast enhancement methods. The purpose of the qualitative analysis is to examine the presence of bothersome distortions and other artifacts in the enhanced images, which can be attributed to subpar functioning of the employed algorithms.

Each figure from Figs. 6.5, 6.6, 6.7, 6.8, 6.9 and 6.10 displays the original grayscale image along with the contrast enhancement results obtained using the ABC, DE, FFA, Gamma-PSO, Fuzzy-IPSO, CLAHE, WTHE approaches, and the MSA-ICE method. Upon analyzing the figures, it can be observed that, in general, the Gamma-PSO and Fuzzy-IPSO methods are effective in enhancing contrast. However, their results often exhibit excessive modification of pixel intensities, leading to extremely

Fig. 6.7 The contrast enhancement outcomes of the image I_3 achieved through the ABC, DE, FFA, Gamma-PSO, Fuzzy-IPSO, CLAHE, WTHE, and MSA-ICE techniques

bright and dark values. On the other hand, the ABC method generates poor enhancement effects, making it difficult to differentiate important details in most of the images. The DE and FFA methods introduce undesired artifacts and lead to a degradation in image quality. The CLAHE scheme produces images with darker intensities, making it challenging to discern details and resulting in an unpleasant visual experience. The WTHE approach performs slightly better than CLAHE in terms of image quality, but the contrast enhancement remains unbalanced. Some regions of the image exhibit an appropriate contrast level, while others still suffer from intensity saturation issues. Although the enhanced images by Gamma-PSO and Fuzzy-IPSO methods are superior to those produced by ABC, DE, FFA, and CLAHE, they also encounter challenges with intensity saturation. These methods generate enhanced images that tend to be slightly overexposed or slightly underexposed, depending on whether it is Gamma-PSO or Fuzzy-IPSO, respectively.

Upon careful examination of Figs. 6.5, 6.6, 6.7, 6.8, 6.9 and 6.10, it becomes apparent that the MSA-ICE method generally provides a better visual perception compared to its competitors (with the exception of Fig. 6.8). The enhanced images

Fig. 6.8 The contrast enhancement outcomes of the image I_4 obtained using the ABC, DE, FFA, Gamma-PSO, Fuzzy-IPSO, CLAHE, WTHE, and MSA-ICE methods

obtained with the MSA-ICE technique exhibit clear and distinguishable details and important elements, while the other algorithms tend to preserve interesting hidden structures. This outstanding performance is attributed to the robust mechanism of the MSA-ICE method in eliminating noise and very small structures that do not contribute to the visual effect. This extension enhances the visibility of details in the image for human interpretation. A closer inspection of the enhanced images reveals that the MSA-ICE technique achieves a superior distribution of pixels within the dynamic range, avoiding saturation in both bright and dark regions, unlike other contrast enhancement methods. Conversely, other image contrast enhancement schemes often produce artifacts as a result of their improvement processes or yield deficient distributions of pixel intensities.

To evaluate the computational cost (CC), an experiment was conducted to measure the average computing time required by the eight algorithms to generate an enhanced image (refer to Fig. $I_1 - I_6$). Table 6.4 displays the average times for each enhancement scheme, calculated by processing all six images from Figs. 6.5, 6.6, 6.7, 6.8, 6.9 and 6.10. The results clearly indicate that the MSA-ICE method is the

Fig. 6.9 The contrast enhancement outcomes of the image I_5 achieved using the ABC, DE, FFA, Gamma-PSO, Fuzzy-IPSO, CLAHE, WTHE, and MSA-ICE techniques

fastest approach to obtain an enhanced image, while the Gamma-PSO and Fuzzy-IPSO methods exhibit the highest computational cost in terms of elapsed time. The significant computational time required by Gamma-PSO and Fuzzy-IPSO methods can be attributed to the numerous computational phases involved in their contrast improvement processes. On the other hand, the ABC, DE, and FFA methods have a moderate computational cost. According to existing literature [5, 10], these meta-heuristic methods typically require a higher number of iterations compared to MSA to achieve satisfactory results. While both CLAHE and WTHE methods maintain a relatively low computational burden, their values are still higher than those obtained by the MSA-ICE method. These findings emphasize that MSA-ICE outperforms other methods not only in terms of performance metrics such as $SSIM$, MSE, EPI, E, REC, and RR but also in terms of computational efficiency (CC).

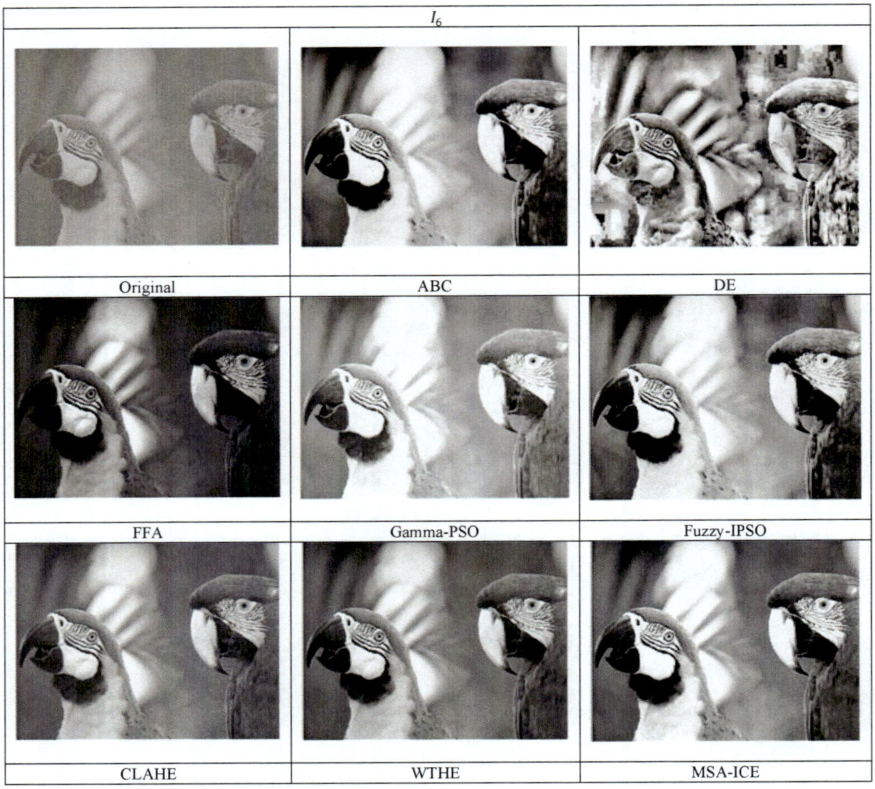

Fig. 6.10 The contrast enhancement outcomes of the image I_6 obtained using the ABC, DE, FFA, Gamma-PSO, Fuzzy-IPSO, CLAHE, WTHE, and MSA-ICE techniques

6.6.2 Performance Comparison Over Distinct Objective Functions

In this section, we compare the contrast enhancement results obtained by the proposed objective function J_P (Eq. 6.16), which is based on the KL-entropy, with the results produced by other objective functions commonly used in the literature for contrast enhancement.

The study considers two representative objective functions. The first objective function, denoted as J_1 and proposed in [11], evaluates the quality of the enhanced image **I** based on the number of edges and the entropy. The formulation of this objective function is as follows:

$$J_1 = \log\left(\sum_{i=0}^{L-1} P(q) \cdot \log(P(q))\right) \cdot NE(\mathbf{I}) \tag{6.28}$$

Table 6.3 Performance outcomes of ABC, DE, FA, Gamma-PSO, Fuzzy-IPSO, CLAHE, WTHE, and MSA-ICE algorithms, considering the indexes $SSIM$, MSE, EPI, E, REC, and RR, for the selected images $I_1 - I_6$

Image	Index	ABC	DE	FFA	Gamma-PSO	Fuzzy-IPSO	CLAHE	WTHE	MSA-ICE
I_1	$SSIM$	0.8745	0.6751	0.7574	0.8733	0.8824	0.7874	0.8014	0.9121
	MSE	7.32E+03	8.25E+03	8.12E+03	7.41E+03	7.35E+03	7.26E+03	7.26E+03	7.22E+03
	EPI	1.1974	0.9874	1.0074	1.1899	1.2008	1.0567	1.0987	1.3412
	E	6.1721	5.1047	5.8741	6.0987	6.1894	5.9740	6.0041	6.8714
	REC	1.1007	0.9421	0.9815	1.0965	1.1101	1.0010	1.0657	1.2101
	RR	43.1271	36.4197	39.4108	43.0014	44.2101	38.7410	41.7131	47.1240
I_2	$SSIM$	0.7422	0.6657	0.6871	0.7398	0.7592	0.7011	0.7127	0.7941
	MSE	5.02E+03	6.11E+03	5.87E+03	5.03E+03	4.98E+03	5.67E+03	5.41E+03	4.71E+03
	EPI	1.0041	0.8432	0.8741	1.0004	1.0940	0.9642	0.9857	1.0874
	E	5.5124	3.8297	4.1702	5.3897	5.6721	4.5787	4.8974	5.8471
	REC	0.9841	0.8714	0.8897	0.8897	0.9900	0.9057	0.9657	1.0074
	RR	36.4170	28.9840	30.7425	36.0120	37.5107	31.7421	33.8721	39.7487
I_3	$SSIM$	0.5621	0.4721	0.4897	0.5502	0.5800	0.5074	0.5214	0.60721
	MSE	3.69E+03	4.22E+03	4.10E+03	3.71E+03	3.68E+03	3.98E+03	3.87E+03	3.41E+03
	EPI	0.9721	0.8014	0.8971	0.9604	0.9740	0.9101	0.9510	0.9814
	E	4.4101	3.6512	3.7841	4.3821	4.5997	3.9740	4.0147	4.8721
	REC	0.9410	0.8011	0.8214	0.9304	0.9604	0.8714	0.8921	0.9941
	RR	39.1781	31.1487	33.1470	39.0011	40.0087	35.7840	37.8710	42.3180
I_4	$SSIM$	0.6321	0.5374	0.5847	0.62987	0.6501	0.6024	0.6121	0.6921
	MSE	6.18E+03	7.34E+03	6.68E+03	6.27E+03	6.25E+03	6.36E+03	6.24E+03	6.21E+03
	EPI	1.1014	0.9014	0.9874	1.0984	1.1001	1.0009	1.0874	1.2101

(continued)

Table 6.3 (continued)

Image	Index	ABC	DE	FFA	Gamma-PSO	Fuzzy-IPSO	CLAHE	WTHE	MSA-ICE
	E	4.8714	3.6874	3.8712	4.8004	4.9001	4.0874	4.2471	5.1401
	REC	1.0874	0.9514	0.9741	1.0074	1.0970	0.9872	1.0214	1.1921
	RR	31.0142	24.0142	26.1424	30.5784	32.0120	27.1024	29.4710	35.8724
I_5	$SSIM$	0.9611	0.8011	0.8724	0.9501	0.9604	0.9417	0.9510	0.9714
	MSE	7.31E+03	8.57E+03	8.25E+03	7.42E+03	7.10E+03	8.01E+03	7.81E+03	7.02E+03
	EPI	0.9841	0.8011	0.8741	0.9704	0.9004	0.9014	0.9410	1.0254
	E	6.1411	5.1472	5.5741	6.0874	6.1009	5.8914	6.0014	7.1101
	REC	0.9747	0.7547	0.8154	0.9667	0.9698	0.8874	0.9211	1.0074
	RR	35.1574	28.3210	30.4178	34.0098	34.9870	31.0440	33.1420	37.1421
I_6	$SSIM$	0.6187	0.4874	0.5017	0.6087	0.6211	0.5514	0.5784	0.6421
	MSE	3.43E+03	4.31E+03	4.10E+03	3.50E+03	3.40E+03	3.87E+03	3.58E+03	3.10E+03
	EPI	1.0874	0.8524	0.8974	0.9007	1.0047	0.9821	1.0044	1.1121
	E	5.0142	4.0020	4.4421	5.0071	5.0989	4.7784	4.8974	5.2140
	REC	1.0478	0.8745	0.90142	0.9870	0.9921	0.9974	1.0009	1.1987
	RR	43.2014	35.1470	37.2180	40.3710	43.3100	39.5870	41.0187	45.1470

Table 6.4 Averaged computational times of each enhancement scheme after processing images $I_1 - I_6$

Algorithm	Execution time [s]
ABC	260.1478
DE	190.2410
FFA	204.2132
Gamma-PSO	340.6623
Fuzzy-IPSO	315.9021
CLAHE	169.4721
WTHE	158.2870
MSA-ICE	139.4101

Here, $P(q)$ represents the probability density function at the intensity level q (where q belongs to the range of 0 to $L - 1$) of the image **I**. L denotes the total number of grayscale levels present in **I**. $NE(\mathbf{I})$ represents the count of edge pixels in **I**, which is calculated using the Sobel operator.

The second objective function J_2, as proposed in [22], incorporates three distinct indexes: Entropy (En), Contrast evaluation (CE), and the count of edges (NE). As a result, J_2 can be expressed in the following manner:

$$J_2 = \sqrt{\frac{En + CE(\mathbf{I}) + NE(\mathbf{I})}{3}} \tag{6.29}$$

The calculations for Entropy (En) and Contrast evaluation (CE) are performed using the following equations:

$$En = \sum_{i=0}^{L-1} P(q) \cdot \log(P(q)) \quad CE(\mathbf{I}) = \frac{1}{M \times N} \sum_{i=1}^{M} \sum_{j=1}^{N} \left(\mathbf{I}(i, j) - \overline{I}\right)^2 \tag{6.30}$$

where $\overline{I}$ represents the intensity mean of the complete enhanced image **I**.

The experiment utilized the MSA metaheuristic algorithm to optimize the redistribution process, while employing different objective functions (J_1, J_2 and J_P) to assess the quality of the solutions. The evaluation was performed on the entire set of images from the CSIQ dataset. The population size was fixed at $N = 30$, and the maximum number of iterations was set to $k_{\max} = 100$. This choice of stopping criterion was made to ensure consistency with other related studies documented in the literature [15].

Table 6.5 presents the average values of each performance index calculated across the set of 116 images from the CSIQ dataset. Upon careful examination of the table, it becomes evident that the proposed objective function J_P outperforms the other considered objective functions, namely J_1 and J_2, in terms of the averaged values. This outcome implies that J_P not only achieves the optimal distribution of data in relation to the contained information but also preserves the highest similarity to the

Table 6.5 The performance of the proposed objective function J_P is compared to that of the other objective functions, J_1 and J_2, using the 116 images from the CSIQ dataset

Index	J_1	J_2	J_P
$SSIM$	0.4010	0.4521	0.8620
MSE	7.22E+03	6.41E+03	5.22E+03
EPI	0.9047	1.0991	1.2491
E	4.8874	5.4570	6.3871
REC	0.8514	0.9011	1.1272
RR	36.7814	39.2410	43.0121

original image data, thereby avoiding distortions and artifacts. The second objective function, J_2, obtains the second-best results in terms of the performance indexes, while the objective function J_1 produces the least desirable outcomes. Additionally, it is noteworthy that the proposed objective function J_P, which is based on the KL-entropy, achieves the highest Range Redistribution (RR) index. This finding demonstrates that J_P enables a superior redistribution of information, effectively preventing the concentration of pixel intensities observed with J_1 and J_2.

6.6.3 Comparison Over Grayscale Images Considering Other Metaheuristic Techniques

Metaheuristic schemes, including the MSA algorithm, are utilized to address complex optimization problems. These schemes are designed to be suitable for specific problems, as no single metaheuristic method can efficiently solve all problems. Therefore, it is crucial to evaluate the performance of a metaheuristic scheme in an appropriate manner. Existing comparisons among metaheuristic techniques in the literature suffer from a significant limitation: they are conducted using a set of synthetic functions with known behaviors and exact solutions, without considering their real-world application scenarios.

In this section, we examine the performance of the MSA method in the context of contrast enhancement, utilizing the KL-entropy-based objective function (Eq. 6.16). We compare its results with those obtained from other metaheuristic optimization techniques, namely the Artificial Bee Colony (ABC) algorithm [17], Differential Evolution (DE) [28], Gravitational Search Algorithm (GSA) [29], and Firefly Algorithm (FA) [37]. The parameter configuration for the ABC, DE, and FFA schemes remains the same as described in Sect. 6.6.1. For the GSA, we set the initial value of the gravitation constant as $G_o = 100$, and the constant parameter alpha is set to $\alpha = 20$ [29].

In the conducted experiments, the focus was on a set of representative images $I_1 - I_6$ extracted from Figs. 6.5, 6.6, 6.7, 6.8, 6.9 and 6.10. The objective of these experiments was to demonstrate the exceptional performance of the MSA method in achieving optimal redistribution during the contrast enhancement process, thereby avoiding local optima. To compare the different methods, five indicators were considered: the best solution f_{best}, the worst solution f_{worst}, the mean solution f_{mean}, the median solution f_{mean}, and the standard deviation f_{std}. For each contrast enhancement method, 30 independent executions were performed on each image, resulting in 30 different values of $\text{Fit}(H_R)$ (Eq. 6.16). Consequently, f_{best} represents the best solution found among these 30 $\text{Fit}(H_R)$ values, while f_{worst} represents the worst value. The mean solution f_{mean} is calculated by averaging the 30 outcomes, and the median solution (f_{median}) is obtained from the median values. Furthermore, the standard deviation f_{std} is computed to assess the robustness of the contrast enhancement method. These indexes, namely f_{best}, f_{worst}, f_{mean} and f_{median}, collectively evaluate the accuracy and robustness of the produced solutions.

Table 6.6 presents the results obtained from 30 independent executions for each method, and the performance indexes f_{best}, f_{worst}, f_{mean}, f_{median} and f_{std} are considered for the analysis. The experimental findings, as shown in Table 6.6, provide compelling evidence that the MSA method outperforms the other methods by delivering significantly superior results, as indicated by the maximal values. The results clearly indicate that the GSA scheme exhibits premature convergence to a local maximum.

6.6.4 Performance in Color Images

In this section, we have extended the evaluation of the MSA-ICE method to include color images. We compare the results obtained by our method with those of other existing schemes, such as the Artificial Bee Colony (ABC) algorithm, Differential Evolution (DE), Firefly Algorithm (FFA), Gamma-PSO algorithm, Fuzzy-IPSO method, Local and Adaptive Histogram Equalization (CLAHE), and Weighted Thresholded Histogram Equalization (WTHE).

In the color enhancement scheme, we start with an original input color image $I_{RGB} = \{RGB\}$ consisting of Red, Green, and Blue (RGB) channels. This I_{RGB} image is then transformed into the HSI color space using the transformation I_{RGB}. From the resulting I_{HSI} image, we focus on enhancing the intensity channel, similar to grayscale images. Once the intensity channel is enhanced, it is combined with the hue (H) and saturation (S) channels, and the image is transformed back to the RGB color space.

Table 6.6 The objective function performance results for each image $I_1 - I_6$ were evaluated for the following metaheuristic optimization techniques: ABC, DE, FA, GSA, and MSA

		ABC	DE	FA	MSA-ICE	GSA
I_1	f_{best}	8.3021E−05	−9.3946E−11	−83.3445799	239.768911	−27.5748866
	f_{worst}	1.7807E−05	−1.5681E−09	−157.605614	120.396818	−62.7913761
	f_{mean}	3.2351E−05	−2.5627E−10	−124.982729	156.979902	−46.9371447
	f_{median}	2.8065E−05	−1.5559E−10	−127.791923	153.180999	−47.6883123
	f_{std}	1.4815E−05	3.281E−10	18.003993	22.5155784	8.42339971
I_2	f_{best}	0.0040786	−7.7514E−11	−200.706366	304.79439	−39.0419775
	f_{worst}	7.5212E−05	−1.2561E−10	−273.437793	263.339792	−150.332676
	f_{mean}	0.00032857	−9.1745E−11	−242.402576	284.670928	−87.3715575
	f_{median}	0.0001596	−9.1515E−11	−237.01505	283.576551	−88.8296675
	f_{std}	0.00071902	1.1513E−11	22.9687993	10.7332911	26.5575174
I_3	f_{best}	0.0035828	−8.781E−11	−89.6788823	191.258436	−11.5687404
	f_{worst}	0.0001412	−1.9929E−10	−169.945826	174.287047	−83.5914617
	f_{mean}	0.00087394	−1.0814E−10	−123.078405	180.490427	−56.3344625
	f_{median}	0.00056547	−1.03E−10	−100.358516	180.417433	−69.0447586
	f_{std}	0.00093826	2.1262E−11	33.2097933	4.14439612	23.6166717
I_4	f_{best}	16.7490028	−8.919E−11	−17.0997514	33.9544188	−4.99519205
	f_{worst}	3.5058E−05	−1.2773E−10	−29.9979348	22.9055029	−10.6570464
	f_{mean}	0.74256915	−1.0718E−10	−25.6800729	29.6958407	−8.2317065
	f_{median}	8.6057E−05	−1.0506E−10	−25.7403874	30.1263737	−8.0785434
	f_{std}	3.18674734	9.5663E−12	3.50303534	2.72610241	1.49040519
I_5	f_{best}	0.71729222	−7.2286E−11	−0.63298755	0.91735293	−0.41317112
	f_{worst}	8.9332E−06	−1.0993E−10	−0.7232744	0.73117568	−0.50635223
	f_{mean}	0.33422472	−9.1787E−11	−0.65835117	0.79974294	−0.46670659
	f_{median}	0.55615539	−9.2211E−11	−0.65579256	0.79697052	−0.46752599
	f_{std}	0.32018918	8.4288E−12	0.0193367	0.03761757	0.02402692
I_6	f_{best}	0.61679468	−6.2086E−11	−0.60573131	0.79490854	−0.44279736
	f_{worst}	2.5523E−06	−9.4708E−11	−0.67033251	0.71277209	−0.5178735
	f_{mean}	0.13467604	−7.5153E−11	−0.6417174	0.75247277	−0.48489713
	f_{median}	1.2197E−05	−7.4366E−11	−0.64253346	0.75207066	−0.48364784
	f_{std}	0.24874496	6.589E−12	0.01479577	0.02204985	0.01842073

To assess the effectiveness of each color enhancement scheme, we have utilized the same set of six performance indexes employed for grayscale images. These indexes include the Structural Similarity Index Measurement ($SSIM$), Mean Squared Error (MSE), Edge Preserve Index (EPI), Entropy (E), Relative Enhancement Contrast (REC), and Range Redistribution (RR). By analyzing these indexes, we can evaluate and compare the performance of each enhancement scheme for color images.

In addition to the aforementioned indexes, we have also included the Colorfulness (C) index in our evaluation. The Colorfulness index, proposed by Hasler in 2003 [12], measures the level of color content in an image from a visual standpoint. A higher value of C indicates a more vibrant and visually appealing color perception. To calculate the Colorfulness index, we consider the color channels R, G, and B of an I_{RGB} image, and it is computed using the following formula:

$$C = \sigma_{RGYB} + (0.3 \times \mu_{RGYB}) \tag{6.31}$$

where

$$\sigma_{RGYB} = \sqrt{\sigma_{RG}^2 + \sigma_{YB}^2} \quad \mu_{RGYB} = \sqrt{\mu_{RG}^2 + \mu_{YB}^2} \tag{6.32}$$

This metric assesses color perception by considering the mean μ and standard deviation σ of the pixel values in the channel differences RG and YB. These channel differences are defined as follows:

$$RG = (R - G) \quad YB = 0.5 \times (G + R) - B \tag{6.33}$$

In the experiments, a variety of color images from the CSIQ and TID2013 databases were used. However, for the purpose of brevity, a subset of three representative images was considered. These images are shown in Fig. 6.11 and represent different levels of complexity. The first three images (IC_1, IC_2 and IC_3) display a high color content, while the last three images (IC_4, IC_5 and IC_6) exhibit properties related to low-light conditions.

Fig. 6.11 The set of representative color images used in the experiments were selected from the CSIQ and TID2013 databases. These images were chosen to provide a diverse representation of color characteristics and were specifically selected for their relevance to the study

The results obtained for each image from Fig. 6.11 in terms of the performance indexes, including Structural Similarity Index Measurement (SSIM), Mean Squared Error (MSE), Edge Preserve Index (EPI), Entropy (E), Relative Enhancement Contrast (REC), Range Redistribution (RR), and Colorfulness (C), are recorded in Table 6.4. From Table 6.7, it is evident that the MSA-ICE method achieves the best performance indexes compared to other existing methods. Specifically, the presented method shows the highest E value, indicating better preservation of the original information content. Moreover, the RR values obtained by the presented method are comparatively higher, indicating a more effective redistribution of the enhanced image. The MSA-ICE method also achieves the highest C values, indicating enhanced color content that appears more natural compared to the original input image. These results indicate that the enhanced images produced by the presented method are visually appealing. Upon analyzing Table 6.7, it is clear that the MSA-ICE method consistently outperforms the other algorithms in terms of the performance indexes. The Fuzzy-IPSO, Gamma-PSO, and ABC approaches rank second in most of the performance indexes. The CLAHE and WTHE techniques represent the third category, while the DE and FFA methods exhibit the least favorable results.

The visual outcomes of the eight contrast enhancement methods can be observed in Figs. 6.12, 6.13, 6.14, 6.15, 6.16 and 6.17. Upon qualitative analysis of these figures, it is apparent that the ABC, DE, and FFA methods only provide slight enhancement to the images, resulting in a loss of fine details and the presence of artifacts. These methods are limited in their ability to effectively operate with the objective function and constraints outlined in Eq. (6.16). As a result, they only identify sub-optimal solutions (local solutions) instead of the global solution. These sub-optimal solutions tend to produce images with an overall brightness but lack detail preservation.

On the other hand, the Gamma-PSO, Fuzzy-IPSO, CLAHE, and WTHE approaches exhibit a relative level of success in contrast enhancement. However, the images produced by these methods often suffer from over-brightening or over-darkening effects, hindering the observation of important details. Furthermore, these effects result in a lack of color perception. In contrast, the MSA-ICE method stands out with the best visual results. It is capable of highlighting details and important characteristics while mitigating undesired effects. The pixel intensities are modified in a manner that avoids detrimental impacts. Although all approaches, including ABC, DE, FFA, Gamma-PSO, Fuzzy-IPSO, CLAHE, WTHE, and MSA-ICE, exhibit some degree of color distortion in the enhanced images, MSA-ICE produces the least color distortion.

Table 6.7 The performance of ABC, DE, FFA, Gamma-PSO, Fuzzy-IPSO, CLAHE, WTHE, and MSA-ICE algorithms is evaluated in terms of various indexes, including $SSIM$, MSE, EPI, E, REC, RR, and C, for each image shown in Fig. 6.11

Image	Index	ABC	DE	FFA	Gamma-PSO	Fuzzy-IPSO	CLAHE	WTHE	MSA-ICE
IC_1	$SSIM$	0.8014	0.6574	0.7487	0.7921	0.8110	0.7651	0.7821	0.8854
	MSE	8.24E+03	9.34E+03	9.10E+03	8.34E+03	8.15E+03	8.87E+03	8.65E+03	8.01E+03
	EPI	1.0121	0.9012	0.9651	0.9032	0.9891	0.9874	1.091	1.1247
	E	6.1041	4.9870	5.1274	5.8341	6.1421	5.7428	5.8797	6.7421
	REC	1.0241	0.9754	1.0004	0.9841	0.9951	1.0094	1.0100	1.1241
	RR	42.8710	34.7481	36.4107	40.2251	42.7840	38.2170	40.8941	44.2781
	C	0.31014	0.2574	0.2841	0.3006	0.2981	0.2978	0.3041	0.3341
IC_2	$SSIM$	0.6014	0.5412	0.5621	0.5902	0.5891	0.5832	0.5900	0.7021
	MSE	6.34E+03	7.02E+03	6.88E+03	6.50E+03	6.82E+03	6.75E+03	6.46E+03	6.17E+03
	EPI	0.9621	0.8210	0.8411	0.9231	0.8892	0.8941	0.9241	0.9874
	E	4.1074	3.2740	3.4174	3.8562	3.7831	3.7024	3.9875	4.8821
	REC	0.9014	0.8289	0.8421	0.8841	0.7961	0.8741	0.8832	0.9441
	RR	36.2701	28.2278	30.7704	32.7241	30.7231	32.0147	34.2839	38.1420
	C	0.2874	0.2141	0.2214	0.2522	0.2401	0.2417	0.26471	0.3011
IC_3	$SSIM$	0.7512	0.6697	0.6874	0.7681	0.7751	0.7104	0.7247	0.8421
	MSE	7.57E+03	8.21E+03	8.07E+03	7.41E+03	7.38E+03	7.93E+03	7.81E+03	7.32E+03
	EPI	0.9074	0.7924	0.8154	0.9142	0.9201	0.8621	0.8874	0.9421
	E	3.8721	2.6270	2.8974	3.8803	3.9002	3.1421	3.6421	4.1014
	REC	0.8040	0.7454	0.7528	0.8124	0.8241	0.7721	0.7974	0.8941
	RR	34.2210	27.2170	28.2247	34.2341	34.4581	30.8740	32.8740	36.2140
	C	0.2741	0.1987	0.2274	0.2781	0.2790	0.2422	0.2574	0.2987

(continued)

Table 6.7 (continued)

Image	Index	ABC	DE	FFA	Gamma-PSO	Fuzzy-IPSO	CLAHE	WTHE	MSA-ICE
IC_4	$SSIM$	0.7231	0.6532	0.6482	0.7421	0.7632	0.6931	0.7020	0.8021
	MSE	8.65E+03	9.11E+03	9.20E+03	8.54E+03	8.23E+03	8.84E+03	8.81E+03	7.89E+03
	EPI	0.8843	0.8042	0.8162	0.9321	0.9362	0.8654	0.8670	0.9765
	E	3.9821	3.5672	3.6640	4.1023	4.2103	3.7892	3.8124	4.5672
	REC	0.8450	0.7853	0.8004	0.8572	0.8631	0.8021	0.8164	0.9023
	RR	33.0051	28.7489	29.8743	33.4561	35.5630	30.2385	32.6739	37.9632
	C	0.2084	0.1672	0.1899	0.2234	0.2373	0.1892	0.1903	0.2893
IC_5	$SSIM$	0.8842	0.7892	0.7994	0.8856	0.8870	0.8009	0.8182	0.9061
	MSE	7.93E+03	8.45E+03	8.37E+03	7.86E+03	7.80E+03	8.20E+03	8.18E+03	7.53E+03
	EPI	0.9320	0.8563	0.8753	0.9662	0.9872	0.9032	0.9125	1.1261
	E	4.4676	4.0021	4.1292	4.6673	4.7032	4.2193	4.3389	5.9632
	REC	0.9537	0.8283	0.8772	0.9732	0.9862	0.9123	0.9245	1.2809
	RR	30.9850	22.7893	25.8303	31.6743	34.7620	27.8692	29.7630	38.9032
	C	0.2345	0.1782	0.2893	0.2253	0.2567	0.1945	0.2008	0.3892
IC_6	$SSIM$	0.8302	0.7984	0.8032	0.8573	0.8730	0.8122	0.8237	0.9032
	MSE	8.22E+03	8.88E+03	9.00E+03	7.90E+03	8.03E+03	8.58E+03	8.40E+03	7.81E+03
	EPI	0.9243	0.8245	0.8641	0.9421	0.9643	0.8840	0.9003	0.9902
	E	4.1032	3.0037	3.2743	4.4052	4.7004	3.6783	3.8830	5.0393
	REC	0.9435	0.8673	0.8870	0.9356	0.9560	0.9054	0.9180	1.0342
	RR	32.6703	25.9730	28.9803	33.6704	35.9982	29.9045	30.9850	39.8701
	C	0.2560	0.1905	0.2076	0.2483	0.2589	0.2076	0.2167	0.3678

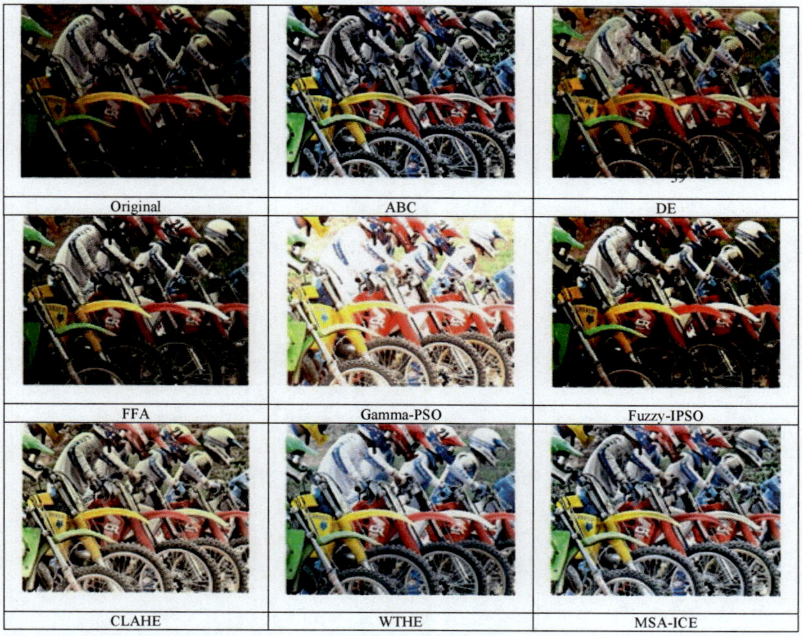

Fig. 6.12 Results for contrast enhancement of the color image IC_1 considering the ABC, DE, FFA, Gamma-PSO, Fuzzy-IPSO, CLAHE, WTHE and MSA-ICE methods

It is worth noting that Fig. 6.13 represents an image with the highest degree of color distortion produced by the MSA-ICE method. While some distortion is present, it is still relatively lower compared to the other methods. The last three images pose a challenge as they were captured under low-light conditions. Most of the ICE methods struggle to perform well under such circumstances due to significant intensity value changes required. Consequently, there is a tendency to generate intensity saturation and amplify noise in the enhanced image. Despite these challenges, the MSA-ICE method exhibits the best performance compared to its competitors.

6.7 Conclusions

The presence of noise and small sets of pixel intensities has a detrimental effect on the image enhancement process. These elements represent pixels that are insufficient to accurately represent objects or visible details in the image. Their existence prevents the adjustment of the original image distribution across the entire range. In this chapter, we introduce an Image Contrast Enhancement (ICE) algorithm for images. Our approach consists of two steps: the elimination of irrelevant information and the improvement of contrast.

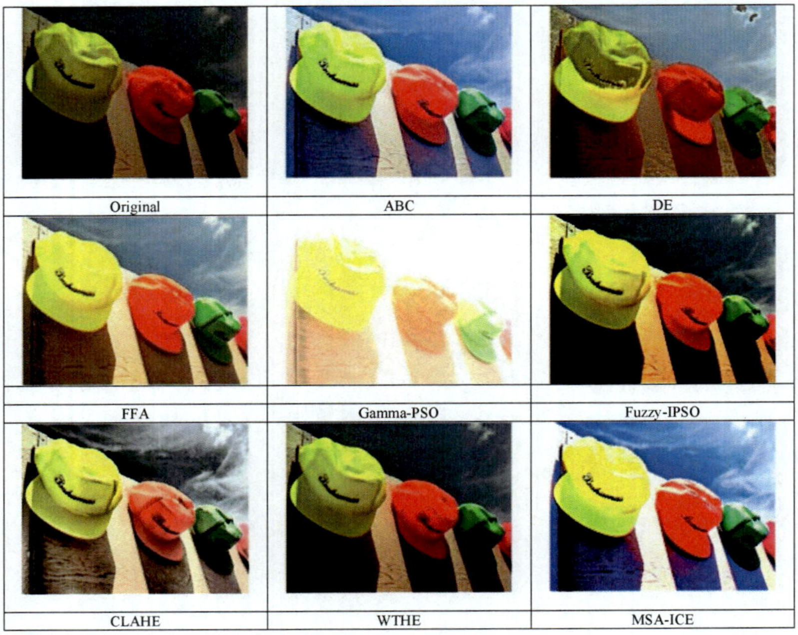

Fig. 6.13 Results for contrast enhancement of the color image IC_2 considering the ABC, DE, FFA, Gamma-PSO, Fuzzy-IPSO, CLAHE, WTHE and MSA-ICE methods

In the first step, we utilize the Mean-shift method to remove insignificant small sets of pixels. The goal is to replace these sporadic characteristics, which are present in the original histogram, with meaningful pixel densities represented by local maxima. This helps to refine the histogram and reduce the impact of irrelevant pixel values. In the second step, we focus on adjusting the contrast of the refined histogram. To accomplish this, we employ the Moth Swarm Algorithm (MSA). The MSA redistributes the pixel intensities within the reduced histogram, maximizing the value of the Kullback–Leibler Entropy (KL-entropy) between a candidate distribution and the best possible configuration, as determined by our algorithm.

We evaluate the performance of the presented approach using a representative set of grayscale and color images commonly found in the literature. We compare the results with other similar techniques, including artificial bee colony (ABC), differential evolution (DE), firefly algorithm (FFA), the Gamma-PSO algorithm, the Fuzzy-IPSO method, Local and Adaptive Histogram Equalization (CLAHE), and Weighted Thresholded Histogram Equalization (WTHE). Experimental results indicate that the presented method outperforms other schemes in terms of various performance indexes that assess the quality of enhancement.

In future work, we plan to validate the algorithm using different contrast metrics and conduct a comparative analysis between them. Additionally, we aim to explore modifications to further enhance the MSA algorithm and apply the ICE technique to other domains of interest, such as medical images, satellite images, and more.

Fig. 6.14 Results for contrast enhancement of the color image IC_3 considering the ABC, DE, FFA, Gamma-PSO, Fuzzy-IPSO, CLAHE, WTHE and MSA-ICE methods

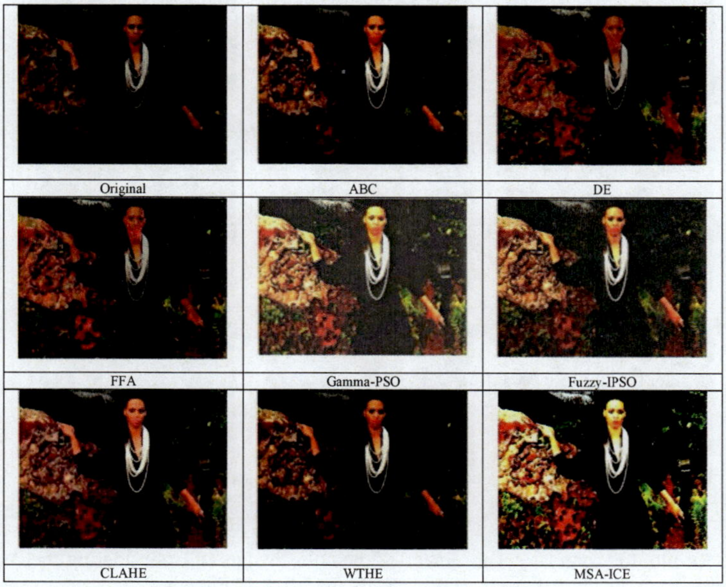

Fig. 6.15 Results for contrast enhancement of the color image IC_4 considering the ABC, DE, FFA, Gamma-PSO, Fuzzy-IPSO, CLAHE, WTHE and MSA-ICE methods

Fig. 6.16 Results for contrast enhancement of the color image IC_5 considering the ABC, DE, FFA, Gamma-PSO, Fuzzy-IPSO, CLAHE, WTHE and MSA-ICE methods

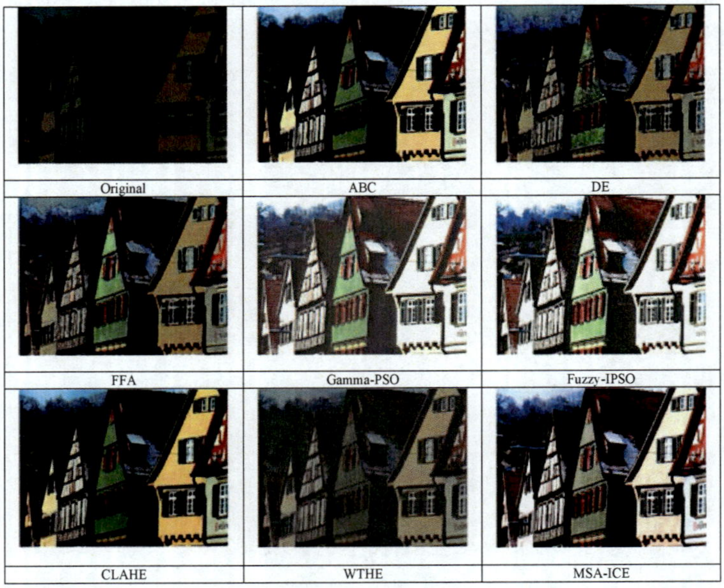

Fig. 6.17 Results for contrast enhancement of the color image IC_6 considering the ABC, DE, FFA, Gamma-PSO, Fuzzy-IPSO, CLAHE, WTHE and MSA-ICE methods

References

1. Agarwal, M., Mahajan, R.: Medical images contrast enhancement using quad weighted histogram equalization with adaptive Gama correction and homomorphic filtering. Procedia Comput. Sci. **115**, 509–517 (2017)
2. Agarwal, M., Mahajan, R.: Medical image contrast enhancement using range limited weighted histogram equalization. Procedia Comput. Sci. **125**, 149–156 (2018). https://doi.org/10.1016/J.PROCS.2017.12.021
3. AliyariGhassabeh, Y., Rudzicz, F.: The mean shift algorithm and its relation to kernel regression. Inf. Sci. **348**, 198–208 (2016). https://doi.org/10.1016/J.INS.2016.02.020
4. Bhandari, A.K., Soni, V., Kumar, A., Singh, G.K.: Cuckoo search algorithm based satellite image contrast and brightness enhancement using DWT-SVD. ISA Trans. **53**(4), 1286–1296 (2014)
5. Bhandari, A.K., Rahul, K.: A context sensitive Masi entropy for multilevel image segmentation using moth swarm algorithm. Infrared Phys. Technol. **98**, 132–154 (2019). https://doi.org/10.1016/J.INFRARED.2019.03.010
6. Cabella, B.C., Sturzbecher, M.J., De Araujo, D.B., Neves, U.P.: (2009), Generalized relative entropy in functional magnetic resonance imaging. Phys. A Stat. Mech. Appl. **388**(1), 41–50 (2009)
7. Chen, Z., Abidi, B.R., Page, D.L., Abidi, M.A.: Gray-level grouping (GLG): an automatic method for optimized image contrast enhancement—part I: the basic method. IEEE Trans. Image Process. **15**(8), 2290–2302 (2006)
8. dol Coelho, L.S., Sauer, J.G., Rudek, M.: Differential evolution optimization combined with chaotic sequences for image contrast enhancement. Chaos Solitons Fractals **42**(1), 522–529 (2009). https://doi.org/10.1016/j.chaos.2009.01.012
9. Cuevas, E., Cienfuegos, M.: A new algorithm inspired in the behavior of the social-spider for constrained optimization. Expert Syst. Appl. **41**(2), 412–425 (2014). https://doi.org/10.1016/j.eswa.2013.07.067
10. Ebrahim, M.A., Becherif, M., Abdelaziz, A.Y.: Dynamic performance enhancement for wind energy conversion system using moth-flame optimization based blade pitch controller. Sustain. Energy Technol. Assess. **27**, 206–212 (2018). https://doi.org/10.1016/J.SETA.2018.04.012
11. Hashemi, S., Kiani, S., Noroozi, N., Moghaddam, M.E.: An image contrast enhancement method based on genetic algorithm. Pattern Recogn. Lett. **31**(13), 1816–1824 (2010). https://doi.org/10.1016/j.patrec.2009.12.006
12. Hasler, Suesstrunk, S.E.: Measuring colorfulness in natural images. In: Electronic Imaging. Int. Soc. Opt. Photon., pp. 87–95.G (2003)
13. Horová, I., Koláček, J., Zelinka, J.: Kernel Smoothing in Matlab (2012). https://doi.org/10.1142/8468
14. Jamil, M., & Zepernick, H.J. (2013). Lévy flights and global optimization. In: Swarm Intelligence and Bio-Inspired Computation, pp. 49–72. https://doi.org/10.1016/B978-0-12-405163-8.00003-X
15. Jayanthi, K., Sudha, L.R.: Optimal Gray level mapping for satellite image contrast enhancement using grey wolf optimization algorithm. Int. J. Eng. Sci. Invent. **45**, 38–44 (2018)
16. Joseph, J., Periyasamy, R.: A fully customized enhancement scheme for controlling brightness error and contrast in magnetic resonance images. Biomed. Signal Process. Control **39**, 271–283 (2018)
17. Karaboga, D., Basturk, B.: A powerful and efficient algorithm for numerical function optimization: artificial bee colony (ABC) algorithm. J. Glob. Optim. **39**(3), 459–471 (2007). https://doi.org/10.1007/s10898-007-9149-x

18. Kennedy, J., Eberhart, R.: Particle swarm optimization. In: Proceedings of International Conference On, Neural Networks, 1995, vol. 4, pp. 1942–1948. IEEE (1995). https://doi.org/10.1109/ICNN.1995.488968
19. Kullback, S., Leibler, R.A.: On information and sufficiency. Ann. Math. Stat. **22**(1), 79–86 (1951). https://doi.org/10.1214/aoms/1177729694.JSTOR2236703
20. Larson, E.C., Chandler, D.M.: Most apparent distortion: full-reference image quality assessment and the role of strategy. J. Electron. Imag. **19**(1), 11006–11021 (2010)
21. Lewin, J.: Comparison of contrast-enhanced mammography and contrast-enhanced breast MR imaging. Magn. Reson. Imaging Clin. N. Am. **26**(2), 259–263 (2018). https://doi.org/10.1016/J.MRIC.2017.12.005
22. Mahmood, A., Khan, S., Hussain, S., Almaghayreh, E.: An adaptive image contrast enhancement technique for low-contrast images. IEEE Access **7**, 161584–161592 (2019)
23. Mantegna, R.N.: Fast, accurate algorithm for numerical simulation of Lévy stable stochastic processes. Phys. Rev. E **49**(5), 4677–4683 (1994). https://doi.org/10.1103/PhysRevE.49.4677
24. Manpreet, K., Jasdeep, K., Jappreet, K.: Survey of contrast enhancement techniques based on histogram equalization. Int. J. Adv. Comput. Sci. Appl. **2**(7), 2011137 (2015)
25. Maurya, L., Mahapatra, P.K., Kumar, A.: A social spider optimized image fusion approach for contrast enhancement and brightness preservation. Appl. Soft Comput. J. **52**, 575–592 (2017). https://doi.org/10.1016/j.asoc.2016.10.012
26. Mohamed, A.-A.A., Mohamed, Y.S., El-Gaafary, A.A.M., Hemeida, A.M.: Optimal power flow using moth swarm algorithm. Electr. Power Syst. Res. **142**, 190–206 (2017). https://doi.org/10.1016/J.EPSR.2016.09.025
27. Ponomarenko, N., Jin, L., Ieremeiev, O., Lukin, V., Egiazarian, K., Astola, J., Vozel, B., Chehdi, K., Carli, M., Battisti, F., Kuo, C.C.: Image database TID2013: peculiarities, results and perspectives. Signal Process Image Commun. **30**, 57–77 (2015)
28. Qin, A.K., Huang, V.L., Suganthan, P.N.: Differential evolution algorithm with strategy adaptation for global numerical optimization. IEEE Trans. Evol. Comput. **13**(2), 398–417 (2009). https://doi.org/10.1109/TEVC.2008.927706
29. Rashedi, E., Nezamabadi-pour, H., Saryazdi, S.: GSA: a gravitational search algorithm. Inf. Sci. **179**(13), 2232–2248 (2009). https://doi.org/10.1016/j.ins.2009.03.004
30. Reza, A.M.: Realization of the contrast limited adaptive histogram equalization (clahe) for real-time image enhancement. J. VLSI Signal Process Syst. Signal Image Video Technol. **38**, 35–44 (2004)
31. Sahu, S., Singh, A.K., Ghrera, S.P., Elhoseny, M.: An approach for de-noising and contrast enhancement of retinal fundus image using CLAHE. Opt. Laser Technol. https://doi.org/10.1016/J.OPTLASTEC.2018.06.061
32. Subrahmanyeswara-Rao, B.: Dynamic histogram equalization for contrast enhancement for digital images. Appl. Soft Comput. J. **89**, 106114 (2020)
33. Wang, X., Chen, L.: An effective histogram modification scheme for image contrast enhancement. Signal Process. Image Commun. **58**, 187–198 (2017). https://doi.org/10.1016/J.IMAGE.2017.07.009
34. Wang, Q., Tan, R.K.: Fast image/video contrast enhancement based on weighted thresholded histogram equalization. IEEE Trans. Consum. Electron. **53**(2), 757–764 (2007)
35. Wu, H.-T., Tang, S., Huang, J., Shi, Y.-Q.: A novel reversible data hiding method with image contrast enhancement. Signal Process. Image Commun. **62**, 64–73 (2018). https://doi.org/10.1016/J.IMAGE.2017.12.006
36. Wu, K.-L., Yang, M.-S.: Mean shift-based clustering. Pattern Recogn. **40**(11), 3035–3052 (2007). https://doi.org/10.1016/J.PATCOG.2007.02.006

37. Yang, X.-S.: Firefly algorithm, Lévy flights and global optimization. In: Research and Development in Intelligent Systems XXVI, pp. 209–218 (2010). https://doi.org/10.1007/978-1-84882-983-1_15
38. Zhao, C., Wang, Z., Li, H., Wu, X., Qiao, S., Sun, J.: A new approach for medical image enhancement based on luminance-level modulation and gradient modulation. Biomed. Signal Process. Control **48**, 189–196 (2019)

Chapter 7
Multi-objective Optimization of Anisotropic Diffusion Parameters for Enhanced Image Denoising

Noise often corrupts images during their acquisition, storage, and transmission processes, resulting in a degradation of image quality. The main challenge in image denoising is to remove the corrupted information while preserving the fine details of the original image. Anisotropic Diffusion (AD) is a well-established method for noise removal in digital images that effectively preserves the image edges. However, selecting the appropriate parameters for the AD operation significantly influences the filtering outcomes. Despite its importance, the automatic determination of AD parameters based on image requirements has received limited attention in the literature. This chapter introduces a novel multi-objective approach for obtaining optimal AD parameters that yield effective filtering results. The presented methodology aims to strike the best possible balance between two conflicting objectives: minimizing noise content in the image while maximizing contrast. Improving one objective without compromising the other is not feasible. To address this, the presented approach employs the Non-dominated Sorting Genetic Algorithm based on reference points (NSGA-III), which is known as a robust and powerful algorithm for multi-objective optimization. The final solution is determined by analyzing the Optimal Pareto front. Experimental results demonstrate that the presented method outperforms existing filtering algorithms in terms of both visual quality and standard performance metrics.

7.1 Introduction

Several computer vision applications require noise reduction and image enhancement to achieve more reliable outcomes [1]. Anisotropic Diffusion (AD) is widely recognized as an effective technique for eliminating noise in digital images while preserving their edges [1]. AD introduces a novel concept of scale-space representation in image processing [2]. It employs a non-linear partial differential equation to model a diffusion process. By utilizing AD, the adverse effects caused by linear

E. Cuevas et al., *New Metaheuristic Schemes: Mechanisms and Applications*, Intelligent Systems Reference Library 246, https://doi.org/10.1007/978-3-031-45561-2_7

filtering, such as blurring or edge attenuation, are significantly mitigated. Consequently, AD has been extensively applied in various image processing tasks, including image enhancement, edge detection, image smoothing, and image segmentation.

AD filtering allows noise elimination in an image while maintaining object boundaries and small details intact. However, the efficacy of this operation heavily depends on appropriately configuring its essential parameters. The conductance function, gradient threshold, and stopping parameter form a parameter set that determines the magnitude of diffusion and the behavior of the filtering process. Modifying any of these parameters beyond a certain threshold can result in undesired blurring effects in the image. Conversely, using excessively small values may render the operation ineffective, leaving the image with residual noise. Hence, it is crucial to determine optimal parameter values that align with the specific requirements of a given image in terms of noise magnitude and edge quality [1].

In recent years, numerous studies have been conducted to explore the filtering capabilities of both discrete and continuous versions of Anisotropic Diffusion (AD) [2]. The continuous version of AD has been the subject of several investigations [3–12], focusing on analyzing its filtering behavior and addressing its ill-posed formulation. Many of these works propose modifications to the ill-posed AD definition through regularization methods to establish well-posed equations. For instance, one study [13] introduces a discrete implementation of the AD equation to achieve improved isotropy, while other works [14, 15] present semi-implicit schemes for reformulating the AD definition. A novel well-posed AD definition is proposed in [12], employing a multi-grid numerical scheme that exhibits interesting steady-state properties. Additionally, a time-dependent solution to the AD formulation is presented in Yu et al. [16] using a different numerical method. Reference [17] introduces a new AD implementation that incorporates the standard deviation of grayscale values in the neighborhood of each pixel. Furthermore, Chao and Tsai [18] introduces a diffusion model suitable for low-contrast images and irregular illumination.

Despite the significance of the AD operation, the automatic estimation of its parameters based on image requirements has received limited attention in the literature. Exceptions include the Edge Preserving AD filter (EPAD) [19] and the Robust AD filter (RAD) [20] approaches, which incorporate preprocessing stages to estimate the noise level in the image and calibrate the AD filter accordingly. Given the numerous parameter values that can be used in configuring the AD scheme, the filtering results can vary significantly. Therefore, an appropriate definition of the AD configuration is essential to effectively reduce noise in the image while minimizing edge attenuation.

The determination of appropriate parameters for AD filtering has received limited attention in the literature [21]. Some estimation techniques involve determining the gradient threshold parameter using a decreasing function [22]. In this approach, the parameter is chosen based on the noise characteristics present in the image. Hence, the magnitude of the gradient threshold corresponds to the value at which the noise is eliminated while preserving the edges. Other methods utilize alternative approaches such as morphological operators [23] or statistical properties of the image [2, 24, 25] to estimate the gradient threshold parameter. These methods have been considered in comparison to the presented algorithm, which achieves superior filtering results in terms of accuracy and robustness.

The effect of the filter on an image is determined during the iterative solution of the AD equations. It is crucial to determine the optimal stopping point for the iterations to prevent the image from being excessively smoothed. Several approaches have addressed the estimation of the stopping parameter [26–30] to achieve satisfactory AD filtering outcomes. These methods incorporate schemes such as [9, 31, 32], where a fidelity factor is introduced in the computation of the AD equations. The AD computation is halted when the fidelity factor indicates that the filtered image deviates significantly from the original image. While these approaches yield interesting results, they may not effectively remove the noise. In Reference [25], a novel formulation of the AD equations is proposed, which is insensitive to the number of iterations used in the computation. Additionally, in Chen et al. [12], a multi-grid scheme is introduced, utilizing a normalized cumulative periodogram to automatically determine the stopping point of the AD computation.

However, these methods estimate the stopping parameter solely based on the criterion of noise elimination by the AD filter, without considering the preservation of edges. The attenuation of existing edges in the final image is closely tied to the results of the AD filter. Thus, the exclusion of this information significantly limits its filtering potential.

A recent trend in formulating image processing problems involves considering approaches capable of addressing multiple objectives in their decision-making processes [33–38]. In formulations with multiple objectives, the optimization functions defined often conflict with each other, making it challenging to individually optimize each objective. In fact, many image processing applications can be framed as multi-objective problems. For example, in segmentation tasks, it is necessary to minimize the distance between homogeneous elements while maximizing the distance between heterogeneous elements [35, 36].

Over the past decade, multi-objective algorithms based on evolutionary principles have emerged as efficient techniques for solving problems that involve conflicting conditions. Several novel multi-objective approaches have been proposed in the literature, such as the non-dominated sorting genetic algorithm (NSGA-III) [39, 40], the strength Pareto evolutionary algorithm (SPEA2) [41], and the indicator-based selection in multi-objective search (IBEA) [42]. Among these, NSGA-III has gained popularity due to its appealing characteristics. Unlike other approaches, NSGA-III enables the generation of a more accurate and evenly distributed Pareto front by utilizing a strategy based on reference points. These properties have motivated its application in various multi-objective scenarios, including feature selection [43], economic dispatch problems [44], and wind energy optimization [45], among others.

Images often experience noise corruption during the processes of acquisition, storage, and transmission [46, 47]. This alteration typically degrades the perceptual quality of the image. The primary challenge in image denoising is to remove the corrupted information while preserving the fine details of the original image. Contrast enhancement aims to enhance the visual quality of images. Image contrast refers to the distinguishable differences in visible characteristics that allow objects to stand out from other elements in the image. In visual perception, contrast is determined by the existing variations among the key attributes of the original image. However, intensifying these differences results in an increase in the noise content of the resulting image.

This chapter introduces a novel multi-objective methodology for determining the AD parameters to achieve effective filtering outcomes [48]. The presented approach seeks to find appropriate values for the AD parameters that strike the best possible balance between two conflicting objectives: eliminating the corrupted information while enhancing the key characteristics of the original image. Improving one objective without compromising the other is not feasible. To tackle this problem, the presented scheme employs the Non-dominated Sorting Genetic Algorithm based on Reference Points (NSGA-III) [39, 40], which is recognized as one of the most robust and powerful algorithms for multi-objective optimization. By analyzing the Optimal Pareto front, the final solution is obtained [48]. Experimental results demonstrate that the presented method outperforms existing filtering algorithms in terms of visual quality and standard performance metrics. The superiority of the presented approach is validated through empirical evidence [48].

The subsequent sections of this chapter are structured as follows: Sect. 7.2 provides a concise overview of the anisotropic diffusion scheme. Section 7.3 delves into the fundamental aspects of multi-objective optimization and discusses the NSGA-III algorithm. Section 7.4 defines the objective functions employed in our multi-objective scheme. Section 7.5 presents a comprehensive description of the presented approach. Section 7.6 showcases the experimental results obtained through our approach. Lastly, Sect. 7.7 concludes the chapter by summarizing the key findings and implications.

7.2 Main Concepts of Anisotropic Diffusion

The basic formulation of the anisotropic diffusion (AD) model was originally presented in [1] as follows:

$$\frac{\partial I(x, y, t)}{\partial t} = div(g(\|\nabla I(x, y, t)\|)\nabla I(x, y, t)) \tag{7.1}$$

In this formulation, the parameter t represents time, $I(x, y, 0)$ represents the original image, and $\nabla I(x, y, t)$ denotes the gradient of the image $I(x, y)$ at time t. The conductance model, denoted as $g(\cdot)$, is chosen in such a way that the diffusion has maximum effect within homogeneous regions ($h \rightarrow 0$), while it has no impact on the borders ($h \rightarrow 0$). The formulation can be mathematically expressed as follows:

$$\lim_{h\to 0} g(h) = 1, \quad \lim_{h\to\infty} g(h) = 0, \tag{7.2}$$

The commonly used conductance functions are represented by the following equations:

$$g_1(h) = \exp\left(-\left(\frac{h}{K}\right)^2\right) \tag{7.3}$$

$$g_2(h) = \frac{1}{1 + \left(\frac{h}{K}\right)^2}, \tag{7.4}$$

The parameter K represents the gradient threshold value, which determines the extent of the diffusion effect. This threshold acts as a dividing point between the gradient components associated with noise information and those associated with the actual image boundaries. In the context of robust statistics, Black et al. [24] proposed an alternative conductance function known as the Tukey biweight function. The Tukey biweight function can be mathematically expressed as:

$$g_3(h) = \begin{cases} \frac{1}{2}\left(1 - \left(\frac{h}{S}\right)^2\right)^2 & h \leq S \\ 0 & \text{otherwise} \end{cases}, \tag{7.5}$$

where $S = K\sqrt{2}$.

Perona and Malik presented in [1] a discretized variant of the anisotropic diffusion model described in Eq. (7.1). The discretized formulation is defined as follows:

$$I_{T+1} = I_t + \frac{\lambda}{|\eta_s|}\sum_{p\in\eta_s} g_d\left(\left|\nabla I_{s,p}\right|\right)\nabla I_{s,p}, \tag{7.6}$$

In the proposed formulation, I represents a specific image, s denotes the pixel position in a two-dimensional lattice, T represents the iteration number, and g_d represents the conductance model with d taking values from 1 to 3. The parameter λ determines the diffusion magnitude and lies within the range (0,1). η_s represents the local spatial 4-pixel neighborhood centered around pixel s, consisting of the North, South, East, and West neighboring pixels. Thus, $\eta_s = \{N, S, E, W\}$ represents the set of these neighbors. The cardinality of η_s, denoted as $|\eta_s|$, is typically 4, except for edge pixels. The gradient operation, represented by ∇, is computed as the difference between neighboring pixels. Consequently, the gradient value $\nabla I_{s,p}$ is computed as follows:

$$\nabla I_{s,p} = I_T(p) - I_T(s), \quad p \in \eta_s \tag{7.7}$$

The discretized version of the anisotropic diffusion model presented in Eq. (7.6) is not an exact representation of the continuous model described in Eq. (7.1). Nevertheless, this discrete model retains essential characteristics of the continuous model while offering a lower computational complexity [1].

Estimating the suitable parameters for AD filtering requires determining two elements: the gradient threshold parameter K and the optimal iteration stopping time T.

7.3 NSGA-III and Multi-objective Optimization

7.3.1 Multi-objective Optimization

A general definition of a multi-objective formulation (MOP) typically involves a minimization statement, which can be expressed mathematically as shown in previous studies [48, 49].

$$\begin{aligned} \text{Minimize} \quad & F(\mathbf{x}) = (f_1(\mathbf{x}), \ldots, f_M(\mathbf{x}))^T \mathbf{x} = (x_1, \ldots, x_n) \\ \text{Subject to:} \quad & g_i(\mathbf{x}) \leq 0, i = 1, \ldots, p; \\ & h_j(\mathbf{x}) \leq 0, j = 1, \ldots, q; \\ & x_d \in [a_d^{min}, a_d^{max}], d \\ & = 1, \ldots, n; \end{aligned} \tag{7.8}$$

Here, $\Omega = \prod_{i=1}^{n}[a_d^{min}, a_d^{max}] \subseteq R^n$ represents the decision space, where $\mathbf{x} = (x_1, \ldots, x_n) \in \Omega$ denotes the d-dimensional design variable. The mapping from Ω to R^M represents the transformation from the design variable space to the objective space. The inequality constraints $g_i(\mathbf{x}) \leq 0$ (where $i \in 1, \ldots, p$) and the equality constraints $h_j(\mathbf{x})$ (where $j \in 1, \ldots, q$) illustrate the p inequality constraints and q equality constraints, respectively.

Pareto dominance: A candidate solution $\mathbf{x}_a$ is said to dominate another candidate solution $\mathbf{x}_b$ (denoted as $\mathbf{x}_a \prec \mathbf{x}_b$) if and only if $f_i(\mathbf{x}_a)$ is better than or equal to $f_i(\mathbf{x}_b)$ for every objective function i from 1 to M. Additionally, there must be at least one objective j (where j belongs to the range 1 to M) for which $f_j(\mathbf{x}_a)$ is strictly better than $f_i(\mathbf{x}_b)$).

Pareto optimal solution: A candidate solution $\mathbf{x}^*$ in the decision space Ω is considered Pareto optimal if there is no other solution $\mathbf{x}$ that dominates $\mathbf{x}^*$ ($\mathbf{x} \prec \mathbf{x}^*$).

Pareto optimal set: The set of all Pareto optimal solutions is denoted as P_t and can be defined as $P_t = \{\mathbf{x} \in \Omega | \ \exists \mathbf{x}^* \in \Omega, \mathbf{x}^* \prec \mathbf{x}\}$.

7.3.2 The NSGA-III

Several methods have been documented in the literature to tackle multi-objective problems [43–45, 50, 51]. Among these methods, NSGA-III (The non-dominated sorting genetic algorithm) has become popular due to its notable characteristics. NSGA-III was developed by K. Deb et al. [39] as an extension of the original NSGA-II for multi-objective optimization. While NSGA-III shares a similar structure with NSGA-II, it introduces a selection mechanism based on reference points. These reference points represent trade-offs between candidate solutions and objective function values. By incorporating reference points, NSGA-III enhances the diversity and spatial distribution of solutions in the Pareto front.

In the NSGA-III process, an initial set of H reference points is generated on a hyperplane. To initialize the algorithm, a population P_t consisting of N individuals is randomly created. If the termination criterion is not met, the following stages are iteratively applied. At the tth iteration, N individuals from the population P_t are selected as parent agents. These parent agents undergo simulated binary crossover (SBX) and polynomial mutation operations to generate a new population Q_t, which also consists of N individuals. The populations Q_t and P_t are then merged to form a resulting population R_t, which contains $2N$ individuals ($R_t = P_t \cup Q_t$). The individuals in R_t are classified based on their non-dominance rank. Once Rt has been classified, a new population St of N individuals is generated by adding solutions from different non-dominated groups, one group at a time. The generation of S_t starts by including the best non-dominated group F_1 and continues with elements from the second non-dominated group F_2, followed by the third non-dominated group F_3, and so on. Since the size of R_t is $2N$, not all fronts can be included in the available positions of St. Thus, the front Fi where $|S_t \cup F_i| \geq N$ is considered the splitting front. If $|S_t| + |F_i| = N$, all elements of F_i are included in St without any selection mechanism. However, if there is a shortfall, a niching scheme is employed to select those elements from Fi that have fewer connections with the reference directions already associated with individuals in S_t.

To assign individuals to the reference directions H, the individuals in S_t are normalized using the minimum point $\hat{z}^*$ and the nadir point $\hat{z}^{nad}$ as the lower and upper limits, respectively. The association between agents and reference points is determined based on their perpendicular distance. A niching-preservation process is then employed to select the $N - |S_t|$ missing elements. The niching count ρj represents the number of candidate solutions from F_i that are connected to the jth reference point. In this context, two different cases can occur:

(I) If there is any element from F_i that is connected to the jth reference point, two scenarios are taken into account. (A) When $\rho_j = 0$, the element with the smallest perpendicular distance to the jth reference point is selected and included in S_t. (B) When $\rho_j > 0$, an element from F_i that is associated with the jth reference point is randomly chosen and added to S_t. After this step, the value of ρ_j is increased.

(II) If none of the elements from F_i are connected to the jth reference point, the reference point is not taken into consideration during the selection process.

The process of selecting elements for S_t is repeated until a stop criterion is met, and once all the elements of S_t are determined, it becomes the new parent population P_{t+1}.

7.4 Function Cost

Multi-objective optimization tackles the challenge of simultaneously optimizing multiple conflicting objective functions that cannot be improved independently.

In this approach, we aim to find the optimal parameter values for AD filtering by striking the best possible balance between two conflicting objectives. The optimization process involves minimizing the noise present in the image while maximizing the contrast of the key features in the original image. In this section, we outline the objective functions that capture these procedures.

7.4.1 Estimation of Noise

The primary task in image denoising is to remove the corrupted information while preserving the essential features of the original image. In this scheme, we utilize the evaluation of the noise variance, denoted as $\sigma(I)$, in the image I as a means to estimate the level of noise present in the image.

The noise estimator [52] should not only be capable of determining the magnitude of noise in the image but also be insensitive to the presence of edge features. Edge structures in an image typically exhibit prominent second-order differential characteristics. As edge points can be identified using the Laplacian operator, the noise estimation operator N is defined as the mask obtained through the relationship $N = 2(L_2 - L_1)$. Figure 7.1 illustrates the Laplacian masks L_1 (Fig. 7.1a) and L_2 (Fig. 7.1b), while Fig. 7.1c displays the noise operator N.

The noise estimation component N has a mean value of zero and a variance of 36 times the squared standard deviation of the noise at each pixel, denoted as σ_p. Considering an image I of size $W \times H$, the noise variance σ_I can be calculated using the following formula:

$$\sigma(I) = \sqrt{\frac{1}{36(W-2)(H-2)} \sum_{\substack{1 \le x \le W \\ 1 \le y \le H}} (I(x, y) * N)^2} \tag{7.9}$$

The expression $I(x, y) * N$ represents the value obtained by applying the mask N at the pixel location (x, y) in the image I.

$$L_1 = \begin{bmatrix} 0 & 1 & 0 \\ 1 & -4 & 1 \\ 0 & 1 & 0 \end{bmatrix} \quad L_2 = \frac{1}{2} \begin{bmatrix} 1 & 0 & 1 \\ 0 & -4 & 0 \\ 1 & 0 & 1 \end{bmatrix} \quad N = \begin{bmatrix} 1 & -2 & 1 \\ -2 & 4 & -2 \\ 1 & -2 & 1 \end{bmatrix}$$

Fig. 7.1 Laplacian windows as noise operators

7.4.2 *Improvement of Contrast*

In the context of visual perception, contrast refers to the noticeable distinctions among the key features of the original image. To assess the contrast $CE(I)$ [53] of an image I, a model is employed that takes into account the intensity of edge pixels, the quantity of edge pixels, and the entropy of the entire image. This formulation is expressed as follows:

$$CE(I) = \log(\log(E(I))) \cdot \frac{NE(I)}{W \cdot H} \cdot H(I), \tag{7.10}$$

In the given equation, $E(I)$ represents the sum of edge intensities obtained through the Sobel filter. $NE(I)$ denotes the count of edges detected in image $I \cdot H(I)$ represents the entropy of image I. Equation (7.10) clearly shows that the contrast value CE increases when there are greater differences among the main characteristics of the original image, which can be achieved by either increasing the number of edges or enhancing the overall entropy.

7.5 Multi-objective Methodology

Estimating the suitable parameters for AD filtering involves determining two elements: the gradient threshold parameter K and the optimal stopping time for iterations, denoted as t. In this approach, these parameter values are obtained by solving a multi-objective problem that aims to achieve the best trade-off between two conflicting objective functions, namely f^1 and f^2. The optimization process focuses on minimizing the noise present in the image (f^1) while simultaneously maximizing the contrast of the key features of the original image (f^2).

7.5.1 *Process of NSGA-III*

During its operation, the anisotropic diffusion (AD) filtering process takes three inputs: the initial corrupted image O, the gradient threshold parameter K, and the optimal stopping time for iterations T. The output of the AD scheme is the resulting filtered image I. If we represent the AD filtering process as $I = \mathbf{AD}(K, T)$, the multi-objective optimization problem can be formulated in the following manner:

$$\text{Minimize} \;\; f^1 = \sigma(\mathbf{AD}(\mathbf{x})) = \sqrt{\frac{1}{36(W-2)(H-2)} \sum_{\substack{1 \le x \le W \\ 1 \le y \le H}} (\mathbf{AD}(\mathbf{x}) * N)^2}$$

$$\text{Maximize } f^2 = CE(\mathbf{AD}(\mathbf{x})) = \log(\log(E(\mathbf{AD}(\mathbf{x})))) \cdot \frac{NE(\mathbf{AD}(\mathbf{x}))}{W \cdot H}$$
$$\mathbf{x} = (x_1, x_2) = (K, T) \tag{7.11}$$

The NSGA-III method is employed to solve this multi-objective formulation.

Hence, the NSGA-III operates by evolving a population X_t ($\{\mathbf{x}_1^t, \ldots, \mathbf{x}_N^t\}$) of N individuals from the initial stage ($t = 1$) to a maximum number of iterations ($t = maxiter$). Each element $\mathbf{x}_i^t$ ($i = 1, \ldots, N$) in the population corresponds to a two-dimensional vector, where each dimension represents a variable of the optimization problem, $\mathbf{x}_i^t = \left(K_i^t, T_i^t\right)$. The quality of each candidate solution $\mathbf{x}_i^t$ is evaluated based on its non-dominance with respect to the two objective functions ($M = 2$). Throughout the optimization process, NSGA-III applies its operators (described in Sect. 7.3) iteratively to the population X_t. The best Pareto front P_t discovered so far is retained as it represents the current best solution to the multi-objective problem. After conducting extensive experiments, the configuration parameters of NSGA-III have been set as follows: $N = 200$, $H = 2$, and 500 generations.

7.5.2 Analysis of the Optimal Pareto Front

Following the execution of NSGA-III, a final Pareto optimal front P_F is generated, consisting of c distinct solutions: $P_F = \{\mathbf{f}_1, \ldots, \mathbf{f}_c\}$. Each position $\mathbf{f}_j = (f_j^1, f_j^2)$ is associated with a two-dimensional vector $\mathbf{x}_j = (K_j, T_j)$, representing the gradient threshold parameter K and the optimal time to stop the iterations T for solution j. The evaluations of each solution $\mathbf{x}_j$ with respect to the objective functions are represented by the elements $\left(f_j^1, f_j^2\right)$, where f_j^i denotes the evaluation of $\mathbf{x}_j$ for the ith objective function ($j = 1, \ldots, c$; $i = 1, 2$). Each element in P_F signifies an optimal solution that balances both objective functions f^1 and f^2. Thus, all solutions hold significance in terms of the trade-off between these objectives. However, from an implementation perspective, only one solution from $\{\mathbf{f}_1, \ldots, \mathbf{f}_c\}$ is required.

The task of determining the optimal trade-off between the objective functions f^1 and f^2 can be likened to identifying the knee point in system engineering. The knee point represents the "right decision point" where the relative value of a variable becomes less significant in terms of its final contribution. Various approaches for detecting knee points have been proposed in the literature [54, 55]. In this chapter, we adopt the method introduced in [54] due to its simplicity. Under this approach, the set P_F consisting of c pairs $\{\mathbf{f}_1, \ldots, \mathbf{f}_c\}$ is normalized, scaling their values $\left(f_j^1, f_j^2\right)$ to range from 0 to 1. Using the distribution of ($l = 1, \ldots, c$), a new cost function $\mathbf{v}_l = (v_l^1, v_l^2)$ is constructed for each $\mathbf{f}_l$, which relates the noise increment f_l^1 to the significance of its contrast in capturing the main characteristics f_l^2. The computation of this function is performed as follows:

$$v_l^1 = f_l^1 + f_l^2; \; v_l^2 = f_l^2; \tag{7.12}$$

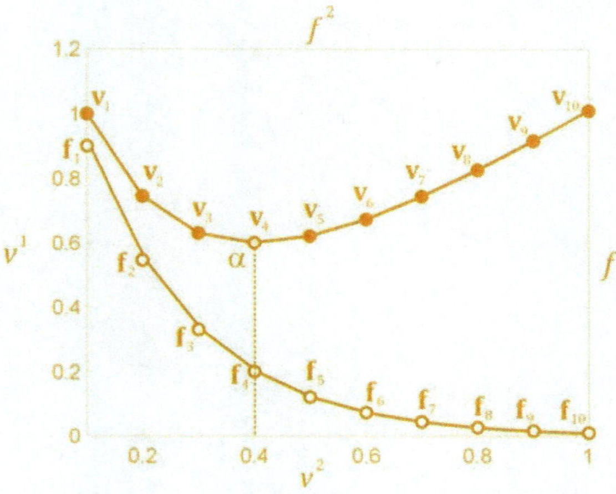

Fig. 7.2 The process to determine the best possible trade-off between both objective functions f^1 and f^2

An essential property of the function v_l^1 is its singular global minimum α, which represents the knee point. The knee point is defined as follows:

$$\alpha = \underset{0 \leq l \leq c}{\operatorname{argmin}}\, v_l^1 \tag{7.13}$$

In the approach, the solution $\mathbf{f}_\alpha$ corresponds to the optimal trade-off between the objective functions f^1 and f^2. This means that the parameters $\mathbf{x}_\alpha = (K_\alpha, T_\alpha)$ yield an image with the highest visual quality while minimizing the amount of noise. Figure 7.2 illustrates the process of determining the best trade-off between f^1 and f^2. From the figure, it can be observed that the fourth element $\mathbf{f}_4$ ($\alpha = 4$) has the lowest value. Hence, the parameters $\mathbf{x}_4 = (K_4, T_4)$ result in the best filtered image.

Figure 7.3 depicts a visual representation of the analysis of the optimal Pareto front. The plot shows the final Pareto optimal front P_F, and specific points A, B, and C have been selected for further examination. Point A represents a solution with the lowest noise level (lowest f^1) and minimal contrast of the main characteristics (lowest f^2). This results in an over-filtered image, as evident in Image A shown in Fig. 7.3. On the other hand, point C corresponds to a solution with the highest noise content (highest f^1) and maximal contrast among its characteristics (highest f^2). Image C in Fig. 7.3 illustrates this condition, where there is a significant amount of noise present along with high contrast. Lastly, point B represents the solution $\mathbf{f}_\alpha$, which corresponds to the optimal trade-off between both objective functions f^1 and f^2. Image B in Fig. 7.3 showcases this balance, where the noise has been effectively eliminated while the main characteristics are still preserved.

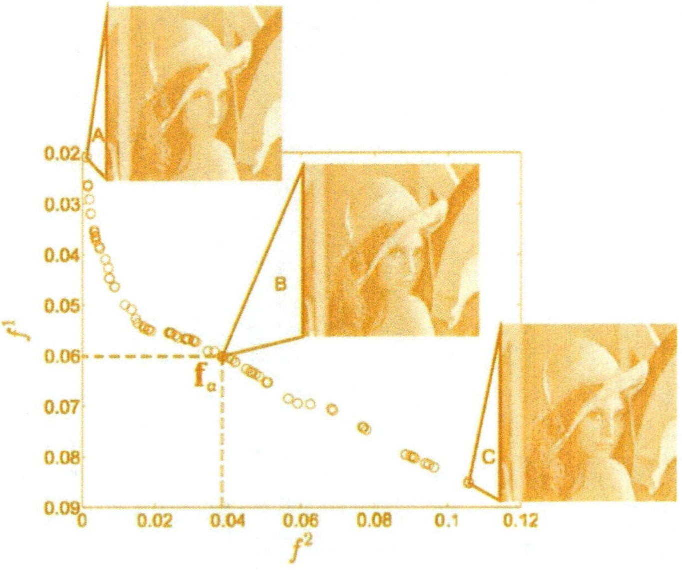

Fig. 7.3 An example of a Pareto front analysis

7.6 Experiments

7.6.1 Configuration of the Tests

To evaluate the performance of our filtering approach, a set of twenty images was used in the experiments. These images were collected from various datasets commonly used in image processing applications, ensuring compatibility with similar methods. The size of all the images in the set was fixed at 481 × 321 pixels. The original images used in the analysis can be seen in Fig. 7.4.

In the experiments, the presented algorithm, Multi-objective AD (MAD), was applied to five images from the dataset. The results obtained from MAD were compared to those produced by two other state-of-the-art filtering techniques based on AD principles: Edge Preserving AD filter (EPAD) [19] and Robust AD filter (RAD) [20]. These methods, EPAD and RAD, are known to deliver competitive results in complex filtering scenarios [56–59]. The configurations of each algorithm used in the comparisons were set based on their reported values, representing the best performance achievable according to their respective references.

Fig. 7.4 Images considered in the experiments

7.6.2 Indexes of Evaluation

In our study, an original image **R** from the dataset was deliberately contaminated with noise to create a noisy image **C**. This noisy image **C** served as the input for an AD scheme, which resulted in the generation of a filtered image **I**.

To assess the performance of the filtering approach, six performance indexes were considered: Structural Similarity Index Measurement (*SSIM*), Mean Squared Error (*MSE*), Peak Signal-to-Noise Ratio (*PSNR*), Edge Preservation Index (EPI), Normalized Cross-Correlation (*NCC*), and Computational Complexity (*CC*). These indexes were chosen to ensure compatibility with other studies [19, 20]. The first five indexes evaluate the quality of the segmentation achieved by the solution, while the last index measures the computational effort required by the filtering process. These indexes collectively assess the quality of the filtered image **I** by comparing it to the reference image **R**, which represents the original image without noise.

The Structural Similarity Index Measurement (*SSIM*) quantifies the similarity between a filtered image and a reference result, typically considered as the ground truth. Let $\mathbf{I} = \{p_1, \ldots, p_{M \times N}\}$ denote the pixel data of the denoised image, and $\mathbf{R} = \{r_1, \ldots, r_{M \times N}\}$ represent the reference data. The computation of *SSIM* is performed as follows:

$$SSIM = \frac{(2\mu_{\mathbf{I}}\mu_{\mathbf{R}} + Q_1)(2\sigma_{\mathbf{IR}} + Q_2)}{\left(\mu_{\mathbf{I}}^2 + \mu_{\mathbf{R}}^2 + Q_1\right)\left(\sigma_{\mathbf{I}}^2 + \sigma_{\mathbf{R}}^2 + Q_2\right)} \tag{7.14}$$

Here, Q_1 and Q_2 represent two small positive constants, typically set to 0.01. $\mu_{\mathbf{I}}$ and $\mu_{\mathbf{R}}$ denote the mean values of the filtered and reference data, respectively. $\sigma_{\mathbf{I}}$ and $\sigma_{\mathbf{R}}$ represent the variances of the filtered and reference data, respectively. $\sigma_{\mathbf{IR}}$ corresponds to the covariance between the elements of **I** and **R**. The values computed by *SSIM* range from 0 to 1.

Mean squared error (MSE) measures the dissimilarity between a filtered image **I** and its reference **R** by calculating the average squared difference between the pixel values of the filtered image and the corresponding pixels in the reference image. The MSE is computed using Eq. (7.15). A lower MSE value indicates a higher similarity or resemblance between the image **I** and its ground truth **R**.

$$MSE = \frac{1}{MN}\sum_{i=1}^{M}\sum_{j=1}^{N}(\mathbf{I}(i,j) - \mathbf{R}(i,j))^2 \tag{7.15}$$

Peak Signal to Noise Ratio (*PSNR*) quantifies the relationship between the maximum possible pixel value of an image (*MAXI*) and its similarity to the reference image in terms of signal magnitude. A lower PSNR value indicates better performance, indicating a higher similarity between the images. The *PSNR* is calculated using the following equation:

$$PSNR = 20\log_{10}\left(\frac{MAXI}{MSE}\right) \tag{7.16}$$

The Edge Preservation Index (*EPI*) is used to assess the preservation of edge features, which are the main characteristics of an image. A higher value of this index indicates better performance of the filtering algorithm in terms of edge preservation. The *EPI* is calculated using the following equation:

$$EPI = \frac{\sum_{i=1}^{M}\sum_{j=1}^{N}\left[\Delta\mathbf{I}(i,j) - \overline{\Delta\mathbf{I}}\right]\cdot\left[\Delta\mathbf{R}(i,j) - \overline{\Delta\mathbf{R}}\right]}{\sqrt{\sum_{i=1}^{M}\sum_{j=1}^{N}\left[\Delta\mathbf{I}(i,j) - \overline{\Delta\mathbf{I}}\right]^2\cdot\left[\Delta\mathbf{R}(i,j) - \overline{\Delta\mathbf{R}}\right]^2}}, \tag{7.17}$$

The *EPI* is determined by comparing the results of a 3 × 3 Laplacian filter applied to both the filtered image **I** and its original reference **R** at each pixel (i, j). The values $\Delta\mathbf{I}(i,j)$ and $\Delta\mathbf{R}(i,j)$ represent the outcomes of the Laplacian filter for the filtered image **I** and the reference image **R**, respectively, at pixel (i, j). Additionally, the values of ΔI and ΔR are computed as averages using the following formulas:

$$\overline{\Delta\mathbf{I}} = \frac{1}{MN}\sum_{i=1}^{M}\sum_{j=1}^{N}\Delta\mathbf{I}(i,j);\ \overline{\Delta\mathbf{R}} = \frac{1}{MN}\sum_{i=1}^{M}\sum_{j=1}^{N}\Delta\mathbf{R}(i,j) \tag{7.18}$$

The *NCC* measures the correlation between a filtered image **I** and its reference **R**, indicating the level of similarity between them. A lower *NCC* value indicates better performance of the tested method. The *NCC* is computed using the formula provided in Eq. (7.19).

$$NCC = \frac{\sum_{i=1}^{M}\sum_{j=1}^{N}\mathbf{I}(i, j)\cdot\mathbf{R}(i, j)}{\sum_{i=1}^{M}\sum_{j=1}^{N}\mathbf{R}(i, j)} \tag{7.19}$$

Due to the complex nature of the considered segmentation methods, which involve random operations and stochastic sub-routines, conducting a deterministic complexity analysis is impractical. Therefore, the computational complexity (*CC*) is evaluated based on the computational effort in seconds required for each algorithm.

7.6.3 Results of the Comparison

This subsection presents a comparative analysis of different filtering AD schemes when applied to eliminate distortions caused by various noise models. The objective is to assess the performance of each method using the performance indexes defined in Sect. 6.2. The comparison of denoising methods follows the following methodology:

1. Each image from the dataset (Fig. 7.4) is intentionally distorted using a specific noise model.
2. The AD schemes (MAD, EPAD, and RAD) are individually executed to generate the filtered image **I**.
3. Performance indexes are then computed based on the filtered image **I**.
4. The above steps are repeated for all images in the dataset.
5. The entire framework is implemented in MATLAB® and executed on a PC with an i5 processor running at 3 GHz and equipped with 12 GB of RAM.

The study comprises five experiments. The first and second tests compare the performance of each filtering AD technique in the presence of Gaussian noise with magnitudes of 25 and 75% of the variance, respectively. The third experiment focuses on evaluating all methods against speckle noise. The fourth test investigates the performance of all methods when subjected to impulsive noise. Finally, the fifth experiment presents the results obtained under conditions of salt & pepper noise. These tests adhere to the standard analysis commonly used to evaluate the performance of filtering schemes [46, 47, 56–59].

Gaussian Noise Characterized by a Variance of 25%

Noise refers to the unwanted random fluctuations in pixel intensity. In the case of Gaussian noise, each pixel in the noisy image p_N^G is obtained by adding a random value $N(p_A, \sigma)$. following a Gaussian distribution to the original pixel value p_A. Thus, the distortion can be mathematically represented as follows:

$$p_N^G = p_A + N(p_A, \sigma), \tag{7.20}$$

In the equation above, the variables p_A and σ within the function N represent the mean and variance, respectively. The initial experiment focuses on evaluating the performance of the filtering methods when subjected to Gaussian noise with a variance of 25%.

Table 7.1 presents the average values of each performance index, considering the set of twenty images from Fig. 7.4. The table highlights the best values obtained for each index. Based on the table, the MAD method demonstrates the highest performance across all indexes. EPAD follows as the second-best approach, while the RAD scheme performs the poorest. The MAD method achieves the highest SSIM value, indicating a strong resemblance between the filtered image **I** and the original noise-free reference image **R**. Additionally, MAD exhibits the best PSNR value, indicating a significant reduction in image noise. Moreover, MAD shows the highest EPI value, suggesting excellent preservation of the main characteristics of the original image **R**. This behavior reflects an optimal balance between minimal noise and maximal preservation of key image features. In terms of computational complexity (CC), the RAD method demonstrates the lowest value, while MAD performs favorably, and EPAD yields the poorest result.

To provide a detailed analysis of the visual outcomes, a subset of four representative images has been chosen from the dataset of twenty images. These selected images are randomly sampled to showcase the visual results. Figure 7.5 illustrates the filtered images produced by each filtering scheme. It is evident that the presented method exhibits the highest visual quality performance. The algorithm effectively reduces the noise in the image while preserving the main characteristics. On the other hand, the other methods exhibit varying levels of performance, but they are unable to completely eliminate all the noise present.

Gaussian Noise with 75% of the Variance

The second test evaluates the performance of the filtering methods using a Gaussian noise model, specifically p_N^G, with a variance of 75%. This level of noise represents a significant level of contamination in the original images.

Table 7.2 displays the average values of each performance index across the set of twenty images from Fig. 7.6. The best-performing values in the table are highlighted. The results are similar to those obtained in the case of a 25% variance. In this scenario of high noise contamination, the MAD method achieves the highest performance

Table 7.1 The average performance indexes for the scenario involving Gaussian noise with a variance of 25%. The highlighted values in the table represent the best results obtained

	SSIM	PSNR	EPI	MSE	NCC	CC
MAD	**0.7259**	**28.4861**	**0.5525**	**96.0224**	**0.9895**	128.12
EPAD	0.6435	27.4831	0.5001	118.0187	0.9886	215.10
RAD	0.5669	27.2096	0.4791	125.5855	0.9874	**94.78**

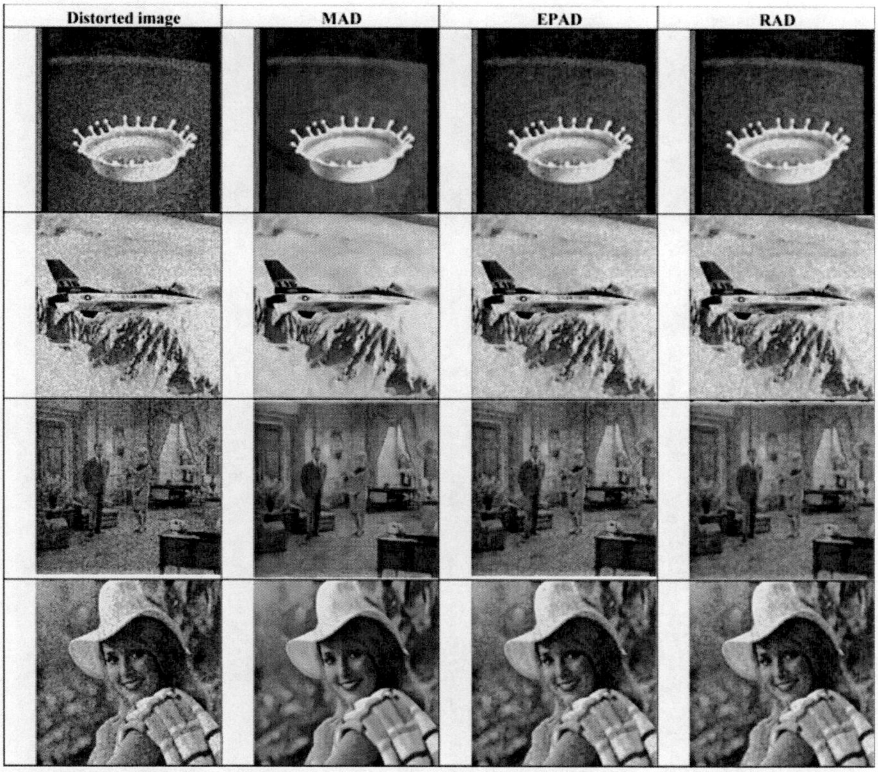

Fig. 7.5 Filtered images obtained when dealing with Gaussian noise with a variance of 25%

Table 7.2 The average performance indexes for the scenario involving Gaussian noise with a variance of 75%. The highlighted values in the table represent the best results obtained

	SSIM	PSNR	EPI	MSE	NCC	CC
MAD	**0.4781**	**22.7763**	**0.3884**	**352.6269**	**0.9783**	129.75
EPAD	0.1679	16.7083	0.3504	1394.9027	0.9767	288.12
RAD	0.1104	13.9937	0.3527	2647.6856	0.9752	**98.02**

scores in terms of SSIM, PSNR, EPI, MSE, and NCC. The EPAD approach performs second best, while the RAD scheme yields the lowest performance numbers. Overall, all the indexes for each AD filtering scheme are significantly affected by the high noise levels.

In the case of the MAD method, it obtains the highest SSIM score, indicating that its filtered image **I** closely resembles the original reference image **R** (noise-free image). The MAD algorithm also achieves the best PSNR value, indicating substantial reduction in noise within the image. Additionally, the MAD scheme demonstrates the highest EPI value, signifying its close resemblance to the characteristics of the

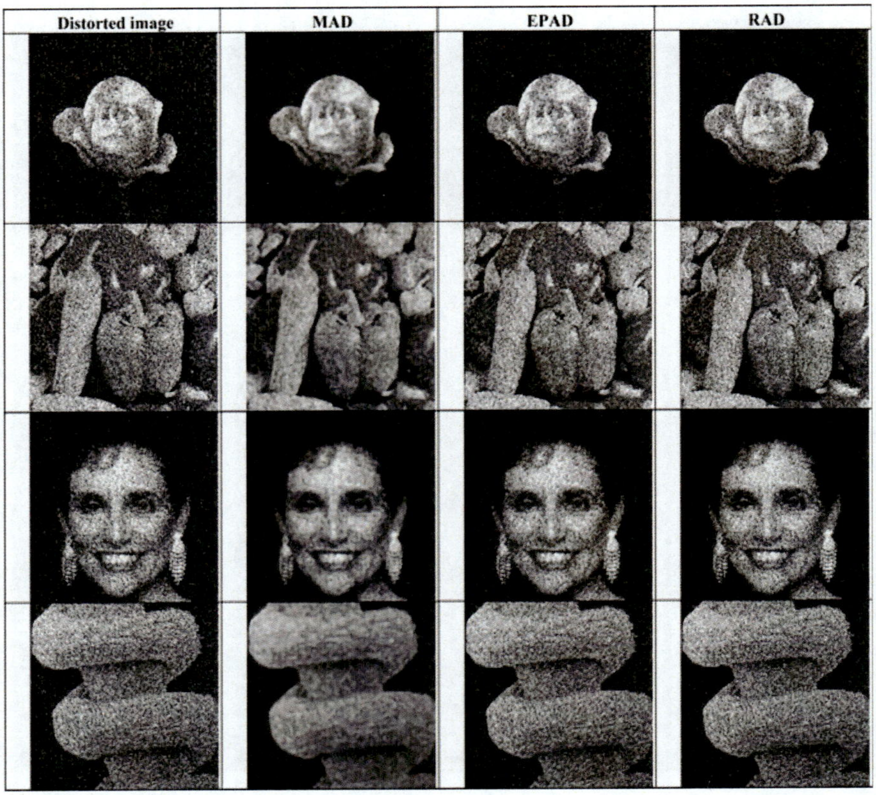

Fig. 7.6 Filtered images obtained when dealing with Gaussian noise with a variance of 75%

original image **R**. This behavior represents an optimal trade-off between minimal noise content and maximal preservation of the main characteristics of the original image.

Regarding computational time, the RAD method exhibits the lowest value. Although the MAD method has a competitive computational time, its results are slightly higher. On the other hand, the EPAD method shows inferior computational times. Based on these findings, it can be concluded that the MAD approach offers the best balance between denoising accuracy and processing speed.

Figure 7.6 showcases the resulting images obtained from the filtering process using each scheme, considering a subset of four images from the complete dataset of twenty. Similar to the first experiment, the MAD method demonstrates the highest visual quality performance. An examination of Fig. 7.6 reveals that the MAD algorithm effectively filters out the noise within the images while preserving their main characteristics. The other methods exhibit varying levels of performance. However, compared to the MAD approach, they are unable to completely eliminate the excessive noise present in the images. Figure 7.6 clearly demonstrates that no method is capable of completely eliminating all the noise contained in the images. Nevertheless, the MAD approach exhibits the least visual distortion among the methods.

Speckle Noise Experiment

Unlike the additive nature of Gaussian noise, speckle noise introduces a multiplicative effect. In the case of speckle noise, each pixel in the noisy image p_N^S is generated by multiplying the actual pixel value p_A by a random weight. Thus, this type of distortion is represented by the following expression:

$$p_N^S = p_A + r \cdot p_A, \tag{7.21}$$

where r represents a uniformly distributed number between 0 and 1.

The averaged results for various performance metrics, including Structural Similarity Index Measurement (SSIM), Mean Squared Error (MSE), Peak Signal to Noise Ratio (PSNR), Edge Preservation Index (EPI), Normalized Cross-Correlation (NCC), and Computational Complexity (CC), are presented in Table 7.3 for all 20 images in the selected dataset. Higher values indicate better performance for all metrics, except for MSE and CC, where lower values are desirable. The best results are highlighted in the table.

According to Table 7.3, the MAD method achieves the highest SSIM and EPI values, indicating its superior ability to preserve edge information and important features present in the original image. This makes it suitable for tasks involving feature-related processing. The MAD method also produces the lowest MSE values, indicating minimal distortion or error in the processed image. In terms of NCC, which evaluates the resemblance of the filtered image to the original noise-free image, the method performs comparatively well.

In terms of computational complexity, the RAD method is the fastest in producing a filtered image, while the EPAD method has the highest computational cost due to requiring multiple function evaluations to achieve competitive results. On the other hand, the MAD approach strikes a good balance between computational time and performance quality across metrics such as SSIM, MSE, PSNR, EPI, and NCC.

For a detailed analysis of the visual results, a set of four representative images has been selected from the dataset of twenty. These images are used to showcase the visual outcomes of the three AD methods, as shown in Fig. 7.7. The MAD method exhibits superior visual quality compared to its competitors. In all images, important details and elements are clearly visible, while the other algorithms tend to obscure

Table 7.3 The average performance indexes for the case of speckle noise are presented in the table. The best results are highlighted

	SSIM	PSNR	EPI	MSE	NCC	CC
MAD	**0.6997**	**27.3096**	**0.5902**	**134.5720**	**0.9785**	127.23
EPAD	0.5681	24.8010	0.4128	249.0135	0.9776	280.72
RAD	0.6983	23.7060	0.5238	324.9990	0.9741	**96.32**

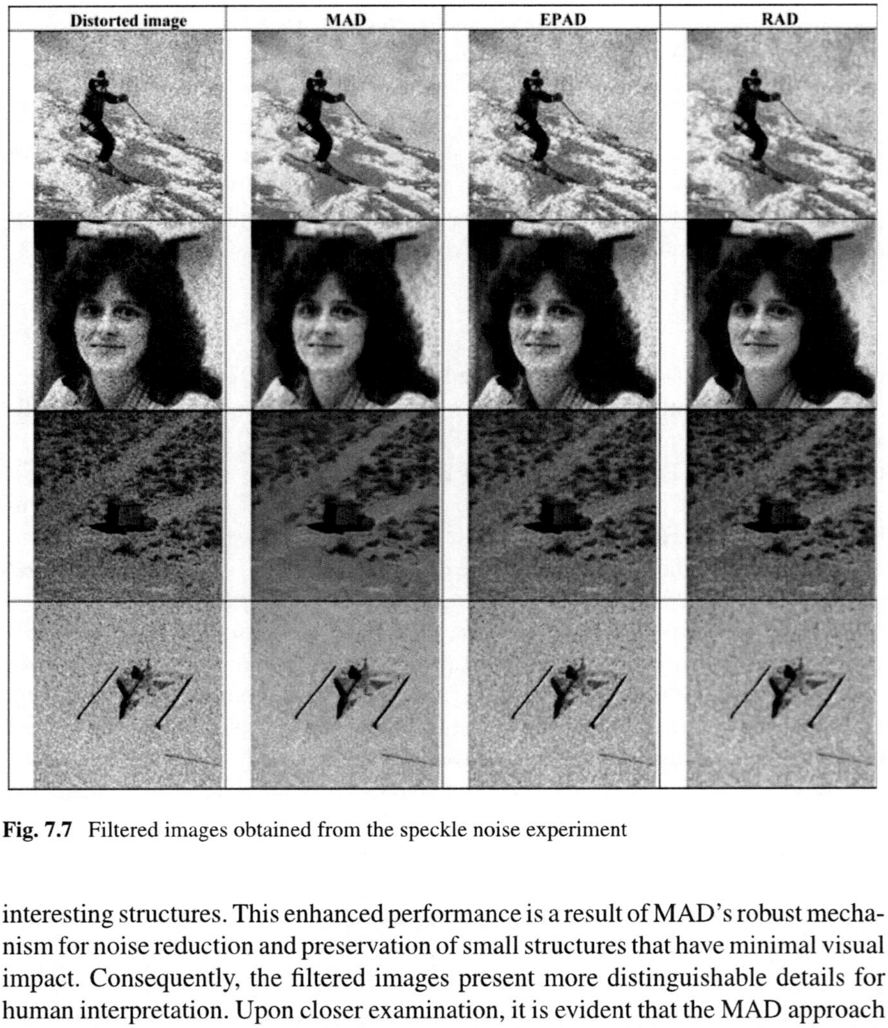

Fig. 7.7 Filtered images obtained from the speckle noise experiment

interesting structures. This enhanced performance is a result of MAD's robust mechanism for noise reduction and preservation of small structures that have minimal visual impact. Consequently, the filtered images present more distinguishable details for human interpretation. Upon closer examination, it is evident that the MAD approach effectively retains various details from the noise-free image **R**. Conversely, other AD methods introduce artifacts as a byproduct of their enhancement processes.

Impulsive Noise Experiment

Impulse noise is a type of distortion that occurs in images as a result of signal loss during transmission. In the case of impulsive noise, each corrupted pixel p_N^I is independently generated through a probabilistic selection based on the following model:

$$p_N^I = \begin{cases} ImV & \text{with probability} P \\ p_A & \text{with probability } (1 - P) \end{cases} \tag{7.22}$$

Based on this formulation, if the probability value P is met, the pixel p_N^I is set to ImV (impulsive value); otherwise, the actual pixel p_A remains unaffected. In the fourth experiment, the performance of the filtering approaches is evaluated using an impulsive noise model p_N^I with a probability value P of 0.3 and an impulsive value of $ImV = 255$.

Table 7.4 displays the averaged results of the three algorithms using all 20 images. The MAD method achieved the highest performance indexes, while the EPAD approach ranked second in most of the metrics. Conversely, the RAD scheme produced the least favorable results. The best values from the table have been highlighted. The success of the presented method can be attributed to two key factors. Firstly, the adoption of a multi-objective scheme allows for the identification of the optimal filtering solution that effectively eliminates noise while preserving the main image characteristics. Secondly, the integration of the NSGA-III algorithm enhances the exploration–exploitation capabilities, enabling the method to avoid getting trapped in local minima and obtain the best possible filtered image across various performance indexes. Figure 7.8 illustrates the visual outcomes of the three AD methods when applied to impulsive noise. A subjective assessment of the enhanced images generated by all algorithms is possible. The results clearly demonstrate that the MAD scheme produces the most visually appealing results, preserving image details without excessive blurring.

Salt and Pepper Noise

Salt and pepper noise belongs to the category of impulse noise. Unlike other impulsive noise models, salt and pepper noise introduces two distinct impulsive values, denoted as $ImV_1 = 255$ and $ImV_2 = 0$. Consequently, eliminating this type of noise poses more challenges. In the case of salt and pepper noise, a pixel p_N^{SP} is affected according to the following expression:

$$p_N^{SP} = \begin{cases} ImV_1 = 255 & \text{with probability}\, P_S \\ ImV_2 = 0 & \text{with probability}\, P_P \\ p_A & \text{with probability}\, 1 - (P_S + P_P) \end{cases} \tag{7.23}$$

The probabilities P_S, P_P and $1 - (P_S + P_P)$ represent the chances that a pixel p_N^{SP} is set to 255, 0, or p_A (unchanged), respectively. In this model, the overall probability P that a pixel p_A is affected by the values ImV_1 and ImV_2 is calculated as $P = P_S + P_P$. The term P is commonly referred to as noise density. For instance, if $P_S = 0.03$

Table 7.4 The average performance indexes for the case of impulsive noise are presented in the table. The best results are highlighted

	SSIM	PSNR	EPI	MSE	NCC	CC
MAD	**0.7613**	**18.2932**	**0.5901**	**966.8723**	**0.8049**	122.12
EPAD	0.6535	18.0858	18.0858	1013.9736	0.7987	270.89
RAD	0.6647	18.2296	0.5440	980.8551	0.8011	**97.24**

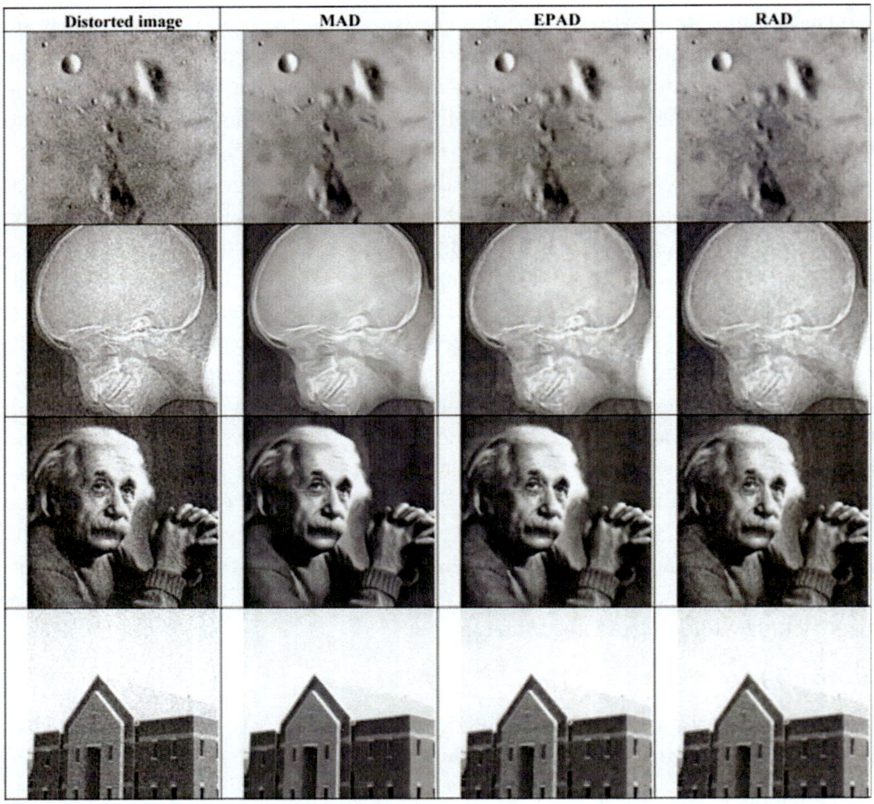

Fig. 7.8 Filtered images obtained from the impulsive noise experiment

and $P_P = 0.03$, then $P = 0.06$. This noise density indicates that approximately 6% of the image pixels are contaminated by salt and pepper noise. To further expand the comparative analysis, the fifth experiment examines the performance of the filtering methods under a salt and pepper noise model, assuming $P_S = 0.04$ and $P_P = 0.04$.

The averaged values of each performance index for the images in Fig. 7.4 are presented in Table 7.5. When dealing with salt and pepper noise, the MAD method achieves the highest performance indexes in terms of SSIM, PSNR, MSE, NCC, and CC. The MAD scheme effectively reduces the impact of impulse noise, resulting in smoother images where the affected pixels are not considered as edge points. As a result, the EPI values for the MAD method are slightly lower compared to the other approaches.

The filtering results of each scheme are displayed in Fig. 7.9, showcasing the impact of their respective effects. It is widely acknowledged that salt and pepper noise poses challenges for the anisotropic diffusion mechanism, making it difficult to eliminate. However, there are specific parameter configurations that can effectively smooth out this noise. From the figure, it is evident that the MAD method

Table 7.5 The average performance indexes for the case of Salt and pepper noise are presented in the table. The best results are highlighted

	SSIM	PSNR	EPI	MSE	NCC	CC
MAD	**0.7281**	**27.3263**	**0.6394**	**155.8670**	**0.9864**	124.53
EPAD	0.4060	20.7921	0.6281	249. 552.5865	0.9842	275.21
RAD	0.5727	24.4018	0.5760	246.1730	0.9812	**95.31**

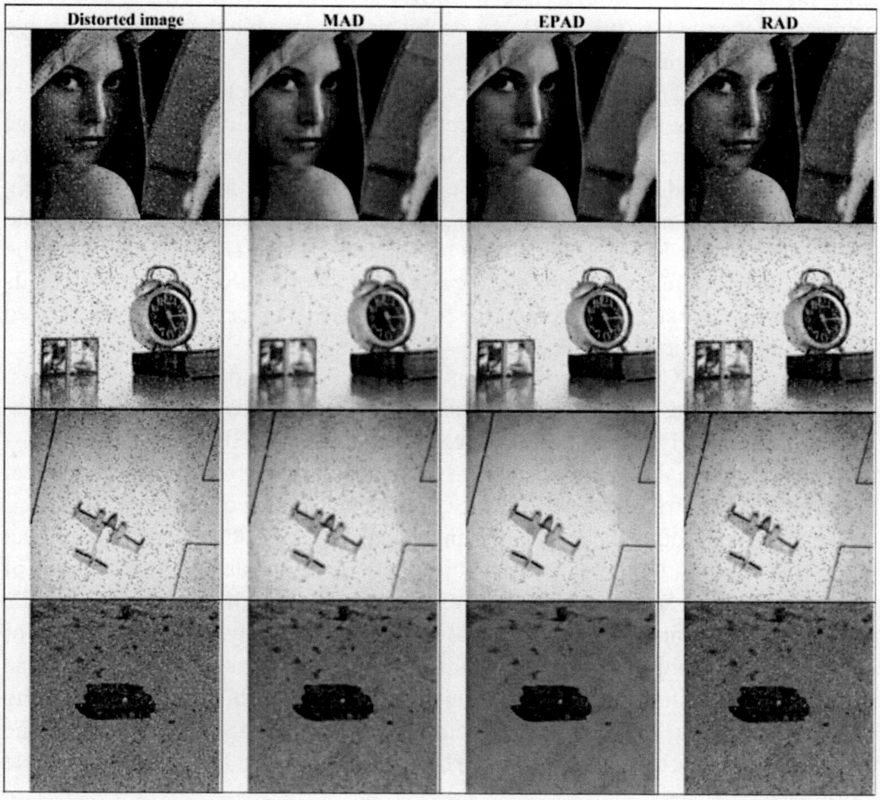

Fig. 7.9 Filtered images obtained from the salt and pepper noise experiment

outperforms the others in terms of visual quality. The algorithm successfully filters out the impulsive noise and mitigates its destructive effects. While the other methods exhibit varying levels of performance, they are unable to completely eliminate all the noise present.

7.6.4 Discussion of the Results

In this subsection, we provide a brief discussion on the strengths and limitations of the MAD scheme based on the evaluation conducted through various experiments with different noise models. Six performance measures were used to assess the results by comparing the filtered image (I) with the reference noise-free image (R).

The experiments clearly indicate that the MAD algorithm outperforms other denoising schemes in terms of PSNR and MSE values. This signifies that the denoised images (I) generated by the MAD approach closely resemble the noise-free image (R) to a greater extent.

While PSNR and MSE are straightforward metrics with a clear physical interpretation, they may not always accurately reflect the visual quality of a filtered image from a human perspective. Hence, the SSIM, EPI, and NCC criteria are considered more closely aligned with human perception of visual quality in images. Comparing the proposed algorithm with other approaches, it is evident that the SSIM, EPI, and NCC values of the MAD scheme are consistently higher. This indicates that the denoised images effectively preserve the main characteristics and important details from the original images, surpassing the performance of other methods when faced with various distortions [60–64].

Regarding the CC metric, it is evident that the RAD method achieves the lowest value. Although the MAD method demonstrates competitive computational time, its results are slightly higher in this aspect. Conversely, the EPAD method exhibits poorer computational times. Considering these conditions, it can be concluded that the MAD scheme strikes the best balance between denoising quality and speed.

The remarkable outcomes of the presented method can be attributed to two crucial factors. Firstly, the adoption of a multi-objective scheme enables the identification of the optimal filtering solution that effectively removes most of the existing noise while preserving the essential characteristics of the image. Secondly, the incorporation of the NSGA-III algorithm significantly enhances the search strategy for obtaining the best Pareto front, improving exploration–exploitation capabilities and preventing the algorithm from getting trapped in local minima [65–68]. Consequently, the presented method excels in achieving the best possible filtered image across multiple performance indexes.

Despite its favorable results, the MAD scheme does have certain drawbacks and limitations as a filtering approach. Under the MAD scheme, the filtering process is formulated as a multi-objective optimization problem. Thus, the denoised image corresponds to a solution that minimizes the noise content in the image (objective one) while maximizing the contrast of the main characteristics of the original image (objective two). However, the use of only two objectives does not guarantee the best solution in terms of all six performance indexes. As part of future work, there are plans to introduce a third objective function that integrates the SSIM, EPI, and NCC metrics. By incorporating these metrics, the performance of the MAD scheme could be significantly improved.

7.7 Conclusions

Images often suffer from noise contamination during the acquisition, storage, and transmission processes. This interference significantly degrades the perceived quality of the image. The primary objective of image denoising is to remove the corrupted information while preserving the essential features of the original image.

In this chapter, a novel approach is introduced to address the denoising problem using a multi-objective methodology. The presented approach aims to determine the optimal parameters for anisotropic diffusion (AD) filters that strike a balance between two conflicting objectives: eliminating the corrupted information and enhancing the main characteristics of the original image. These objectives cannot be individually improved without negatively impacting the other. To solve this optimization problem, the Non-dominated Sorting Genetic Algorithm based on Reference Points (NSGA-III) is employed, as it is known for its robustness and effectiveness in multi-objective optimization. The final solution is obtained by analyzing the Optimal Pareto front, which represents the trade-off between the two objectives.

To assess the performance of the presented scheme, various experiments were conducted. The presented method, called Multi-objective AD (MAD), was applied to a dataset consisting of five representative images, and its results were compared to those obtained by the Edge Preserving AD filter (EPAD) and the Robust AD filter (RAD). The experimental findings demonstrate that the presented method outperforms existing filtering algorithms in terms of both visual quality and standard performance metrics.

References

1. Perona, P., Malik, J.: Scale space and edge detection using anisotropic diffusion. IEEE Trans. Image Process. **12**(8), 629–639 (1990)
2. Witkin, A.: Scale-space filtering. In: International Joint Conference on Artificial Intelligence, Karlsruhe, West Germany, pp. 1019–1021 (1983)
3. Mazzia, A.: A numerical study of the virtual element method in anisotropic diffusion problems. Math. Comput. Simul. **177**, 63–85 (2020)
4. Alvarez, L., Lions, P.L., Morel, J.M.: Image selective smoothing and edge detection by nonlinear diffusion II. SIAM J. Numer. Anal. **29**(3), 845–867 (1992)
5. Catté, F., Lions, P.L., Morel, J.M., Coll, T.: Image selective smoothing and edge detection by nonlinear diffusion. SIAM J. Numer. Anal. **29**(1), 182–193 (1992)
6. Chen, Y., Barcelos, C., Mair, B.: Smoothing and edge detection by time-varying coupled nonlinear diffusion equations. Comput. Vis. Image Underst. **82**(2), 85–100 (2001)
7. Kichenssamy, S.: The Perona-Malik paradox. SIAM J. Appl. Math. **57**, 1328–1342 (1997)
8. Nitzberg, M., Shiota, T.: Nonlinear image filtering with edge and corner enhancement. IEEE Trans. Pattern Anal. Mach. Intell. **14**(8), 826–833 (1992)
9. Rudin, L.I., Osher, S., Fatemi, E.: Nonlinear total variation based noise removal algorithms. Physica D **60**, 259–268 (1992)
10. Weickert, J.: A review of nonlinear diffusion filtering. In: Scale-Space Theory in Computer Vision. Lecture Notes in Computer Science, vol. 1252, pp. 3–28 (1997)

11. You, Y.L., Xu, W., Tannenbaum, A., Kaveh, M.: Behavioral analysis of anisotropic diffusion in image processing. IEEE Trans. Image Process. **5**(11), 1539–1553 (1996)
12. Chen, D, MacLachlan, S., Kilmer, M.: Iterative parameter-choice and multigrid methods for anisotropic diffusion denoising. SIAM J. Sci. Comput. **33**, 2972–2994
13. Gerig, G., Kubler, O., Kikinis, R., Jolesz, F.: Nonlinear anisotropic filtering of MRI data. IEEE Trans. Med. Imaging **11**(2), 221–232 (1992)
14. Weickert, J.: Anisotropic diffusion in image processing. In: Series ECMI, Teubner, Stuttgart, Germany (1998)
15. Weickert, J., Romeny, B.M.T.H., Viergever, M.A.: Efficient and reliable schemes for nonlinear diffusion filtering. IEEE Trans. Image Processing **7**, 398–410 (1998)
16. Yu, X., Wu, C., Jia, T., Chen, S.: A time-dependent anisotropic diffusion image smoothing method. In: 2nd International Conference on Intelligent Control and Information Processing, vol. 2, July 2011, pp. 859– 862
17. Chao, S.M., Tsai, D.M.: An improved anisotropic diffusion model for detail and edge-preserving smoothing. Pattern Recogn. Lett. **31**(October), 2012–2023 (2010)
18. Chao, S.M., Tsai, D.: Anisotropic diffusion with generalized diffusion coefficient function for defect detection in low-contrast surface images. Pattern Recogn. **43**(5), 1917–1931 (2010)
19. Nair, R.R., David, E., Rajagopal, S.: A robust anisotropic diffusion filter with low arithmetic complexity for images. EURASIP J. Image Video Process. **48**, 1–14
20. Veerakumar, T., Esakkirajan, S., Vennila, I.: Edge preserving adaptive anisotropic diffusion filter approach for the suppression of impulse noise in images. Int. J. Electron. Commun. (AEÜ) **68**, 442–452 (2014)
21. Tsiotsios, C., Petrou, M.: On the choice of the parameters for anisotropic diffusion in image processing. Pattern Recogn. **46**(5), 1369–1381 (2013)
22. Li, X., Chen, T.: Nonlinear diffusion with multiple edginess thresholds. Pattern Recogn. **27**(8), 1029–1037 (1994)
23. Voci, F., Eiho, S., Sugimoto, N., Sekiguchi, H.: Estimating the gradient threshold in the Perona-Malik equation. IEEE Signal Process. Mag. **21**(3), 39–46 (2004)
24. Black, M.J., Sapiro, G., Marimont, D., Heeger, D.: Robust anisotropic diffusion. IEEE Trans. Image Process. **7**(3), 421–432 (1998)
25. Monteil, J., Beghdadi, A.: New interpretation and improvement of the nonlinear anisotropic diffusion for image enhancement. IEEE Trans. Pattern Anal. Mach. Intell. **21**(9), 940–946 (1999)
26. Ilyevsky, A., Turkel, E.: Stopping criteria for anisotropic PDEs in image processing. J. Sci. Comput. **45**, 333–347 (2010)
27. Gilboa, G., Sochen, N., Zeevi, Y.Y.: Estimation of optimal PDE-based denoising in the SNR sense. IEEE Trans. Image Process. **15**(8), 2269–2280 (2006)
28. Mrázek, P., Navara, M.: Selection of optimal stopping time for nonlinear diffusion filtering. Int. J. Comput. Vis. **52**(2/3), 89–203 (2003)
29. Papandreou, G., Maragos, P.: A cross-validatory statistical approach to scale selection for image denoising by nonlinear diffusion. In: IEEE Conference on Computer Vision and Pattern Recognition, pp. 625–630 (2005)
30. Solo, V.: Automatic stopping criterion for anisotropic diffusion. In: International Conference on Acoustics, Speech, and Signal Processing, vol. 6, pp. 3929–3932 (2001)
31. Gilboa, G., Sochen, N., Zeevi, Y.Y.: Forward-and-backward diffusion processes for adaptive image enhancement and denoising. IEEE Trans. Image Process. **11**(7), 689–703 (2002)
32. Weickert, J.: Applications of nonlinear diffusion in image processing and computer vision. Acta Math. Univ. Comenian. **70**, 33–50 (2001)
33. Maulik, U., Saha, I.: Modified differential evolution based fuzzy clustering for pixel classification in remote sensing imagery. Pattern Recogn. 42(9) (2009)
34. Saha, I., Maulik, U., Bandyopadhyay, S.: An improved multi-objective technique for fuzzy clustering with application to IRS image segmentation. In: Proceedings of EvoWorkshops 2009 on Applications of Evolutionary Computing (2009)

35. Shirakawa, S., Nagao, T.: Evolutionary image segmentation based on multiobjective clustering. In: Proceedings of Congress on Evolutionary Computation (CEC '09), Trondheim, Norway (2009)
36. Mukhopadhyay, A., Maulik, U., Bandyopadhyay, S.: Multiobjective genetic clustering with ensemble among pareto front solutions: application to MRI brain image segmentation. In: Proceedings of 7th International Conference on Advances in Pattern Recognition (2009)
37. Nakib, A., Oulhadj, H., Siarry, P.: Image thresholding based on Pareto multiobjective optimization. Eng. Appl. Artif. Intell. **23**(3), 313–320 (2010)
38. Saha, S., Bandyopadhyay, S.: A new symmetry based multiobjective clustering technique for automatic evolution of clusters. Pattern Recogn. **43**(4), 738–751 (2010)
39. Deb, K., Pratap, A., Agarwal, S., Meyarivan, T.: A fast and elitist multi-objective genetic algorithm: NSGA-II. IEEE Trans. Evol. Comput. **6**(2), 182–197 (2002)
40. Wang, Y., Shen, Y., Zhang, X., Cui, G., Sun, J.: An Improved non-dominated sorting genetic algorithm-II (INSGA-II) applied to the design of DNA codewords. Math. Comput. Simul. **151**, 131–139 (2018)
41. Zitzler, E., Laumanns, M., Thiele, L.: SPEA2: improving the strength pareto evolutionary algorithm. In: Proceedings of Evolutionary Methods for Design, Optimization and Control with Application to Industrial Problems, pp. 95–100 (2001)
42. Zitzler, E., Kunzli, S.: Indicator-based selection in multi-objective search. In: 8th International Conference on Parallel Problem Solving from Nature, pp. 832–842 (2004)
43. Yingying, Z., Junwei, L., Jianyong, C., Ming, Z.: An improved NSGA-III algorithm for feature selection used in intrusion detection. Knowl.-Based Syst. **116**, 74–85 (2017)
44. Bhesdadiya, R.H., Trivedi, I.N., Jangir, P., Jangir, N., Kumar, A.: An NSGA-III algorithm for solving multi-objective economic/environmental dispatch problem. Cogent Eng. **3**, 1269383 (2016)
45. Yuan, X., Tian, H., Yuan, Y., Huang, Y., Ikram, R.M.: An extended NSGA-III for solution multi-objective hydro-thermal-wind scheduling considering wind power cost. Energy Convers. Manage. **96**, 568–578 (2015)
46. Karunanayake, N., Aimmanee, P., Lohitvisate, W., Makhanov, S.S.: Particle method for segmentation of breast tumors in ultrasound images. Math. Comput. Simul. **170**, 257–284 (2020)
47. Alghafisd, A., Firdousiac, F., Syeda, MajidKhanab, Batoolae, I., Amin, M.: An efficient image encryption scheme based on chaotic and Deoxyribonucleic acid sequencing. Math. Comput. Simul. **177**, 441–466 (2020)
48. Srinivas, N., Deb, K.: Multi-objective optimization using nondominated sorting in genetic algorithms. Evol. Comput. **2**(3), 221–248 (1994)
49. Miettinen, K.: Nonlinear Multi-objective Optimization. Kluwer, Norwell, MA: (1999)
50. Wei, X., Hao Liu, Z., Ping, L., Zhao, T.: An efficient multi-objective optimization algorithm based on level swarm optimizer. Math. Comput. Simul. **177**, 588–602 (2020)
51. Abdolrazzagh-Nezhad, M., Radgohar, H., Salimian, S.N.: Enhanced cultural algorithm to solve multi-objective attribute reduction based on rough set theory. Math. Comput. Simul. **170**, 332–350 (2020)
52. Immerkær, J., Estimation, F.N.V.: Comput. Vis. Image Underst. **64**(2), 300–302 (1996)
53. Draa, A., Bouaziz, A.: An artificial bee colony algorithm for image contrast enhancement. Swarm Evol. Comput. **16**, 69–84 (2014)
54. Satopa, V., Albrecht, J., Irwin, D., Raghavan, B.: Finding a "Kneedle" in a haystack: detecting knee points in system behavior. In: 31st International Conference on Distributed Computing Systems Workshops, pp. 166–171 (2011)
55. Zhao, Q., Xu, M., Fränti, P.: Knee point detection on bayesian information criterion. In: 20th IEEE International Conference on Tools with Artificial Intelligence, pp. 431–438 (2008)
56. Zhu, R., Li, X., Zhang, X., Ma, M.: MRI and CT medical image fusion based on synchronized-anisotropic diffusion model. IEEE Access **8**, 91336–91350 (2020)
57. Khalil, H.A., Darwish, S., Ibrahim, Y.M., Hassan, O.F.: 3D-MRI brain tumor detection model using modified version of level set segmentation based on dragonfly algorithm. Symmetry **12**, 1256 (2020)

58. Veerakumara, T., Narayan, B., Esakkirajan, S.S.: Empirical mode decomposition and adaptive bilateral filter approach for impulse noise removal. Expert Syst. Appl. **121**, 18–27 (2019)
59. Yu, W., Yuanjun, W.: Anisotropic diffusion filtering method with weighted directional structure tensor. Biomed. Signal Process. Control **53**, 101590 (2019)
60. Ramírez-Ortegón, M.A., Tapia, E., Ramírez-Ramírez, L.L., Rojas, R., Cuevas, E.: Transition pixel: a concept for binarization based on edge detection and gray-intensity histograms. Pattern Recogn. **43**(4), 1233–1243 (2010)
61. Zaldivar, D., Morales, B., Rodríguez, A., Valdivia-G, A., Cuevas, E., Pérez-Cisneros, M.: A novel bio-inspired optimization model based on Yellow Saddle Goatfish behavior. Biosystems **174**, 1–21 (2018)
62. Cuevas, E., Gálvez, J., Hinojosa, S., Avalos, O., Zaldívar, D., Pérez-Cisneros, M.: A comparison of evolutionary computation techniques for IIR model identification. J. Appl. Math. (2014)
63. Ibrahim, R.A., Elaziz, M.A., Oliva, D., Cuevas, E., Lu, S.: An opposition-based social spider optimization for feature selection. Soft. Comput. **23**, 13547–13567 (2019)
64. Cuevas, E., Wario, F., Osuna-Enciso, V., Zaldivar, D., Pérez-Cisneros, M.: Fast algorithm for multiple-circle detection on images using learning automata. IET Image Proc. **6**(8), 1124–1135 (2012)
65. Cuevas, E., González, A., Fausto, F., Zaldívar, D., Pérez-Cisneros, M.: Multithreshold segmentation by using an algorithm based on the behavior of locust swarms. Math. Prob. Eng. (2015)
66. Cuevas, E., Zaldivar, D., Pérez-Cisneros, M.: Seeking multi-thresholds for image segmentation with Learning Automata. Mach. Vis. Appl. **22**, 805–818 (2011)
67. Bandyopadhyay, R., Basu, A., Cuevas, E., Sarkar, R.: Harris Hawks optimisation with simulated annealing as a deep feature selection method for screening of COVID-19 CT-scans. Appl. Soft Comput. **111**, 107698 (2021)
68. Hinojosa, S., Dhal, K.G., Abd Elaziz, M., Oliva, D., Cuevas, E.: Entropy-based imagery segmentation for breast histology using the stochastic fractal search. Neurocomputing **321**, 201–215 (2018)